How to Do Things with Books in Victorian Britain

How to Do Things with Books in Victorian Britain

Leah Price

PRINCETON UNIVERSITY PRESS

Princeton and Oxford

Second printing, and first paperback printing, 2013
Paperback ISBN 978-0-691-15954-6

The Library of Congress has cataloged the cloth edition of this book as follows

Price, Leah.
    How to do things with books in Victorian Britain / Leah Price.
        p. cm.
    Includes bibliographical references and index.
    ISBN 978-0-691-11417-0   1. Books and reading—Great Britain—
History—19th century. 2. Books—Great Britain—Psychological aspects—History—19th
century. 3. Books—Social aspects—Great Britain—History—19th century. 4. Book
industries and trade—Great Britain—History—19th century. 5. English fiction—19th
century—History and criticism. 6. Books and reading in literature. 7. Books in
literature.   I. Title.
    Z1003.5.G7P75 2012
    028'.9094109034—dc23   2011037436

British Library Cataloging-in-Publication Data is available

This book has been composed in Sabon

10 9 8 7 6 5 4 3 2

Contents

Illustrations

Acknowledgments

MANY READERS HELPED me write. Thanks especially to Srinivas Aravamudan, Margaret Beetham, Peter de Bolla, Larry Buell, Amanda Claybaugh, Nancy Cott, Patricia Crain, Nicholas Dames, Robert Darnton, Ian Duncan, Drew Faust, William Flesch, John Forrester, Elaine Freedgood, Debra Gettelman, Lisa Gitelman, Simon Goldhill, David Hall, Susan Halpert, Richard Hardack, Barbara Hochman, Isabel Hofmeyr, Hansun Hsiung, Virginia Jackson, Melissa Jenkins, Jane Kamensky, Sol Kim-Bentley, Michèle Lamont, Michael Ledger-Lomas, Yoon Sun Lee, Spencer Lenfield, Lauren Lepow, Seth Lerer, Deidre Lynch, Alison MacKeen, Peter Mandler, Jane Mansbridge, Sharon Marcus, Maia McAleavey, Deborah Nord, Geoff Nunberg, Alexander Parker, Clare Pettitt, John Plotz, Christopher Prendergast, Peter Pruyn, Harriet Ritvo, Catherine Robson, Jan Schramm, Jason Scott-Warren, James Secord, Sharmila Sen, Stuart Shieber, James Simpson, Diana Sorensen, Peter Stallybrass, William St Clair, Christopher Stray, Michael Suarez, Ramie Targoff, Pam Thurschwell, Katie Trumpener, Judy Vichniac, David Vincent, Michael Warner, Hanne Winarsky, and Ruth Yeazell. Thanks, too, to King's College, Cambridge, the Radcliffe Institute, the Stanford Humanities Center, and the National Endowment for the Humanities.

Earlier versions of material from chapters 2, 4, and 7 appeared in *Reading Victorian Feeling*, edited by Rachel Ablow (Ann Arbor: University of Michigan Press, 2010), 47–68; *Representations* 108 (Fall 2009): 120–38; and *Bookish Histories*, edited by Ina Ferris and Paul Keen (New York: Palgrave, 2009), 148–68. Thanks to all three for permission to reprint.

Ann Blair has been the best possible coconspirator in all things book-historical. Natalka Freeland's friendship remains as strong as her misreadings. Nir Eyal's love and example teach me daily how to do things with words and without.

Introduction

UPON COMING INTO his master's fortune, Dickens's illiterate dustman Mr. Boffin immediately hires a ballad-seller to entertain him by reading aloud. Only one detail remains to be checked: "You are provided with the needful implement—a book, sir?"

> 'Bought him at a sale,' said Mr. Boffin. 'Eight wollumes. Red and gold. Purple ribbon in every wollume, to keep the place where you leave off. Do you know him?'
>
> 'The book's name, sir?' inquired Silas.
>
> 'I thought you might have know'd him without it,' said Mr. Boffin slightly disappointed. 'His name is Decline-And-Fall-Off-The-Rooshan-Empire.' (Dickens, *Our Mutual Friend* 59)

Because no one reading this passage shares Mr. Boffin's illiteracy, and because few readers of late Dickens have not read at least the spines of Gibbon, we smile. But what if the geographical confusion made bibliographical sense? As a waste-dealer familiar with tanners, Mr. Boffin would have heard of "Russia" as a metonymy for a leather produced in that country, calfskin (often dyed red) tanned with birch oil that imparted a characteristic smell. In this hypothesis, the hope that "you might have know'd him" would look perfectly reasonable: cannier than Silas, Mr. Boffin does recognize the book "without," if not within. "In what I did know," David Copperfield reflects upon leaving warehouse for school, "I was much farther removed from my companions than in what I did not" (Dickens, *David Copperfield* 218). If we took Russia to refer to container rather than contents, then the dustman's class position would reflect less a deficiency of interpretive skill than an excess of sensitivity to color, texture, and smell. His ignorance of the history in the book would throw into relief how much he knows about the history of the book. "Bought him at a sale": Boffin knows not only how the "wollumes" were manufactured, but whether he is their first owner. Once endowed with a life story, even a book judged by "his" cover can elicit affection.

When Silas later arrives to take up his task, it remains unclear whether the "gorging Lord-Mayor's-Show of wollumes (probably meaning gorgeous, but misled by association of ideas)" will end up on Mrs. Boffin's side of the room (whose shelves display stuffed birds) or Mr. Boffin's (lined with cold joints). As binding is to text, so "gorgeous" to "gorging":

do books resemble decorative outsides or functional insides? Should the volumes that Boffin has "ranged flat, in a row, like a galvanic battery" be treated as an implement or a show?

In short, what meanings do books make even, or especially, when they go unread? And why did Victorian novelists care? That books function both as trophies and as tools, that their use engages bodies as well as minds, and that printed matter connects readers not just with authors but with other owners and handlers—these facts troubled a genre busy puzzling out the proper relation of thoughts to things, in an age where more volumes entered into circulation (or gathered dust on more shelves) than ever before.

It's not that they hated books. But the great realists did loathe anyone who loved the look of books—who displayed "a great, large handsome Bible, all grand and golden, with its leaves adhering together from the book-binder's press," or whose "splendidly bound books furnished the heavily carved rosewood table" (Gaskell, *North and South* 79; Jewsbury 13, 37). One wellborn narrator remarks, in the house of a wealthy tradesman, that "the round rosewood table was in a painfully high state of polish; the morocco-bound picture books that lay on it, looked as if they had never been moved or opened since they had been bought; not one leaf even of the music on the piano was dogs-eared or worn" (W. Collins, *Basil* 61). Book against text, new money against old money—and secondary characters against protagonist. The opening scene of *Ranthorpe* establishes the hero's depth by describing what aspects of books he *fails* to notice. "He cared not for rare editions, large paper copies, or sumptuous bindings . . . he cared not even whether they had covers at all" (Lewes 4–5).

A moral test doubles as a political stance: the post-Gutenberg consensus that makes differently priced editions of a text functionally equivalent becomes a proxy for the more controversial demand to value human souls alike, whatever the color of their money or their skin. Or was the problem, on the contrary, that literacy was spreading too widely to remain a reliable marker of rank or gender? To use books no longer proved anything; to refrain from misusing them did. The *Gentleman's Magazine*'s lament that "too many women value a book solely for its binding" (Watkins 102) is dramatized in a joke about a lady complaining to the librarian: "Look what an atrocious cover it has; haven't you one bound in saxe-blue to match my costume?" (Coutts 147). In 1851, an Evangelical magazine contrasts the good child who "puts books into his head" with the dunce whose books are "only on your shelves" ("How to Read Tracts").

Nothing against books, then, but something against the eyeing and pricing of books imagined to compete with internalizing them. *The Oxford English Dictionary* dates to 1847 the use of "reading copy" as a euphemism for a book so battered that the only value left lies in the words that it contains. "Books are now so dear," Southey had reported at the dawn of the Regency phenomenon known as "the bibliomania," "that they are becoming rather fashionable articles of *furniture* more than anything else; they who buy them do not read them, and they who read them do not buy them. I have seen a Wiltshire clothier, who gives his bookseller no other instructions than the *width* of his shelves."[1] Made to be seen through, books find themselves seen. By 1887, an article titled "Literary Voluptuaries" could declare that "the collector is curious about margins, typography, and casings, but comparatively indifferent to contents" (805). Cover and content, authenticity and appearance: the language of insides and outsides makes any consciousness of the book's material qualities signify moral shallowness. Leather bindings rub off on their skin-deep owners.

Commission reinforces omission. Not content to ignore the outsides of books, a good reader actively scorns them. "Due attention to the inside of books, and due contempt for the outside," Chesterfield had pronounced in 1749, "is the proper relation between a man of sense and his books" (1291). One dictionary defined bibliomania as the fact of being "rather seduced by the exterior than the interior" (Dibdin 58). An article titled "Furniture Books" compared loving one's "handsomely dressed" volumes to "thinking more of the jewels of one's mistress than of her native charms" (97). Reciprocally, Wilde could shock by comparing a woman wearing a "smart gown" to "an édition de luxe of a bad French novel" (178, 37).[2]

No cheaper cue for our sympathies, no surer predictor of the plot: a character who sells his father-in-law's library can't be trusted not to buy a mistress; a character who wants his books bound in leather will marry the blonde; a character who manhandles books will abuse children. The great nineteenth-century novels of individual development domesticate Heine's 1821 prediction that "when they have burned books, they will end in burning human beings." In liberal democracies, the traditional state prosecution of books whose content is judged treasonous gives way to homegrown persecution of persons whose reading is judged antisocial. After Julien Sorel's father catches his attention by knocking a book out of his hands, the book is drowned; when Hugh Trevor's master beats him for being "deeply engaged in my book," the book is burnt (Holcroft 41). The public hangman burned books in place of their author, but domestic tyrants made books a proxy for the readers under their control. When John Reed reduces books to projectiles or Tom Tulliver asks why a bankrupt's books shouldn't be auctioned off along with his chairs, their refusal to treat the book as a protected category signals their blindness to what's special about Jane or Maggie.

A Note on Language

One of the dark red volumes of the first edition of the *New Oxford English Dictionary* defines "book" as

> a. . . . a treatise occupying numerous sheets or leaves fastened together at one edge called the *back* . . . But, since either the form of the book or its subject may be mainly or exclusively the object of attention, this passes on either side into
> b. The material article so made up, without regard to the nature of its contents, even though its pages are occupied otherwise than with writing or printing, or are entirely blank . . .
> c. A literary composition such as would occupy one or more volumes, without regard to the material form or forms in which it actually exists . . .

> In sense b every volume is a 'book'; whilst in sense c one 'book' may occupy several volumes; and on the other hand one large volume may contain several 'books,' i.e. literary works originally published as distinct books.

The minute the contributor pictures the material container, the textual contents empty out: the example imagined is "entirely blank." Charles Chestnutt's 1904 story "Baxter's Procrustes" makes that zero-sum logic a plot twist, imagining a club of book-collectors tricked into accepting a blank book for their collection. "The true collector loves wide margins, and the Procrustes, being all margin, merely touches the vanishing point of the perspective" (830). A thumbed-to-death "reading copy" stands opposite an illegible collectible clean not only of smudges and underlinings, but of print.

You'll have noticed my contortions attempting to distinguish "text"—a string of words—from "book" or "book-object": a physical thing. In an everyday language incapable of even deciding what preposition should link the two—the text "of" a book, the text "in" a book?—one term appears sometimes as contained within the other, sometimes as antithetical to it.[3] If "book" really connoted materiality, there would be no need to affix the pleonastic "object"; if "text" really provided an adequate term for a linguistic structure, I would refer to what you're now reading as "this text." Only the ambiguity of sentence openings prevented me from generalizing the distinction between the Bible (a text) and the bible (an object) to Books and books.[4]

The Victorians cathected the text in proportion as they disowned the book. More specifically, they identified themselves as text-lovers in proportion as they distinguished themselves from book-lovers. To take in a text is to tune out its raw materials: a newspaper isn't called a "rag" if the speaker thinks it worth reading. More surprisingly, in 1818 William Hazlitt could ridicule a book by pointing out the *high* cost of the paper it was printed on: "Mr. Campbell always seems to me to be thinking how his poetry will look when it comes to be hot-pressed on superfine wove paper" (295). Whenever a review mentions the price or appearance of a book, we know that its textual contents will be either ridiculed or dismissed as beneath contempt. Even in the digital age, to name the ingredients of a book is to insult it—as when an MIT professor refers to "tree flakes encased in dead cow" or a Microsoft researcher to "sooty marks on shredded trees."[5]

Conversely, the best texts eclipse the book. When Amazon launched its first e-reader, Jeff Bezos boasted that the Kindle emulated the way in which "the physical book is so elegant that the artifact itself disappears into the background. The paper, glue, ink and stitching that make up the book vanish, and what remains is the author's world."[6] A successful e-reader, by this logic, would illustrate Marian Evans's contention that "on certain red-letter days of our existence, it happens to us to discover among the *spawn* of the press, a book which, as we read, seems to undergo a sort of transfiguration before us. We no longer hold heavily in our hands an octavo of some hundred pages, over which the eye laboriously travels, hardly able to drag along with it the restive mind; but we seem to be in companionship with a spirit, who is transfusing himself into our souls" (G. Eliot, "J. A. Froude's *The Nemesis of Faith*" 265). The double etymology of "liber" points to the book's Janus-faced potential: some medieval commentators traced it to the word for the "bark" on which texts were inscribed, others to the action ("liberare") that texts were expected to perform.[7] Grounded in a material substance or linked with a lofty abstraction, the same object bound by its medium is credited with the power to free its users.

WHAT USE ARE BOOKS?

The following pages reconstruct nineteenth-century understandings of, and feelings toward, the uses of printed matter. In particular, they excavate the often contentious relation among three operations: reading (doing something with the words), handling (doing something with the object), and circulating (doing something to, or with, other persons by means of the book—whether cementing or severing relationships, whether by giving and receiving books or by withholding and rejecting them). Often pictured as competing, in practice these three modes almost always overlapped. Impossible to read without handling (even if certain genres took pains to suppress any mention of handling), or to get one's hands on a book without its having passed through someone else's hands (even if other genres imagined books as found objects). We might posit, then, that what look like antonyms are in fact subsets: handling without reading is easier to imagine than reading without handling, circulating without reading than reading without circulating. Yet the opposite asymmetry occupies an even more prominent place in certain Victorian literary genres—notably the bildungsroman and the memoir—which represented reading as systematically as they avoided any mention of the social transactions in which the book was enlisted or the material properties with which it was invested. The fact that a few instances of these genres continue to be reprinted and reread, while genres that acknowledged handling now look like repositories of jokes gone flat, and genres that theorized circulation look like depositories for dated didacticism, suggests how much twenty-first century culture values the first use over the second and third.

To ask how one use relates to the other two is also to ask how—even whether—books differ from other kinds of object. Where do books fit in a postal system that mandates different pricing for letters than for freight? What about newspapers, catalogs and advertising circulars, or books that contain nonverbal objects (herbaria, scrapbooks, tradesman's sample books)? When you display a book in your hands or on your shelf, are you implicitly claiming to have read it—and therefore, as often as not, lying? In what operations other than reading can books be enlisted? Is it legitimate to hide behind the newspaper, use an encyclopedia as a doorstop, turn a newspaper into fish wrapping, match the binding of your bible to your dress, fill a study wall with hollowed-out books, decorate a living-room table with intact ones that you have no intention of opening?

Are books likelier to be put to one or another of the three uses if they're free? What about if they're bought, borrowed, inherited, received as a gift from an acquaintance or as a giveaway from an organization? (In some quarters, the price of subsidized bibles was raised in order to prevent their being worth reselling for wastepaper; in others, to inherit, stumble upon, or even steal books was considered morally superior to buying them.) Do traces (verbal or nonverbal) left by past users increase or decrease the value of books (commercial or sentimental)? What should be done with printed matter when its contents go out of date?

Under what circumstances is it acceptable to annotate, extra-illustrate, cut up, disbind, rebind, reprint, recycle, or discard books? And when is it permissible to disperse, sell, or export entire library collections? What should be pulped (and how soon), what should be archived (and how long)? What relation do those persons responsible for interpreting and evaluating texts bear to those responsible for dusting or shelving books? And the formal corollary of that social question: why do Victorian writers develop such a rich language in which to name the manual gestures of holding, turning, and handling, with no matching lexicon to describe the mental act of reading?

Investigating these questions may help us understand the printed "before" against which so many twenty-first-century commentators measure their digital "after." We can learn, in particular, from the Victorians' struggle to articulate how far the power of books (for good and evil) depended on their verbal content, their material form, or the social and antisocial practices that they enabled and even prompted. (In the language introduced a moment ago: on their reading, handling, or circulating.) When we use idealized printed texts as a stick with which to beat real digital ones, we flatten the range of uses to which the book was put before digital media came along to compete with it. If we shift our gaze from the library to the kitchen and the privy, an ethnography that juxtaposes reading with handling and circulating can replace the Manichaean contrast *between* print and digital by distinctions *within* the uses of each. Where nostalgists today conflate the practice of disinterested, linear, sustained attention with the object that is the printed book—equating modular, scattershot, instrumental reading in turn with electronic media—secular novelists like Dickens, Eliot, Brontë, and Trollope assumed that absorption in the text required forgetting its medium. The ideal text was, as we say today, platform-independent; the ideal reader, binding-blind and edition-deaf (see Kirschenbaum). Evangelical tracts, in contrast, showed less interest in the words that the book conveyed than in the interpersonal transactions that served to convey it. Web 2.0 has lent new life to a question that Victorian missionaries first formulated: does the distribution of texts compete with, or piggyback onto, social relationships among human beings?

READER-UNRESPONSE

I was trained in the method known as "reception history." That enterprise shifts literary and intellectual historians' sights from writers to readers, from upstream arguments about a work's sources to downstream speculations about those other works that it influenced or spawned. The chapters that follow form a prototype for what might better be dubbed "rejection history." However much interest books have in being coveted, bought, hoarded, even stolen, a wide range of Victorian genres devote more attention to the energy expended refusing to read or own or touch or even refrain from destroying them.

The umbrella term "nonreading" encompasses an array of practices that have little in common except what they are defined against:

- novelistic narrators replacing the mental act of reading by the manual gesture of holding, in order to repudiate the omniscience that could penetrate characters' thoughts
- writers reducing the term "reading" to a metaphor for activities that involve the interpretation of something other than books, and books to a front for daydreaming or for ignoring others sharing the same physical space
- in the case of free print, refusing to vest time (or shelf space) where you have not chosen to invest money
- a sign of respect for the book—protecting it from wear and tear—or on the contrary an insult to the text: branding it unworthy of your own time and attention or, worse, delegating or relegating it to your social inferiors
- a feeling that you don't belong in its audience, whether your identity doesn't match its implied reader's or because you are too good (or not good enough) to rub elbows with others in its public. Or, more contingently: the sense that it's too soon, or too late, for you to shove your way among them—that a servant, for example, should hand today's news to his master without peeking, contenting himself instead with using last month's paper to wrap food. Or, more comprehensively: the sense that you do not fit into *any* text's audience, either because your place is to handle (or dust or fetch) books rather than to read. Or, more crudely, because you are unable to read at all—or because you *are* able to put the book to humbler uses, such as wrapping groceries in its pages.

"Nonreading" may be too negative a term to encompass one more scenario in which, whether or not a text is worth reading, the book becomes more valuable for some other purpose. The book's material properties trump its textual content when its value (whether for use or for resale) lies in attributes orthogonal to its legibility. This could be for aesthetic reasons, as when a book's textual content is judged particularly worthless *and* its material properties are judged especially valuable: the gap between the two yawns particularly wide, for example, in the case of coffee-table books and their early-nineteenth-century ancestor, the annual. The reason could be that one of those two axes looks more relevant to a particular situation; material value trumps textual value in times and places where paper is particularly scarce, including among the poor, in wartime, at moments when the raw materials fall short, or at times and places in which paper is heavily taxed or imports restricted. Or cultures in which the *idea* of the book signifies more than the content of any particular book: during China's Cultural Revolution, for example, burning formed a sign of hostility not just to a particular text's political message, but also to the social classes that were literate and inherited cultural goods. Or the moment of nonreading could be determined not by the history of a nation, but by that of a book: the point in its life span when its read-by date has passed and its pages are ripe for cutting, wrapping, and even wiping.

As late as 1711, Pope could gibe of a miscellany published by Bernard Lintot (and containing works of his own):

Lintot's for gen'ral Use are fit,
For some Folks read, but all Folks ———.

(Pope 280)

The couplet aligns the gap between the many books that are handled and the few that are read with the gap between the few who read books and the many who use them. To reconstruct the hermeneutics of handling is also to situate the book within a larger social world. Since the nineteenth century, activists and scholars alike have assumed that the place to look for the illiterate classes' relation to printed matter was reading aloud—that is, those moments where the information contained in newspapers overleaps their written medium. Pope directs our attention instead to the converse: those moments where the medium outlives its content.

By the following century, what Pope represents as a subset has become a contrary. At the very moment when the poor are learning to read, the rich are unlearning how to handle—are forgetting, as paper ceases to be taxed and new manufacturing methods substitute cheap wood pulp for expensive linen, how to assess the reuse potential and resale value of pages. Servants continued to eyeball how much animal gelatin had been used to "size" a page (determining whether liquids like ink, and later grease, would sink in or bead up); they knew, therefore, which pages were suitable for sealing food and which for absorbing dirt. Masters, in contrast, now noticed only whether the text was absorbing. Although all folks still ———, not all folks associated that activity with print: memoirists now described Queen Victoria visiting Cambridge "and saying, as she looked over the bridge: 'What are all those pieces of paper floating down the river?' To which, with great presence of mind, [Dr. Whewell, the master of Trinity College] replied: 'These, ma'am, are notices that bathing is forbidden'" (Raverat 34).

Between reading and wiping, a range of uses stretches: the social breadth to which Pope's "gen'ral" alludes is matched by (though not

THE TURF.

Elderly Clergyman (who was passing). "I'm very glad, Cabman, to see you Improving your Mind by reading during your spare time."

Cabby (with a Sporting Paper). "Improvin' my mind! I dunno. I backed this 'ere 'Oss all through last Season, and he never landed me once!—and I've follered 'im up, and now he's dropped me another Dollar on the 'Grand Int'national 'Urdle!'" *(Gloomily.)* "If yer call that Im——" [*The Parson retires!*

Figure I.1. "The Turf," *Punch*, 18 March 1882, 122.

always mapped onto) an equally wide spectrum of practices. If reading can serve different agendas—to save a soul, to form an identity, to do a job, to place a bet, to snub a spouse—handling figures in even more disparate activities.[8] Just as bibliographers have taught us that the changes among successive editions do not necessarily constitute decay, so the Victorian novel can teach us to distinguish absence of reading from absence of use.

Not all uses, however, were created equal. The Victorians plotted the book/text distinction onto every axis imaginable: temporal (new books get read, old books handled), sexual (the text as the province of male thinkers, the book as raw material for women's curlpapers or pie plate liners), generic (the text as the object of piety, the book as the butt of jokes), ethical (the text as an aid to selfhood, the book as a spur to self-ishness), social (the text as the business of intellectuals, the book of filthy rich bibliophiles or literally dirty rag-collectors), even disciplinary (the text as the purview of Skimpoleanly aesthetic sensitivities, the book of Gradgrindianly empirical plodders). All that cuts across these otherwise ill-assorted word-pairs is value: in each case, the text is aligned with which-ever term happens to be considered superior. A higher-order instance of that logic is that the text is associated with moderation, the book with extremes. In social terms, the professional middle classes' rejection of materialism left the book-object in the hands of effete gentry (the owners of country-house libraries as selfish hoarders), rich vulgarians (Manchester manufacturers' wives who chose books to match their color schemes), or poor illiterates (costermongers who priced a book by the absorbency of its pages). And in historical terms, book fetishism looked forward (to new technologies for facsimile reproduction and nouveaux riches furnishing their houses with bran-new bindings) as well as backward (to country-house collectors ignoring the post-1850 public libraries, or superstitious old women eating the pages of their bibles). What was true for users also held for things. Just as the very rich and the very poor, the excessively scholarly and insufficiently literate, were both imagined to be either above or below reading, so books were faulted as too cheap or too expensive. Terms like "penny dreadful" and "shilling shocker" took a low price as metonymic for literary worthlessness; more counterintuitively, mentions of perfumed or hot-pressed paper did the same with high.

CHAPTER SUMMARY

My study starts where Curtius's foundational survey of "the use of writing and the book in figurative language" leaves off: in intellectual terms, at the end of the Enlightenment; in technological, as the handpress era closed (Curtius). It ends with the midcentury legal and technological developments that cheapened paper, shortening its life cycle and narrowing its affordances. The boundaries of my subject, therefore, are at once technological, legal, and literary- (or sometimes intellectual-) historical. Changes in printing and papermaking technology; innovations in distribution systems; institutional changes in schooling, both sacred and secular; legal changes to copyright and to taxes on knowledge—even if these add up to a coherent narrative, they map less neatly onto the time line of literary history, itself complicated by gaps between production and reception. (No argument about the books the Victorians read can confine itself to texts composed by Victorians.)

The proper nouns that appear in this book's table of contents form a grudging concession to the unspoken rule that literary-critical monographs must title each chapter after a different author (or in books about a single author, a different text). Although it fits badly with anonymous texts and worse with those (even thicker on the ground in this study) whose authors are named but whose names command no recognition, the convention humors our own protocols of reception—as well as of selection and rejection. With the exception of professional reviewers, people reading such books often skip straight from the introduction to whichever chapter discusses a text or author that the reader himself happens to be reading or writing about.

In that spirit, some itineraries. Because chapter 1 intertwines an introduction to Victorian debates about media with a survey of (and polemic about) the relation of book history to literary-critical theory and practice, readers interested in methodology should begin at the beginning. Those more interested in the primary texts, however, can easily enough cut straight to the more accessible and more detailed case studies of chapter 2. Husband-wife relations come to the fore there, parent-child in chapter 3, master-servant in chapters 4 through 6. Scholars of reading aloud (and silent listening) may want to skip ahead to chapter 6; bibles figure most prominently in chapters 4, 5, and 6; newspapers in chapters 2, 6, and 7.

Why can realist novels represent the book (the second chapter asks) only at the price of reducing reading, quite literally, to an act? And why does representing reading from the inside (as do the texts discussed in the third) entail abstracting the visible book? What models of causation (the fourth asks) have the nineteenth-century bildungsroman and it-narrative bequeathed to twentieth-century bibliographers and twenty-first-century book historians? The fifth chapter turns to the circulation of free and subsidized print—especially junk mail and religious tracts—among owners and borrowers, givers and receivers, readers and handlers, preservers and destroyers. The sixth asks what relationships the Victorians expected particular copies of a book to establish among their users—whether concurrent, as in reading aloud or subscribing to the same periodical, or sequential, as in secondhand books or association copies. Ending with the end of the book's life, chapter 7 explores the relation between old texts instantiated in new books (reprinting) and new texts transmitted via old books (marginalia, binder's waste, and paper recycling).

Books don't simply mediate a meeting of minds between reader and author. They also broker (or buffer) relationships among the bodies of successive and simultaneous readers—or even between one person who holds the book and others before whose gaze, or over whose dead body, she turns its pages. Ambivalence about circulation runs through these different case studies: untouched books figure as prisoners or wallflowers or clotted blood, but books subjected to too many readers are compared to worn-out prostitutes or knackered horses. The same fictions that credit texts with marking minds blame handlers for marking books. Conservative and radical fiction agree in classifying books as a special category of commodity that can be alienated only at the price of disloyalty. Yet one deplores, and the other celebrates, the intimacies and antagonisms that the book establishes between buyer and seller, lender and borrower, or even between strangers who handle the same piece of paper unbeknownst to one another. Circulation affects not only relations between persons and books, that is, but also between one person and another.

A second tension runs between the book's powers to unite and to divide. Books can link their successive readers, owners, and handlers, whether across classes (as in tracts distributed by the rich to the poor, or papers that find their way from the study to the kitchen) or even (as in the case of "association copies" bought or inherited) across the line that divides the living from the dead. Books could just as easily, however, separate individuals (like the husbands and wives who hide from one another behind books and newspapers in chapter 2, or the children of chapter 3 who hide behind books, and within texts, from the adults who jolt them back to their surroundings by hitting them with a book), or separate classes (like the masters and servants of chapters 6 and 7, who handle the same book or newspaper but for different purposes and at different moments of its life cycle). It's worth emphasizing that this distinction between the book as bridge and the book as wedge does *not* map onto the dichotomy between reception and rejection: on the contrary, withholding a book can assert a relationship (think of a parent denying a child access to a book) as easily as bestowing a book can sever it (the bookseller who gives that child the book he requests is disowning any more personal responsibility for the child's morals).

The tension between commonality and distinction cuts across genres as different as circulating-library triple-deckers representing middle-class couples and didactic tracts written for and about servants. Midcentury middle-class fiction substituted power struggles within the middle-class family (chapters 2 and 3) for more public debates about working-class literacy (chapters 5 and 6). Yet even as the antagonists in these battles of the books shift from master/servant to husband/wife and stepfamily/stepchild, the question of who has "business in the library" (a phrase echoed across these different contexts) continues to determine who stands inside and outside of the "family"—whether in the older sense of an internally stratified economic entity or the newer affective unit divided by age and gender. In both cases, the self-made reader—whether "made" as a middle-class child develops interiority, or as a working-class person climbs the social ladder—may be represented either with empathetic intimacy or with satirical distance, and this generic choice implies an ideological choice between embedding the book within, or counterposing the text to, social structures.

The self-made reader in turn implies a self-propelling text: to acknowledge how books reached one's hands is to recognize one's dependence, in every sense of that now old-fashioned word. Victorian secular fiction deploys two genres of required reading—school textbooks (chapter 3) and religious tracts (chapters 5 and 6)—as foils to its own claim to be freely chosen, even secretly coveted, hoarded, begged, borrowed, or stolen. Tracts are to the mid-Victorian novel what romance was to its predecessors: the inscribed genre against which it defines itself. Institutions like school and church stand opposite the novel's putative market, imagined as an aggregate of independent (even rebellious) individuals. By representing teachers foisting grammars, dictionaries, and prize books upon middle-class children, and tract-distributors doing the same to working-class men, the novel presents itself as a commodity driven more by demand than by supply. A different novelistic subgenre, the Evangelical it-narrative (chapter 4), substitutes divine providence as the motor driving the circulation of books, a logic borrowed, surprisingly enough, by the resolutely anticlerical Henry Mayhew to structure his account of paper recycling (chapter 7).

The subgenres discussed in Part I grope for ways to discuss the circulation and handling of books while bracketing their textual content. Those comedies of manners that I call "behaviorist" perform that substitution lexically (by substituting manual phrases like "turned the page" or spatial phrases like "sat with a book before him" for the mental verb "read") as well as thematically (by representing characters going through the motions of reading or even pretending to read). In novels that more dogmatically prize psychological depth, however, the child who internalizes the content of books at the expense of any awareness of their material or commercial properties stands opposite the adults who throw, display, and sell books with no interest in actually reading them. Seen from the inside, a prompt for absorption; from the outside, a prop for avoidance. Does the book compete with human friendships (as when the metaphorical "companions" that populate a man's library crowd out his wife and children) or enable them (as when the loan of a bible provides a missionary an excuse to enter a home)?

The book as barrier (Part I) gives way to the book as bridge (Part II): reading can create interpersonal bonds (in the sense of constraint as well as of intimacy), but so can using and choosing books—for oneself, or on behalf of others. Over the course of the nineteenth century, new commercial developments (including the introduction of new raw materials for papermaking), new political arrangements (notably the removal of "taxes on knowledge"), and new distribution infrastructures (ranging from the penny post to the missionary press) changed books from a scarce resource to a storage problem. Printed matter came to be figured as a chain, for better and for worse: what linked its users also burdened them. Too much information, too many readers, too much paper: Part II explores the first problem through the rise of junk mail and subsidized tracts (chapter 5), the second through the shift from masters' concern about servants' reading to public library patrons' concern about one another's handling (chapter 6), the third through the fall of paper recycling (chapter 7).

Subsidized Evangelical tracts and middle-class three-volume novels alike shifted their attention from the individual reader to the social and economic transactions that link one user to another (with or without their consent). If hiding behind a book could undercut the compulsory intimacies expected of family members who shared the same domestic space (whether the husbands and wives of chapter 2 or the parents and children of chapter 3), conversely peeking into one's master's bible (chapter 4) or procuring novels on the sly from one's servant (chapter 6) could undermine the distance that unequals were expected to maintain.

Like religious relics, books link us not just to an author but to those who have touched them before: think of Barack Obama's being sworn in at his inauguration on Abraham Lincoln's bible or, two years earlier, Keith Ellison's being sworn in to the House of Representatives on Thomas Jefferson's Koran. Those transitive relationships sound cosy enough: conventionally compared to a friend or companion, the book can also broker friendships, even between the living and the dead. By the nineteenth century, however, the cheapening of both paper and literacy opened the less pleasant possibility of bumping into one's social inferiors within the readership of a particular book, or the handlership of a particular copy of a book. The traditional fear that a text might poison its readers' minds was now joined by a newer anxiety that poor, sick, or dirty fellow handlers might infect their bodies.

The sequential uses of a single copy of a book embodied in "association copies" find their converse in Benedict Anderson's famous analysis of strangers bonding through simultaneous newspaper reading. Yet if we shift our sights from text to book, the relationships enabled by print look more negative—a prop for avoiding persons in the same space, as easily as communicating with strangers at a distance. And if we look beyond reading to handling (an activity that occupies a larger fraction of any newspaper's life cycle), it becomes clear that while the meaning of texts changes as new generations reinterpret them, the relation between page and paper changes as the former ages. In Henry Mayhew's ethnography of the wanderings of books from class to class and hand to hand, we find a media theory that lumps paper together with humbler commodities while insisting on the power of even illiterate users to invest even papers past their read-by date with fresh value.

UNCOMMON READERS

Like most literary-historical arguments, mine has both a corrective and a creative ambition. In negative terms, it seeks historical and critical distance from the heroic myth—whether Protestant, liberal, New Critical, or New Historicist—that makes textuality the source of interiority, authenticity, and selfhood (Raven, *The Business of Books* 132, 377). In more positive terms, it seeks to recover stories that this myth overwrites: stories about women, children, and working-class or non-European men who remained sensitive to the material affordances of books and, therefore, to the stories in which books themselves figured as heroes. Some of the following chapters will trace antibookishness back to a particular time (around 1850) and a particular genre (the secular middle-class bildungsroman). Others excavate Victorian alternatives to a worship of the text that demonizes the book: now-forgotten genres and subcultures whose challenges to that model may be worth fishing out of the dustbin—no: the glass-fronted bookcase—of history.

Within a culture where book is to text as outside to inside, secular middle-class fictions and Evangelical tracts alike make the relation between those terms a surrogate for the relation of the material world to the inner life—whether that life belongs to their characters or to their readers. Printed matter raises ethical questions (how much or little should one care about the look of books?) as much as formal ones (how, and how fully, can a mental act like reading be represented?). Identifying a deep structure underlying different representations of the book, however, doesn't mean lumping "the" Victorians into some monolithic mass. Multiple fault lines separated those narratives and essays that celebrated the spread of ideas from those that mocked the circulation of paper: political and sectarian and economic and educational positions of readers, writers, and publishers; size and format and pricing of books; genre of texts. It's hardly surprising, for example, that Evangelical Protestants produced and consumed texts that figured reading rather differently from those that emerged from Catholic or freethinking subcultures, or that those who favored or feared the social mobility of persons developed different vocabularies in which to discuss books' movement through space and across social ranks, or that proponents of individualistic economic or religious models valued silent reading as highly as others condemned it. What held for discourses applied less evenly to practices: each subculture developed its own ways of showing books off or hiding them away, distributing or hoarding, alienating or personalizing, bequeathing or disposing of, noticing or taking for granted.

Such sectarian and political identities crosscut a second determinant of attitudes toward the book: genre. Cast by circulating-library novels as a buffer between intimates (chapters 2 and 3) and by subsidized tracts as a bridge among strangers (chapters 4–6), the book could figure in the Evangelical press as a picaresque wanderer (chapter 4) or in radical journalism as the protagonist of a providential plan (chapter 7). And a third: narrative mode. First-person accounts showed the individual reader transcending the constraints of space, time, age, and social class—whether that individual was the middle-class child through whom the bildungsroman was focalized, the working-class autodidact of rags-to-riches memoirs, or the narrator of an American slave autobiography. The counternarrative that emphasized the material, social, and commercial properties of paper, in contrast, clustered in third-person comic and anecdotal genres, distanced by the Olympian irony of an omniscient narrator.

The two halves of the book correspond, therefore, not only to different genres and different classes of audience, but also to different models of literacy. Middle-class bildungsromans, like working-class autodidacts' autobiographies, frame reading in terms of individual agency, self-fashioning, even transgression. To read a subsidized tract, in contrast, was to engage in an interpersonal transaction. In that sense, surprisingly, Evangelical tracts (chapters 5 and 6) had less in common with those bildungsromans that secularized the Christian conversion narrative (chapter 3) than with social satires and comedies of manners that cast books as props in etiquette dilemmas (chapter 2).

Yet what divided these genres was ultimately less what powers they ascribed to the book than what value judgments inflected that ascription. All three associated autodidacticism with the text, formal education with the book. If the text guaranteed upward mobility; the book made users placeable. The text signifies individual freedom, the book social determinism; the text generates empathy among different classes and genders, while the book marks differences of rank and age. It's logical enough, in that context, that the religious tracts produced by anti-Jacobin propagandists should celebrate the moment of elders and betters handing books to the young and the poor, while more secular middle-class fictions instead praised texts for propelling themselves into the hands of protagonists (often, again, young and poor) who were badly treated by other human beings. By extension, the protagonists of the great Victorian bildungsromans are characterized less by their love of texts than by their hatred of books—less by immersion in verbal content than by indifference, or even repugnance, to its material container.

Like nonfictional accounts of individual self-improvement and national progress, serious fictions marketed to middle-class circulating-library patrons vest the text with the power to liberate and individuate. They associate the text with mobility, whether through the power of words to move across media (the cheap reprint's claim to be functionally interchangeable with the finest folio models the equality of their respective owners), the power of the author to move through space and time (to be read although dead, to do his work "on the top of a mountain or in the bottom of a pit"), or the power of the text to change the reader's identity (through empathy with fictional characters) and social status (whether by transcending one's social and physical disabilities, or by forging relationships with fellow readers) (Anthony Trollope, *An Autobiography* 209). Whether in the privy or on the sofa table, among collectors or bibliomancers, a book that was placed—either socially or spatially—was always a book not being read.

A fuller ethnography or phenomenology of Victorians' interactions with the book would need to approach a wider range of genres and formats from a wider range of methods. My reliance on a few pieces of printed prose that have survived in twenty-first-century research libraries positions me to offer little more than an account of competing ideologies surrounding the book in a few numerically unrepresentative genres. Yet "ideology" sounds at once too lofty and too dry (or, in a more Victorian language, too coarse) to do justice to the visceral energies driving my subjects to distance themselves from some uses of books and identify themselves with others. In the end, the most interesting question to ask of these hands now quiet may be not what they felt about the book but why they felt so much. To grope our way back into their intellectual and emotional and ethical investment in paper; their urge to cast written matter in etiological narratives and interpersonal dramas; the leaps of faith and logic that pressed trivial decisions about to whom to hand a tract, or on which shelf to stick a volume, or at what angle to hold a newspaper, into the service of hopes and fears and theories and hunches—this exercise may provide a chance to work through the contradictions of our own media theory and practice.

CHAPTER 1

Reader's Block

How to Handle Reading

BOUGHT, SOLD, EXCHANGED, transported, displayed, defaced, stored, ignored, collected, neglected, dispersed, discarded—the transactions that enlist books stretch far beyond the literary or even the linguistic. Frustration first made me wonder where that range begins and ends, for among all those uses, reading elicits the most curiosity and leaves the least evidence. There's a reason that book historians have gravitated toward tearjerkers and pornography: like dolls that cry and wet their pants, past readers come to life through secretion.[1] Yet with the exception of the happy few who work on genres that elicit a measurable somatic response, any reception historian will sooner or later be maddened by the low proportion of traces left in books that are verbal. For every pencil mark in the margin, ten traces of wax or smoke; for every ink stain, ten drink spills.

The book can be used as a napkin for food, a coaster for drink, a device for filing, or (especially in eras where paper was expensive) a surface on which to scribble words only tenuously related to the print they surround. As late as 1897, a manual titled *The Private Library* still needed to sneer that "books are neither card-racks, crumb-baskets, or receptacles for dead leaves" (Humphreys 24). In earlier eras, traces of the hands through which a book had passed formed an expected and even valued part of its meaning; over the course of the nineteenth century, that practice gradually retreated to particular subcultures: botanists using encyclopedias as devices to store and organize pressed flowers; hobbyists "Grangerizing" texts with carte-de-visite daguerreotypes, or, less systematically, books—and not only cookbooks—bearing, Hansel-and-Gretel-like, a trail of crumbs (H. J. Jackson, *Marginalia* 186; Garvey).

Mental actions prove harder to track than manual gestures, human traces that are not intentional, let alone textual, let alone literary. From evidence of reading to nonevidence of reading to evidence of nonreading: those bodily acts that both accompany and replace reading, whether licking a page or turning down a corner, should provide historians of the book with more than a consolation prize. Like the dog that didn't bark in the night, the book with uncut pages constitutes evidence too. As we'll see in chapter 4, such negative evidence can carry forensic weight, as when a bible's pristine condition "bears witness" against its owner. It was precisely in order to stave off such testimony that Flann O'Brien proposed (tongue in cheek) a "*Buchhandlung*" service to break in libraries bought by the yard.

> The spines of the smaller volumes to be damaged in a manner that will give the impression that they have been carried around in pockets, a passage in every volume to be underlined in red pencil with an exclamation or interrogation mark inserted in the margin opposite, an old Gate Theatre programme to be inserted in each volume as a forgotten book-mark . . . , not less than 30 volumes to be treated with old coffee, tea, porter or whiskey stains. (22)

In my nightmares, these are the books I'm studying.

Scholarly populism leads logically enough to inverting the traditional focus on production over use: even outside of textual studies, a historian of technology can axiomize that "the majority have always been mainly concerned with the operation and maintenance of things and processes; with the uses of things, not their invention or development" (Edgerton xv). For scholars as for the secular novelists discussed in the next two chapters, however, reading is harder to document than handling—let alone than writing. If book history began as a supply-side enterprise focused on publishing and printing, it may be because consumption generates less of the hard evidence that can lift a discipline out of humanistic impressionism into social-scientific rigor.[2] Conversely, even as literary critics shifted their focus from the authorial exception to the readerly rule, reader-response theorists and reception historians alike continued to study the text as a linguistic structure, at the expense of the book as a material thing. Only in the past few decades have those developments converged. On the one hand, book historians turned their attention from production to circulation, from printing to reprinting, from genetic criticism of authors' manuscripts to cultural criticism of readers' annotations; on the other, reader-response theorists have followed the rest of the literary-critical profession on its trek from the abstract to the concrete, from the "history" represented or refracted within the text's verbal content to the "history" of the book itself.[3] In both camps, though, an investment in textual interpretation that runs as deep among intellectual historians as among literary critics has distracted both from the wide range of nontextual and sometimes even noninterpretive (which doesn't mean noninterpretable) uses to which the book is put. Is book history a subset of textual interpretation or vice versa?

LITERARY LITERALISM

What exactly would it mean to study books without privileging reading? Any answer remains slippery, even (or especially) for scholars who, by definition, spend their lives surrounded by books. Within literary theory, even as successive aspects of mid-twentieth-century symptomatic reading come under attack—its adversarial stance (Eve Sedgwick's "recuperative reading"), its professional self-differentiation (Michael Warner's "uncritical reading"), its granularity (Franco Moretti's "distant reading"), the wedge it drives between surface and depth (Elaine Freedgood's "metonymic reading" and Sharon Marcus's "just reading")—a familiar noun anchors each new adjective (Sedgwick; Warner, "Uncritical Reading"; Moretti, *Graphs, Maps, Trees*; Freedgood; Sharon Marcus, *Between Women*).[4] It would be a false parallelism to dub the method illustrated in the pages that you're about (I hope) to read "logistical reading," for my target is not a particular kind of reading so much as the primacy of reading itself.

Late twentieth-century literary critics are not alone in overinvesting in reading. Contemporaneous "disciplines from political science to anthropology, and from economics to legal and juridical studies," in Fredric Jameson's words, took "as [their] model a kind of decipherment of which literary and textual criticism is the strong form" (297). Nor is this phenomenon limited to that intellectual-historical moment. Although literary theory lasted barely more than a decade as queen of the disciplines, its reign was both foreshadowed and outlasted by a more diffuse tradition in which interpreting the book of Nature (or, in Clerk Maxwell's metaphor, the magazine of nature) was assimilated to reading—a verb that itself began as a synonym for "interpreting" before it narrowed into its current textual sense.[5] The book of nature, but also the book of culture: in *The Stones of Venice*, Ruskin enjoined us to "read the sculpture. Preparatory to reading it, you will have to discover whether it is legible (and, if legible, it is nearly certain to be worth reading) . . . Thenceforward the criticism of the building is to be conducted precisely on the same principles as that of a book" (230). From 1918 onward, as the *Oxford English Dictionary* reminds us, a shirt can even be "read" for lice.[6]

In 1865, *Our Mutual Friend* already mocked the indiscriminancy of what Eugene Wrayburn called "that very word, Reading, in its critical use": "An actress's Reading of a chambermaid, a dancer's Reading of a hornpipe, a singer's Reading of a song, a marine painter's Reading of the sea, the kettle-drum's Reading of an instrumental passage, are phrases ever youthful and delightful" (Dickens, *Our Mutual Friend* 605). A novel that obsessively plays the symbolic value of literacy against the ubiquity of nonalphabetic "signs" also juxtaposes a reductively material perspective on book-objects (as when the narrator's description of the Veneerings' library stops short at the "backs of the books" in their "bran-new bindings") with the more expansive metaphor that allows Eugene to speak of Mortimer's "reading of my weaknesses" and the narrator to describe Mrs. Lammle "reading" Twemlow or Riah learning to "read" his master's face (605, 263, 605, 281, 636). In the second case, to "read" means not only to translate the twitches of an eyebrow, but also to vocalize the wishes being extrapolated from those signs, as if an expression could be read aloud. The same metaphoric drift erases the difference between book and corpse (as when the narrator compares the morgue to a whitewashed library) or book and flame (as when Charlie compares the way his sister looks at the hearth to the way he looks at printed pages).[7] Dickens expands on Bagehot's remark that in his novels "London is like a newspaper" by adding that "the streets being, for pupils of [Charley's] degree, the great preparatory Establishment in which very much that is never unlearned is learned before and without book" (Bagehot 468).

A century before the rise of cultural studies, then, the alphabetic practices inculcated by formal schooling supplied a template for everyday observation, as well as for what Lorraine Daston has called "other ways of making sense of objects quite different from the manuscript or printed page—the morphology of a plant, the trajectory of a comet, the slide under a microscope, the 'reading' of an instrument. This would have been especially the case for those who—for reasons of class, gender, and the cultural status of literacy—would have learned bookish skills before or to the exclusion of manual ones" (444). Daston achieves more distance than Jameson from the logic that both describe, for the word "before" slyly inverts the received wisdom that positions manual skills as a given, textual operations as a supplement. Yet in casting physical gestures as the alternative to mental operations, she leapfrogs over the manual dimension of reading itself: books handled, pages turned. Like Jameson's "model," moreover, Daston's "template" remains double-edged: both endow written texts with exemplarity at the price of stripping them of specificity.

After the cultural turn, however, that age-old balance of trade shifted. Literary critics now look to other fields not for raw materials but for methodological tools. Where the humanistic social sciences once borrowed literary-critical tricks to interpret nontextual objects ("Reading a Mid-19th-Century, Two-Cylinder Parlor Stove as Text"), literary critics today mine other disciplines—bibliography, history of science, even archaeology—for a vocabulary in which to describe the nontextual aspects of a particular category of material object: books.[8] Instead of "reading" sewer systems, critics now smell leather bindings.[9] Scholars who once "read" the stock market now tabulate paper prices.

In intellectual as much as literary history, the hermeneutics of suspicion has given way to a poetics of deflation.[10] Oxymoronic subfields like "thing theory" and oxymoronic titles like Robert Darnton's *The Business of Enlightenment*, D. F. McKenzie's *Printers of the Mind*, or Elaine Freedgood's *The Ideas in Things* drag ideas into the marketplace, the mind down to the level of the body. In the process, scholars change from the freest of associators into the most slavish of idiots savants. In a discipline that prides itself on discerning hidden depths, superficiality shocks like a purloined letter. Even the repressed is dragged down to earth when Henry Petroski notices that in the list of goods that he brought to Walden Pond, Thoreau omits the one whose trace it constitutes: a pencil.[11] Once, a writing instrument would have stood for something less speakable; now, self-reference finds its home in an everyday tool.

A dogged or even mulish taste for the mundane, the contingent, and the simpleminded finds its only aesthetic outlet in puns. Writing from the "margins" gave way to writing *in* the margin (adversaria provide much of the richest book-historical evidence). The old hermeneutic refrain "it is no accident that" was shunted aside by a new interest in paratextual "accidentals." Isabel Hofmeyr reinvested postcolonial catchwords like "stereotype" and "cliché" with their typographical weight (Hofmeyr 105). Research on the mechanics of writing put the bureau back into bureaucracy; research on the embodied labor of data entry put the digits back into digital; geographers took "space" to refer to the layout of the page, not a concept represented within it.[12] And Peter McDonald retranslated the slogan "il n'y a pas de hors-texte" into a claim about tipped-in pages ("Ideas of the Book" 222–23). By foregrounding the technical sense of Derrida's term, McDonald defines the text by contradistinction to the book, not the world. The dethronement of reading requires an assault upon metaphor.

Bookish Bathos

At one extreme, those professions—from sociologists to linebackers—who dignify their job by claiming to "read" something other than a text; at the other, those populations—from decorators to bibliographers—whom others mock for putting books to some use other than reading. This isn't to say that individuals fall squarely into either of those camps. Even as Dickens's writings (chapter 3 will suggest) aspired to the first position, his decor tilted toward the second. *Our Mutual Friend*'s critique of the extension of "reading" to cover matter other than print inverts its author's interest in unreadable books. Veneering-like, Dickens lined his study with dummy spines, for which he composed titles like *History of a Short Chancery Suit* (in twenty-one volumes) or *Cat's Lives* (in nine).[13] Pointing upward to the aristocratic tradition of trompe l'oeil libraries, the dummy spines also point backward and downward to a working life that began in the pasting of labels, not even onto the spines of books, but only onto blacking-bottles.

Like the twentieth-century use of "reading" to designate nonverbal operations, the literalist backlash against that metaphor has a long history, one that stretches from Dickens's dummy spines to present-day literary theorists' wordplays and present-day software designers' jokes. In a 1995-era user interface dubbed Microsoft Bob, where icons of doors, rolodexes, and typewriters could be clicked on—prefiguring the metaphors of "folder" and "notebook" that now order our virtual desktops—only the Encyclopedia whose bound volumes were displayed on the bookshelf was inert. A click brought up the placeholder message "Note: This is a decorative object. It does not start any programs or do anything special." Once "content" becomes available online, the only place left for its erstwhile containers is the coffee table. Two decades later, one wall in Google's Cambridge office is lined with hollowed-out spines of disbound books, like taxidermists' trophies attesting a successful slaughter. In a nod to the tradition of dummy spines, a cluster of yellow spines has been sliced from Wiley's "For Dummies" series.[14]

To notice that books are things is, literally, for dummies. My corny pun finds its precedent in a midcentury issue of the *Dublin University Magazine*:

> Meeting one day an author newly-fledged, and greatly elated by the hit of his literary first-born, [Daniel O'Connell] shook him heartily by the hand.
>
> "Well, my dear fellow, I congratulate you sincerely on the success of your book; I have seen something extremely good in it."
>
> "What was it—eh?" said the delighted author, rubbing his hands and blushing.
>
> "A mutton pie, my dear fellow," replied the Liberator, chuckling slyly. ("Railway Literature" 280)

Pivoting on conceptual and spatial senses of "in" while reendowing "something" with its literal force, the joke casts the book's content as food, not words—and its users as bodies, not minds. That ethnic slur draws on a long tradition of Irish bulls. The fall from text to book can just as easily, however, be pinned to gender. When the title character of F. E. Paget's neoquixotic *Lucretia; or, The Heroine of the Nineteenth Century* (1868) burns the house down in "volumes of smoke" by reading novels that a male character terms "inflammable trash," the misogynistic joke hinges on the tension between figurative and literal meanings of "volume" and "flame" (18, 22). Or, if not to gender, to rank, as when a cockney clerk remarks that a servant finding the fragments of a letter lying in the grate "must have been very much gratified with the warmth of the epistle" (Crowe 76). Or, if not to rank, to race, as in an American jokebook of 1871:

> ARMY CHAPLAIN. "My young coloured friend, can you read?"
> CONTRABAND. "Yes, sah."

ARMY CHAPLAIN. "Glad to hear it. Shall I give you a paper?"
CONTRABAND. "Sartain, Massa, if you please."
ARMY CHAPLAIN. "Very good. What paper would you choose, now?"
CONTRABAND. "Well, massa, if you chews, I'll take a paper ob terbacker.
    yah! yah!" (*The Railway Anecdote Book* 213)

Or to age: when Mr. Brownlow observes Oliver Twist surveying the books lining the walls of his study, the workhouse boy notices what a propertied adult doesn't.

> "You shall read them, if you behave well," said the old gentleman kindly; "and you will like that, better than looking at the outsides, that is, in some cases; because there are books of which the backs and covers are by far the best parts."
>
> "I suppose they are those heavy ones, sir," said Oliver, pointing to some large quartos, with a good deal of gilding about the binding.
>
> "Not always those," said the old gentleman, patting Oliver on the head, and smiling as he did so; "there are other equally heavy ones, though of a much smaller size. How should you like to grow up a clever man, and write books, eh?"
>
> "I think I would rather read them, sir," replied Oliver.
>
> "What! wouldn't you like to be a book–writer ?" said the old gentleman.
>
> Oliver considered a little while; and at last said, he should think it would be a much better thing to be a bookseller; upon which the old gentleman laughed heartily, and declared he had said a very good thing, which Oliver felt glad to have done, though he by no means knew what it was. (Dickens *Oliver Twist* 107)

Insides and outsides of books, figurative and literal senses of "heavy," writing and selling: the dichotomies mapped onto the difference between gentlemen and paupers or adults and children break down only in laughter.

No use claiming novelty for these jokes: as we'll see at the end of this book, their lineage stretches back to Roman satire. But they cluster in different genres at different moments. In the early nineteenth century, bookish puns migrate from the book review, to the comic press, to the triple-decker realist novel. In the genre that (Jakobson tells us) replaces metaphor by metonymy, the reliteralization of dead metaphors takes on a particular force: the novel's formal partisanship of the literal over the figurative and the concrete over the abstract finds its strongest thematic corollary in the bathetic substitution of material book for verbal text (Jakobson).[15] In *Vanity Fair*, the illustration of George lighting his cigar with a letter from Amelia is captioned "Lieutenant Osborne and

His Ardent Love-Letters." In the figurative sense that the caption leads us to expect, the adjective would have implied romantic passion; in the literal sense that sinks in once we see the illustration, practicality if not coxcombery.

In a culture where page is to paper as ideal is to real, the novel will establish its realism not only by contrasting the content of high-flown literary texts with a more mundane reality (the older quixotic move), but also by replacing textuality *tout court* with a materiality that, like charity, begins at home—that is, that begins with the book in our hands.[16] Generic disillusion traditionally tracks social debasement: romance is to the real as Dulcinea to a cowherd. In the nineteenth century, the lower orders reject abstraction as much as idealism.

The prominence of servants in these jokes reflects their combination of material access to, and intellectual or political unfitness for, literature. The *Yellowplush Papers*, which *Fraser's* commissioned from Thackeray in 1837, are narrated by a "littery" footman who quotes an Irish journalist praising the Cabinet Cyclopaedia as a "litherary Bacon."[17] Thackeray's pun on "beacon" can replace metaphorical enlightenment by literal food, or literature by litter, only because it occurs in the mouth of an Irish hack as quoted by an English servant. From study to kitchen, from lofty brainwork to footmen below-stairs: in ventriloquizing a servant literalizing his masters' language, Thackeray draws on the occupational puns already elaborated in an 1830 pamphlet satirizing the Society for the Diffusion of Useful Knowledge's ambition to publish for the lower orders.

> Shakespeare and Milton they supply
> That those who run may read;
> A circulating library
> It may be call'd indeed.
> . . .
> Meanwhile the butler, worthy man
> So snug o'er his o-port-o,
> Enjoys the 'life of sherry-dan,'
> Appropriately in quarto.
>
> (Moncrieff 16, 19, 32)

Each stanza displaces an intellectual abstraction by its material corollary—whether culinary or bibliographical hardly matters. In overlaying the mind/body distinction with the passage from figurative to literal language, the poem associates lower-class pseudoliteracy with the book's physical format and commercial transmission. It's true enough that the SDUK literalized the metaphor of "enlightenment" when it campaigned against window tariffs as a material obstacle to reading in working-class homes: the pun becomes a political platform.

Mind/body puns proliferate at the moment when those classes who use books for pie lining or sandwich wrapping were beginning to identify themselves as readers. F. B. Doveton makes larder to library as literal to figurative:

> I lost my <u>Bacon</u> t'other day—could anything be harder?
> My cook had taken it by stealth—I found it in the <u>Larder</u>.
>
> (21)

Meat links readers' bodies with books' binding more than would, say, the observation that *Leaves of Grass* is printed on esparto grass. The digital-era metaphor of "spam" can be traced back to the era of pigskin bindings: even more than Lamb's name, Bacon's lent itself to cheap jokes. Irving Brown's "How a Bibliomaniac Binds His Books" ends thus:

> I'd like my favourite books to bind
> So that their outward dress
> To every bibliomaniac's mind
> Their contents should express.
> . . .
> Intestine wars I'd clothe in vellum,
> While pig-skin Bacon grasps . . .
> Crimea's warlike facts and dates
> Of fragrant Russia smell;
> The subjugated Barbary States
> In crushed Morocco dwell.
>
> (G. White 21–22)

When Victorian essayists quote Bacon's aphorism that "some books are to be tasted, others to be swallowed and some few to be chewed and digested," they reverse the logic of the dummy spine that Thomas Hood devised for a library staircase at Chatsworth: "Pygmalion. By Lord Bacon" (W. Jerrold 258). Bacon had changed the tongue from an organ that literally affects the book (licking a finger before it turns the pages, for example) to the vehicle of a metaphor for disembodied mental acts. Hood changed "Bacon" instead from the name of a great mind to the name of an animal's body.

Because puns on bookbinding pit materialist against idealist conceptions of culture, my insistence on belaboring the obvious simply follows the cue of my primary sources. The same could be said of my discipline as a whole: just as Victorian puns prefigure the tension later developed by critics' plays on words like "stereotype" and "hors-texte," so Victorian realist fiction shares its temperamental cast with late twentieth- and early twenty-first-century book history. Both are detail-oriented, business-minded, and petty; both are called upon to integrate descriptions of material details with generalizations about social institutions; both are inclined to privilege the mundane over the ideal, the local over the transcendent, the concrete over the abstract. The overrepresentation of realist fiction among book historians' case studies betrays a craving for role models.

If we recognize twenty-first-century book historians as the heirs to the realist novel, then twenty-first-century literary critics look more like heirs to the sermon. From Protestant theology, secular explicators have learned to prize spirit over matter—and, by extension, the inwardness of selves produced by reading over the outward circumstances of bodies handling books.[18] Where the realist novel found its foil in Evangelical tracts, book historians could find theirs in close reading.

ANIMAL SPIRITS

As abstract is to concrete, common are to proper nouns: Bacon metonymically bound in pigskin or "crushed Morocco" reduced from the name of a country to a piece of leather. No accident that the person who anathematized "things in book's clothing" was named Lamb. He alluded at once to rag paper made from old clothes and to the sheepskin that books as well as wolves were covered in. Puns on animal names remind readers that most European books were bound in animal skins, inscribed on parchment or vellum, or held together with glue made from dead horses. (Cultures with different attitudes toward animal by-products, notably Hinduism, developed quite different methods of manufacture [Trivedi 26].) In 1900 alone, Oxford University Press's binderies used the skins of one hundred thousand animals (Ledger-Lomas 331; Holsinger 619). When La Bruyère sneered in 1688 at a man who "calls a tannery his library," he implied that the out-of-body experience that is reading requires forgetting those animals' corpses (La Bruyère 315). Yet for buyers thumbing a British and Foreign Bible Society catalog that listed bibles bound in calf, sheep, or sheepskin artificially grained to resemble calf, the book remained inseparable from the body (*Fifty-Sixth Report of the British and Foreign Bible Society*).

The book's dependence on animals' bodies would continue to generate black humor as late as the publication of *Animal Farm*, where after the aging workhorse Boxer is taken away—just days shy of the retirement that he plans to spend in pasture "learning the remaining twenty-two letters of the alphabet"—another horse who has gotten further in the alphabet notices something strange:

"Fools! Fools!" shouted Benjamin, prancing round them and stamping the earth with his small hoofs. "Fools! Do you not see what is written on the side of that van?"

That gave the animals pause, and there was a hush. Muriel began to spell out the words. But Benjamin pushed her aside and in the midst of a deadly silence he read:

"'Alfred Simmonds, Horse Slaughterer and Glue Boiler, Willingdon. Dealer in Hides and Bone-Meal. Kennels Supplied.' Do you not understand what that means? They are taking Boxer to the knacker's!" (Orwell 113)

Designed to ensure immortality to authors and disembodiment to readers, the written word here serves as a reminder that books themselves are made from corpses.

The leather binding that linked books metonymically to the animals whose carcasses covered them also linked them metaphorically to other, nontextual leather goods. Whether, in 1860, the *Saturday Review* attacked Trollope for "mak[ing] a novel just as he might make a pair of shoes" ("Review of Castle Richmond"), or on the contrary Bulwer-Lytton defended the genre by urging novelists who declare "I am not going to write a mere novel" to remember that no one "could ever become a good shoemaker if he did not have a profound respect for the art of making shoes" ("On Certain Principles"), they dragged the novel down from head to feet.[19] Trollope, of course, returned the favor, calling writer's block as absurd as if "the shoemaker were to wait for inspiration, or the tallow-chandler for the divine moment of melting" (Anthony Trollope, *An Autobiography* 121).

By the end of the century, the comparison of fiction to shoes was well-enough established for Woolf to assert that "what happened to boots has now happened to books. Books used to be made in small quantities by hand; they are now made in enormous quantities by machinery. Just as hand-made boots fitted better and lasted longer than machine made boots, so hand-made books read better and wore better than do our machine made books" ("Are Too Many Books Written and Published?"). The boot-boy at Claridge's finds his double in the maker of boots. Galsworthy, too, could take that metaphor as the donnée of "Quality," a 1911 story about a craftsman's pride in making shoes by hand, which reads clearly enough as an allegory of the workmanlike in literature.[20] (Here and throughout, I use "literature" to encompass a wider range of genres than nineteenth-century writers themselves did, stretching both to essays as in the eighteenth-century sense of the term, and to the narrative fiction that twentieth-century classifications began to dignify by the name of "literature.") For the producer to devote his earnest attention to any object, no matter how trivial, implies an ethic of service; aesthetic attention on the part of consumers, however, implies self-indulgence. By the same token, the artisanal particularity that sets Galsworthy's bespoke shoemaker apart in an age of machine production is precisely what renders book-collectors ridiculous in the age of the steam-press and the stereotype.

In "The Street Companion; or The Young Man's Guide and the Old Man's Comfort in the Choice of Shoes," De Quincey parodies biblio-philia by the mad-lib-like expedient of replacing the word "book" with "shoes." As Deidre Lynch has shown, by conflating the leather found in both, De Quincey collapses any difference between the high-end consumers and low-end artisans (Lynch, " 'Wedded to Books' " 11). We'll see that Mayhew attacks this distinction from the opposite direction when he describes artisans who resole shoes as "translators." So does the author of *The Missing Link, or, Bible-Women in the Homes of the London Poor* (1860), who describes the need to "translate" Christian books for the benefit of a Jew whose job consists of "translating" old shoes (R. 116). In this context, to conflate novelist with shoemaker is also to confuse text with book. Dressing, tanning, tasting, smelling, excreting: the possibilities that the book resembles a body, or is made from a body, or interacts with a body, or even resembles an object used to clothe a body, can be named only in a comic register.

It's as insulting to imagine the book *resembling* food (in that both consist of slaughtered animals, for example) as it's flattering to imagine the book *replacing* food (as when a literary character starves himself to buy a much-loved text). The narrator whose high-mindedness is established by a preference for books over food—as early as 1791, James Lacking-ton boasts of his wife's dismay that "I have often purchased books with the money that should have been expended in purchasing something to eat"[21]—secularizes the Christian logic that allowed missionaries to re-spond to demands for bread by distributing bibles. When the Shepherd of Salisbury Plain testifies in the anti-Jacobin tract of the same name that "my Bible has been meat, drink, and company to me, as I may say," he proves that reading can defuse the lower orders' demand for food (More, *Tales* 36). Remember that "in early Christian monasticism, reading took its place alongside fasting, prayer, the keeping of vigils, and the making of pilgrimages as an ascetic practice" (I. Hunter, "Literary Theory in Civil Life" 1109). Ruskin turns that political quietism into aesthetic transcen-dence when he praises books "bought out of saved halfpence; and perhaps a day or two's fasting. That's the way to get at the cream of a book."[22] Reducing cream to a metaphor for ideas, the speaker substitutes word for food as decisively as does the hypothetical reader whom he describes.

At the end of the century, "food for the soul" would become the refrain of *The Private Papers of Henry Ryecroft*. Where the writers portrayed in *New Grub Street* produce bad books in order to buy food, here the narrator forgoes food in order to buy good ones.[23] "Books [were] more necessary to me than bodily nourishment," the narrator tells us; reading as a boy, "I was astonished to find that it was four o'clock, and that I had forgotten food since breakfast." Men's forgetfulness of dinner depends, it's true, on women's remembering: "little girls should be taught cooking and baking more assiduously than they are taught to read" (Gissing, *The Private Papers of Henry Ryecroft* 43, 32, 125, 211).[24] Yet as chapter 7 will show, that doesn't imply keeping women from books: on the contrary, they enlist the book enthusiastically in that same "cooking and baking," whether in the form of pie lining or butter wrapping. Exalted when substituted *for* food, paper is degraded when associated *with* it.

The book's status depends on whether it displaces or conjures up its user's body. To convey the peacefulness of his new home, the narrator remarks that "the page scarce rustles as it turns" (Gissing, *The Private Papers of Henry Ryecroft* 75). Hearing joins sight among the senses that books are not supposed to stimulate—more, are supposed to deaden. No looking, listening, touching, tasting, smelling: the sensory deprivation of the post-1850 public library, where food was banned along with talking or even reading aloud, stands opposite the medieval scriptorium where books were voiced, stroked, smelled, and gazed at. Only in a paper-rich and information-saturated society do such acts begin to provoke nervous laughter.

Literary Logistics

Not noticing that the book was made of paper also implied ignoring that others had commissioned, manufactured, and transmitted it, and that other handlings had preceded and would follow one's own. The good reader—himself disembodied and unclassed—forgot what books looked like, weighed, and would fetch on the resale market; he also forgot how books had reached his hands, and through whose, and at what price. (The abstraction of the book thus mimics the abstraction of its readers.)[25] Yet paradoxically, those acts of oblivion themselves became enmeshed in human relationships, since reminders of the book's material attributes got delegated to persons less rich or male or Protestant than oneself.

Once piggybacked onto class and gender, the division between those who refer to "texts" and those who speak of "books"—those who memorize Penguin reprints and those who buy new hardbacks—is now replicated *within* the upper middle class by the difference between English majors and everyone else, from illiterates to book historians. A familiar intellectual-historical narrative tells us that since the New Criticism, literary critics have spearheaded an assault on the book's materiality, elevating the study of literature by demoting bibliographers to a service profession. It's true that if the book has been invisible and intangible even to those literary-critical schools that succeeded them, it isn't only for the negative reason that material culture remains absent from our training; it's also because a commonsense Cartesianism or Platonism more actively numbs us to the look and feel of the printed page. Hence critics' discomfort with purely bibliographic units—the page-break as opposed to the line-break, the volume as opposed to the chapter.

A longer historical view, however, makes it hard to blame or credit literary critics alone for exalting the text to an end and reducing the book to a means. Elaine Scarry frames the status of the book as an aesthetic question when she defines imaginative literature precisely by its power to drown out the significance that would otherwise be attached to its material form. Unlike music, sculpture, or painting, she observes, "verbal art, especially narrative, is almost bereft of any sensuous content. Its visual features . . . consist of monotonous small black marks on a white page." In fact, Scarry argues, what little sensory response the book does provoke is "not only irrelevant but even antagonistic to the mental images that a poem or novel . . . produce[s]" (5).

From a bibliographic perspective, in contrast, the bifurcation that Scarry associates with verbal "art" appears to inhere not in literature, but in print. It holds as true for intellectual history as for literary history: Popper, too, can assume that "*of course* the physical shape of the book is insignificant . . . and frequently even the formulation of an argument does not matter greatly. What do matter are contents, in the logical sense" (45; my emphasis). Carlo Ginzburg has argued that the first humanist printings of the classics set aside sensory data in the process of devaluing all those aspects of documents that vary from one copy to another (95).[26] The difference between a white and a yellowed page, to take Scarry's example, doesn't mean in the same way that color does in an illuminated manuscript.[27] This isn't to say that shades of brown and yellow don't convey useful information about which pages have been most heavily handled, which left untouched—or, indeed, that they don't conjure up "mental images" of the now-dead hands that have turned those pages. (And as my discussion of association copies in chapter 5 will emphasize, those data are the purview of the impassioned amateur at least as much as of the detached historian.) But as far as textual content itself goes, it seems fair to say that as mechanical reproduction stripped away visual and tactile differences among different copies of a single edition—or, at least, downgraded those differences to traces of error, accident, or wear and tear—printed texts in the West came to demand new ways of reading, and of not looking (Ginzburg 95).[28]

"Printed," not "literary": what Scarry claims of "verbal art" applies as well to even the most inartistic printed text, and as badly to even the most aesthetically ambitious of texts produced in a manuscript culture. Historical comparison suggests that what causes readers to bracket sense-data is not (or not only) their status as art, but also their status as reproduced and reproducible. A matter of media, not of aesthetics—with the crucial caveat that aesthetically serious texts and aesthetically intense experience figure as limit cases of the logic that all reading post-Gutenberg is supposed to follow. Far from (or as well as) forming a sphere of heightened attention, the printed as described by Ginzburg—no less than the aesthetic as theorized by Scarry—emerges from refusals to attend.[29] The book so strongly exemplifies the contrast between superficial change and fundamental invariability that the narrator of *Waverley* can use it as a metaphor for the continuity of human character, promising to read aloud a chapter from "the great book of Nature, the same through a thousand editions, whether of black-letter, or wire-wove and hot-pressed" (Scott 36).

By the nineteenth century, what had held since Gutenberg for all readers came to apply especially to good readers, whether that excellence was measured morally (as *Ranthorpe* did) or intellectually (in the manner of the New Critics). In that sense, paradoxically, the new New Bibliography could also be seen as a reductio ad absurdum of the New Criticism against which it appeared to react. The book historians whom literary critics think of as antiformalists have in fact pushed the boundaries of that term to encompass material (along with verbal) form.[30]

The opening of *Ranthorpe* echoes Carlyle's paean—reprinted in several late-Victorian compendia of bibliophilic pieties—to "the most momentous, wonderful and worthy . . . things we call Books! Those poor bits of rag-paper with black ink on them;—from the Daily Newspaper to the sacred Hebrew BOOK" (*On Heroes* 142). Yet the hostility to expensive books that essay shares with the novel suggests that outward form does matter. If what Carlyle calls the "wonder" of books depends on the mismatch between the insignificance of the poor bits of paper and the (metaphorical) richness of the verbal signs that they incarnate, this may be because once object competes with language for attention—as in fine bindings—the former ceases to be available as a foil for the latter.

How to Read Handling

Where the nineteenth-century general-interest press asked what uses of the book were acceptable, twenty-first-century scholars are likelier to ask what uses of the book are legible, and how the skills involved in reading texts (notably those possessed by literary critics and intellectual historians) differ from the skills required to describe objects (notably those possessed by all bibliographers and by some book historians). Closer to home, then, my question is how to situate literary interpretation vis-à-vis the social life of books more broadly understood—and also where different subcultures (from scholarly disciplines to religious traditions to political movements) have drawn the limits of that breadth.

Now that "the history of books and reading" has become a catchphrase, scholars in flight from lexical monotony refer to "the history of the book" interchangeably with "the history of reading."[31] It's true that both demonize the same opponent: the idealism that literary history shares with the history of ideas (which should remind formalist critics that "history" is hardly the opposite term to "literature"). Yet the survey I've just offered of the metaphorization of "reading" and reliteralization of bibliographic terms suggests what gets lost in that lumping. Where late twentieth-century critics insisted that books are not the only thing that can be read, so early twenty-first-century scholars are rediscovering (like so many M. Jourdains) that reading is not the only thing that can be done to books. That some of those other operations can themselves be performed upon objects other than books creates a third methodological problem.

I spoke of a turn away from metaphor, but the opposite case could also be made: that where the old historicism within literary criticism once invoked a metonymic logic to discuss commissioning, writing, editing, printing, and reading—whether upstream as in textual notes or downstream as in reception histories—book historians have substituted something more like metaphor. Reading is compared to other forms of consumption, or writing to other manual practices, or copyright to other forms of property. When Daston brackets the page with a comet, she looks both backward—to the long tradition which exalts reading to an art that other interpretive practices can only hope to emulate—and forward: to new forms of scholarship that reduce the book to one object among many. Where intellectual historians once studied the note-taking habits of individual thinkers, Ann Blair and Peter Stallybrass instead analyze scholarly note taking side by side with commercial record keeping; where an earlier generation of "law and literature" scholarship examined the image of lawyers in Romantic poetry, William St Clair juxtaposes the development of copyright with the changing legal regimes governing the sale of pharmaceuticals; where critics once narrated authors' alcoholism or analyzed the literary figure of the drunkard, Paul Duguid traces the history of authorial signature in parallel to the history of wine branding (Blair, "Note Taking"; Blair and Stallybrass; Duguid, *The Quality of Information*; St Clair). In cutting across different objects (books and ledgers, books and bottles, books and pills) to identify parallel practices, this research topples the text from its taxonomic pedestal.

In some contexts, certainly, verbal content trumps material medium: for someone in search of political information, a newspaper and a radio broadcast have more in common than do a newspaper and a piece of plastic wrap. In others, however, the reverse is true (someone trying to wrap a sandwich can use the newspaper interchangeably with the cling-film more easily than with the broadcast). At some moments, as we'll see in chapter 6, a servant's meddling with her mistress's books looks similar to eavesdropping on conversations, but at others it bears more resemblance to breaking a china vase. What's more, those attributes that set the book apart from other objects need to be disentangled from those that set some books apart from others (for example, literary from nonliterary texts or good works of literature from bad); because even the most un-readable book still differs from nontextual objects in the way it's priced, cataloged, and handled, the exceptionalism of the book should be no less visible to economists than to literary critics. By the same token, few of the issues I've mentioned so far are unique to the book: the logic that exalts reading copies while mocking coffee-table volumes shares its structure with contrasts between showy and serviceable clothing, or even between food addressed to the palate and that designed to please the eye.

If Victorian policy-makers grappled with the special status of the book—should printed matter be mailed at different rates from botanical cuttings? should books be taxed or priced differently from other commodities?—scholars today face analogous questions. Is literary-critical training a help and/or a hindrance to studying the book? How does a library differ from a museum? How should verbal evidence of reception be cross-checked with nonverbal traces? Should textual and material evidence be used to corroborate, to complicate, or even to con-tradict each other?

In marking the gulf separating bibliographic codes from linguistic codes, the pun provides a corrective to the recent strand of book-historical scholarship that set out to dovetail them. It's easy to see why book his-torians trying to find an audience among literary critics have empha-sized those moments where the book's material form converges with its linguistic content: where, for example, the small size or printed form of the novel (and by extension, its susceptibility to being read in private) reinforces its thematics of solitude. Yet the result is a catch-22: when such analyses of the material conditions of production and consumption cor-roborate some formal or thematic analysis of the text itself, they become redundant; and in those rarer cases where they contradict conclusions reached through textual explication, they become irrelevant. Heads you win, tails I lose. It may prove more productive to turn our attention to moments where these two strands of evidence pull apart, when (for ex-ample) an anticolonial manifesto is printed on paper imported from the metropole, or when an oath of revenge is sworn upon the same bible whose text preaches forgiveness.

Book historians are hardly alone in facing this challenge: the evidence that other schools of literary historicism invoke is equally incommensurate with the language in which they frame their claims. William St Clair is right to charge our entire profession with basing historical claims on literary evidence—more specifically, with supporting arguments about early nineteenth-century culture with texts that were produced in that era but found an audience only later (St Clair). By extension, historicist literary critics could be accused of wanting to have their cake and eat it—to make cultural claims about a different time and place while restricting ourselves to evidence that gives a particular set of professional readers aesthetic pleasure in the here and now. I plead guilty to a different bait and switch: I draw my evidence for bibliographical claims from sources chosen at least in part on the basis of their linguistic content. If I had taken my own premises more seriously, the table of contents would have looked rather different. Unlike Dickens, neither Trollope nor Charlotte Brontë employed particularly innovative publishing strategies. More fundamentally, if publishing strategies were my primary focus, authors would not provide the marquee names for my table of contents in the first place. Much less Victorian authors, given that the books most *used* by Victorians were not books *written* by Victorians. In the case of books whose life span exceeded that of human beings (notably the Bible) they were not even books printed by Victorians.

Worse, this study of those uses of the book that exceed or even replace reading is based primarily on the evidence of my own . . . reading. My occasional appeal to material evidence (the traces of wear and tear, of handling or ignoring) is dwarfed by my more regular use of textual proof. Certainly my interpretations are conditioned by the list price and current condition of the books that I study, but they appeal more heavily (to state the obvious once again) to the evidence of words. I confess not only to the absence of a volume of sources or a systematicity of sampling that would satisfy historians—as for most literary critics', my sources are printed (or reprinted or digitized), not manuscript; skewed toward the literary canon; too few in number to support any bibliometric claim—but also to the presence of literary-critical tastes and priorities. My subject is Victorian representations and perceptions of, and fantasies and illusions about, the circulation of books, not the circulation of books itself. And even within representations, my corpus skews heavily toward the literary and journalistic: a search across a few digital repositories for the character string "locked bookcase" yields one kind of result; a search of locksmiths' records to find out how often bookcases tampered with by servants needed to be repaired would have yielded quite another.

To make book-historical claims on the basis of textual evidence is not the same as making cultural-historical claims on the basis of literary evidence: the book's material form can give aesthetic pleasure as easily as its textual content can withhold it. It's a fallacy to assume that analyses of noncanonical or nonliterary texts are somehow more "book-historical" than others: the fact that a text is not aesthetically pleasing does not necessarily make it bibliographically significant, and book historians might do better to analyze the category of the "literary" than to flee it. But for a book historian drawing on a corpus defined by textual parameters (title, keyword, generic form, theme) as much as for a cultural historian drawing on a corpus defined by literary parameters, means will never match perfectly with end.

I mention these inconsistencies and trade-offs as symptoms of dilemmas that face many scholars today—and not just card-carrying literary critics. Even when book historians choose objects that stand outside of the literary, the language in which they describe their own scholarly practices remains parasitic on a literary canon in which reading gets tirelessly thematized. Chapter 4 argues that a particular corner of that canon, the bildungsroman, has both generated and limited the stories scholars tell about reading. I attempt to test those limits not only by finding countermodels in other genres—notably the it-narrative—but also, in the opposite direction, by tracking sweeping generalizations and unspoken pieties back to the tropes and leitmotifs of the particular bildungsromans in which they originate. Like most literary critics (and indeed like most readers), I go to past texts seeking origins for, as well as alternatives to, my own models. I hope this casts my sources as sibyls rather than ventriloquist's dummies.

Or, perhaps, as mirrors. Book history differs from most scholarly disciplines in that its object of study is also its means of transmission—the message is also the medium. For all its interest in marginalia and marginalized persons, the history of books is centrally about ourselves. It asks not only how past readers have made meaning (and therefore, by extension, how others have read differently from us); but also, closer to home, where the conditions of possibility for our own reading came from. Self-referentiality generates self-knowledge at the price of blind spots. The book historian too easily finds herself in the position of Thoreau (the son of a pencil manufacturer) forgetting to list his own pencil; or of the Reverend Alfred Hackman, who died in 1874 after spending thirty-six years as a sublibrarian at the Bodleian:

> During all the time of his service in the Library he had used as a cushion in his plain wooden armchair a certain vellum-bound folio, which by its indented side, worn down by continual pressure, bore testimony to

the use to which it had been put . . . When after Hackman's departure from the Library it was removed from its resting-place of years, some amusement was caused by finding that the chief compiler of the last printed Catalogue had omitted from his Catalogue the volume on which he sat, of which too, although of no special value, there was no other copy in the Library! (Macray 388)

Shortly before Hackman's death, the magazine founded by the original of Mr. Brocklehurst reprinted an American anecdote from half a century earlier:

Some gentlemen of a Bible Association lately calling upon an old woman to see if she had a bible, were severely reproved by a spirited reply. "Do you think, gentlemen, that I am a heathen, that you should ask me such a question?" Then addressing a little girl, she said, "Run and fetch the bible out of my drawer, that I may show it to the gentlemen." They declined giving her the trouble; but she insisted upon giving them ocular demonstration that she was no heathen. Accordingly the bible was brought, nicely covered; and on opening it, the old woman exclaimed, "Well, how glad I am that you have come; here are my spectacles that I have been looking for these three years, and didn't know where to find 'em." . . . My child, which have you: a *dusty* or a *well-worn* bible? ("The Two Bibles")

We think of spectacles as a tool for reading books, but one joke casts books as tools for storing spectacles; we think of a desk chair as a device for reading, but the other presents books as an aid to sitting. Is "furniture book" an oxymoron or a pleonasm?

PEOPLE OF THE TEXT

Two cases mark the limits of my topic: on the one hand, the ephemerality that the newspapers of chapters 2 and 7 share with the middle-class circulating-library novels of chapters 2 and 3 and the didactic fictions of chapters 4 and 6; on the other, the durability of the bibles discussed in chapters 4 and 5, whose verbal text and material form both remain themselves through many adventures. Even the law enforced the distinction between the dated and the timeless. For the first half of the nineteenth century, taxes kept newspaper prices artificially high and bible prices artificially low: the latter benefited from the only exception to the duties levied on all paper used for vernacular publication, to which an extra tax was added from 1819 onward for serials that both contained news and appeared at intervals under twenty-six days (Fyfe, *Science and Salvation* 40; Collet).

The Renaissance scholar James Kearney has described Christianity as "a religion of the book that was always made uneasy by the materiality of the text." To the extent that the Reformation thought of itself as "a return to the book within a religion of the book," "the book became an emblem of the desire to transcend the merely material and irredeemably fallen world of objects," but "at the same time, Reformers were suspicious of all human media . . . [and] distrusted the material dimension of text" (*The Incarnate Text* 8, 3).[32] More specifically, Catholics and heathens alike were accused of subordinating a text to its material container: of violating, whether by idolatry or illiteracy, what Tyndale called a law written "not with ink (as Moses' law) but with the Spirit of the living God: not in tables of stone (as the Ten Commandments) but in the fleshly tables of the heart" (Kearney, "The Book and the Fetish" 436; Tyndale 161). As we'll see in chapters 4 and 5, missionaries boasted that "silent messengers"—that is, religious tracts—would displace the "dumb idols" that natives worship. When a man in Burma proudly shows a missionary his old prayer book, he is told: "You have been ignorantly worshiping the book. I will teach you to worship the God whom this book reveals." Yet the author exults in the enthusiasm of others in Burma who "were all so earnest for tracts, and there not being enough for all those who desired them, they cut the tracts up into bits, that each might have a few words or a few lines of the sacred writings to keep in their houses" (Jones 474, 62, 63).

Catholics, too, were accused of sharing non-Europeans' respect for the material book, as when a bible was kissed during the Mass (Kearney, "The Book and the Fetish" 462). "To evince a belief in the power of the object was to engage in a fundamental category mistake that separated superstitious and credulous others (non-whites, non-Christians, Catholics, the lower classes, and women [and, one might add, children]) from the rational European man" (Kearney, "The Book and the Fetish" 436). That such rationality was not entirely a delusion is perhaps best proven by the it-narratives of chapter 4: while Evangelical publishers systematically produced accounts of bibles being desecrated, torn, and trampled, it would be hard to imagine any equivalent body of literature surrounding the Torah or the Koran.

Protestant missionaries described their goal as spreading literacy; on the ground, however, they often seemed more concerned with limiting bibliolatry. They worried, that is, not only about the *in*ability of the poor and the heathen (and often the Catholic) to read their sacred text, but also about the *ability* of those populations to put the Bible to uses *other* than reading.[33] In colonial contact zones as among working-class populations in Britain, illiteracy made the zero-sum relation between material book and verbal text especially visible. The first English explorers found the natives of Virginia, for example, "glad to touch [a bible], to embrace it, to kisse it, to hold it to their brests and heades, and stroke all over their bodies with it" (Wogan 407). Elsewhere in the present-day United States, native graves have been found to contain a leaf torn from a bible (Amory). Even in Protestant Europe, a bible could be kissed to lend weight to an oath, stuck under invalids' pillows, used as a shield against bullets, and even eaten (Cressy 98). Decisions were made, and the future predicted, by what page fell open. Births and deaths were recorded in blank spaces of bibles (often the only writing surface available in poor households). At the other end of the social scale, the librarian Edward Edwards could compare the unread books in aristocratic private libraries to "idols"—as if bibliophiles were no better than heathens.[34]

Yet the same Protestant clergy who accused others of "idolatry" for investing the book with totemic powers laid themselves open to that charge when they placed their faith in the dissemination of printed matter. People of the book—Protestants or even freethinkers whose faith lay in reading—could be accused in return of fetishizing literacy. Thus Francis Hitchman conflated paganism with a different religion of the Book when he called English social reformers "good people in whose eyes in a book is a species of Fetish, and who look upon printed paper with as much reverence as do the Mahometans" (151). One anticlerical journalist in 1899 complained that "some minds, even in the midst of civilization, retain a sort of heathen, or Arab, reverence for the printed page" (Ogden).

As Patrick Harries points out, Africans who invested books with totemic powers could look uncannily like "Europeans who invested the bible with supernatural powers when taking an oath, or who read the Good Book as divine revelation or self-evident truth" and who "collected books not for the information and ideas they contained, but in order to present a show of knowledge and wealth" (421). When William Carey set up a printing press in Bengal in 1798, "the crowds of natives who flocked to see it, hearing Mr. Carey's description of its wonderful power, pronounced it to be a European idol" (Marshman 80). In West Africa a century later, books were dubbed the "white man's fetish."[35] As Joseph Slaughter has recently documented, post-1945 organizations such as UNESCO have made literacy "the functional boundary—globally as much as locally—between the disenfranchised and the unincorporated" (279–81). The literate world takes the place of Christendom.

In the year when I completed the manuscript that became this book, I sat in on a court-ordered book club in Massachusetts. Changing Lives through Literature, the program sponsoring the group, secures probation for convicted criminals on condition that they meet weekly to read and discuss short stories and poems. Read a book or go to jail: this encounter mixed even more messages than my own task of grading college students on whether they had "done" the reading (Price, "Read a Book, Get out of Jail"). Although elite pupils had long faced the choice between a line of poetry and a stroke of the cane, the spread of formal education made nineteenth-century England one of the first times and places when, for large numbers of children, "school" began to replace "work" as the antonym to "pleasure" (Religious Tract Society).

Cast as a means of rehabilitation—an opportunity for working-class men to practice self-recognition and other-directed empathy in the safely distant imaginative spaces usually reserved for middle-class women— Changing Lives through Literature reminded me how powerful a hold certain Victorian values retain on twenty-first-century Americans (including, as these pages betray, on me). By the middle of the nineteenth century, faith in bible reading had devolved into faith in reading. A Gissing character declares that "every educated person is really a missionary, whose duty it is to go forth and spread the light . . . I couldn't give money—for one thing, I have very little, and then it's so demoralising, and one never knows whether the people will be offended—but I sat down and told the poor woman all about the Prologue to the Canterbury Tales"; she typifies the shift from the first decades of the century, in which the distribution of bibles and tracts came to link social ranks, to its last, when secular texts took over that function (*Our Friend the Charlatan* 155). William McKelvy points out that "literacy was normalized, and illiteracy pathologized, at the point when the state no longer recognized religious heresies" (34)—that is, over the course of the nineteenth century. Over the course of the same century, educational publishing replaced sectarian publishers' share of the market (Ledger-Lomas 328). Small wonder, then, that the Victorians joked about the book in proportion as they took reading seriously. As memoirs of found romances inherited the task of conversion narratives, bibliophilia took over the work of idolatry.

PART I

_____

Selfish Fictions

CHAPTER 2

Anthony Trollope and the Repellent Book

THINK BACK TO THE LANGUAGE in which some Victorian novels establish a character's position in relation to a book. A mouthful, but my periphrasis reflects the thesaurus-sized arsenal of circumlocutions that the Victorian novel itself elaborates in order to avoid coupling its characters' names with the verb "to read."[1]

1. "The quarto Bible was laid open before him at the fly-leaf . . . Mr Tulliver turned his eyes on the page" (G. Eliot, *The Mill on the Floss* 274)
2. "said Mrs Tulliver, going up to his side and looking at the page" (G. Eliot, *The Mill on the Floss* 274)
3. "He sat with a magazine in his hand." (Anthony Trollope, *The Prime Minister* 383)
4. "The gentleman had his head bent over a book" (H. James, "The Middle Years" 335)
5. "the pages of the magazine which he turned." (Anthony Trollope, *The Prime Minister* 514)
6. "with which Mr Osborne spread out the Evening paper" (Thackeray, *Vanity Fair* 134)
7. "Baxter's 'Saints' Rest' was the book [Mrs Glegg] was accustomed to lay open before her on special occasions" (G. Eliot, *The Mill on the Floss* 134)
8. "having always at breakfast a paper or book before him" (Anthony Trollope, *The Claverings* 372)
9. "When he had done yawning over his paper . . . " (A. Brontë 164)

The first two passages describe eye movements, the next four hand gestures. The seventh and especially the eighth go further, invoking spatial terms to describe the position of the printed object in relation to a person's body, but refusing to specify what that person is doing with his hand—let alone with his eye, much less his mind. The last uses "over" in a sense that could be either causal (the book's content inspires a yawn) or purely spatial (in which case the yawn might indicate, on the contrary, that the book is not being read at all).

You could object that these ambiguities are of one reader's own making: that these quotations appear elliptical only because I've wrested them from their original contexts. It's true that in *Vanity Fair*, the description of Mr. Osborne spreading out his newspaper is immediately followed by the punch line that "George knew from this signal that the colloquy was ended and that his Papa was about to take a nap." *The Mill on the Floss* makes us wait longer for that resolution: there, a hundred pages separate the initial refusal to specify whether Mrs. Glegg's laying open of the book forms a prelude to, or a substitute for, the act of reading, from the moment when the narrator remarks, a propos of something quite different, "If, in the maiden days of the Dodson sisters, their bibles opened more easily at some parts than others, it was because of dried tulip petals, which had been distributed quite impartially, without preference for the historical, devotional, or doctrinal" (G. Eliot, *The Mill on the Floss* 284).[2] As "opening" becomes an intransitive verb, and as its agent shifts from Mrs. Glegg to the book itself, a character—as so often in *The Mill on the Floss*—becomes accessory to an inanimate object. At the same time, though, the strategic ambiguity of that "laying open" gives way to a unmistakably broad joke about the precedence that the material book takes over not only its human user, but its verbal content: any reader can tell that the "opening" of the bible doesn't refer to Genesis.

If an individual sentence leaves us to decide for ourselves whether to parse an ambiguous act as "reading," in the long run each novel as a whole makes perfectly explicit the gap between the presence of the book and its user's absence of mind. Yet the time lag between our initial hunch that what we're witnessing will count as reading, and a later moment when that possibility is definitively ruled out, exemplifies a strategy by which Victorian novels spread out over time doubts about the relation of persons to books. Thus *The Mayor of Casterbridge* opens with Henchard and Susan walking "side by side in such a way as to suggest afar off the low, easy confidential chat of people full of reciprocity; but on closer view it could be discerned that the man was reading, or pretending to read, a ballad sheet which he kept before his eyes" (Hardy 1). No sooner does the hypothetical observer congratulate himself on shattering the illusion that a human conversation is going on, than the impression that Henchard is communing only with a printed (and therefore non-"reciprocal") text is punctured, in turn, by the qualification that even physical proximity—whether "side by side" or "before his eyes"—provides no guarantee of mental engagement.[3] The book as obstacle will give way to the book as lens only later, when, after receiving Susan's posthumous letter, "her husband regarded the paper as if it were a window-pane through which he saw for miles" (122).

Characteristically, Hardy fudges the question of whose consciousness that undecidability should be located in: the passive voice leaves unclear exactly who is trying and failing to distinguish bodily actions from cognitive ones. Trollope more often shunts onto his characters his readers' uncertainty about what use a book is being put to. In fact, in the passages that I just quoted from his fiction, the act (in every sense) of reading seems to respond to the presence of such a third party. Thus *The Prime Minister* establishes the breakdown of a marriage by pitting Palliser's newspaper against Glencora's novel: "He busied himself with books and papers,—always turning over those piles of newspapers . . . She engaged herself with the children or pretended to read a novel" (361). In a parallel subplot, Lopez's quarrel with *his* wife leads to the same result: "he sat with a magazine in his hand" (383). The sentence stops there: no need to spell out (as the narrator will when the scene recurs a hundred pages later) that "it may be doubted whether he got much instruction or amusement from the pages of the magazine which he turned" (514).

Along with foxhunting and electioneering, pseudoreading forms one of the set pieces that knit together the loose bagginess of the Palliser series. At its other end, *Can You Forgive Her?* already described the young Palliser "*reading, or pretending to read*, as long as the continuance of the breakfast made it certain that his wife would remain with him" (610; my emphasis). Bibliographic aggression spans lifetimes, bridges subplots, and unites enemies. In *The Claverings*, all that the two feuding brothers share is their use of the book. "At their meals [Sir Hugh] rarely spoke to [his wife],—having always at breakfast a paper or a book before him, and at dinner devoting his attention to a dog at his feet." In a different household, his brother "was reading,—or pretending to read—a review" (372, 114). The book's function as a prop for privacy or prompt for interiority depends less on its being looked at by the character who holds it than on that person's being looked at himself.

BLANK BOOKS

Reading or pretending to read: what's the difference between the two? Not much, Hardy's lack of punctuation suggests. Trollope's pileup of commas with dashes puts more distance between them, but the fact that one rarely appears without the other still reduces the book to a prop. This has more to do with the syntactic construction than with the particular verb chosen: an analogous effect is produced when Elizabeth Sewell's narrator remarks that a character "was reading, or perhaps, more strictly speaking, intending to read; for although a book lay open before her, her eyes wandered chiefly amongst the flowers" (E. M. Sewell 67). By the nineteenth century, the rich repertoire of uses of the book that we saw in the previous chapter—with reading positioned somewhere in the middle of a spectrum running from divining to wrapping—narrows to a binary opposition between authentic reading and its simulacrum.

In the process, new jokes enter the repertoire. *The Small House at Allington*, for example, takes the most economical proof of two honeymooners' hatred for one another to be a game of chicken:

> He had the *Times* newspaper in his dressing-bag. She also had a novel with her. Would she be offended if he took out the paper and read it? The miles seemed to pass by very slowly, and there was still another hour down to Folkestone. He longed for his *Times*, but resolved at last, that he would not read unless she read first. She also had remembered her novel; but by nature she was more patient than he, and she thought that on such a journey any reading might perhaps be almost improper. (Anthony Trollope, *The Small House at Allington* 497)

His and hers, newspaper and novel: the railway carriage echoes the railway platform across which Johnny Eames and Adolphus Crosbie chased one another several chapters earlier, ending up at the W. H. Smith bookstall where Johnny "laid his foe prostrate upon the newspapers, falling himself into the yellow shilling-novel depot by the overt fury of his own energy" (371).[4] Crushing each genre under the weight of a different combatant, the first scene introduces a face-off between newspaper and novel that the second will peg to sexual difference.

By century's end, paired novel and newspaper reading was canonical enough that its interruption could carry a sexual charge: in the 1899 film *The Kiss in the Tunnel*, a woman reading a novel and man reading a newspaper kiss when the train enters the darkness (Hammond 60). *The Small House* shows less interest in analogizing men's newspaper reading to women's novel reading, however, than in pairing unread copies of each. One measure of this is that "remembered her novel" doesn't mean what I'd mean if I wondered, for example, whether you remembered *The Small House at Allington* well enough for me to dispense with plot summary. What Alexandrina remembers isn't the content of the text, but the location of the book. Even when Crosbie can no longer resist taking out his newspaper, "he could not fix his mind upon the politics of the day."

No matter, since what Crosbie needs is less to take his mind off Alexandrina than to get *her* eye off him. Crosbie's conventionally agreed-upon signal for what Erving Goffman calls "civil inattention" is ratified in turn by the novel's refusal to tell us what exactly is in the newspaper going unread. "He could not fix his mind upon the politics of the day"; neither can the narrator, who proceeds without transition to detail the thoughts that crowd out the news: "Had he not made a terrible mistake? Of what use to him in life would be that thing of a woman that sat opposite him?"

and so on, for the space of a paragraph (Anthony Trollope, *The Small House at Allington* 498). We can read (or at least read about) Crosbie's mind more than we can read over his shoulder. His newspaper might as well be one of the fake broadsheets sold by the California-based Earl Hays Press for use as props on film sets throughout most of the twentieth century, the periodical equivalent of a dummy spine. And the symmetry of this scene extends the paper's emptiness to the genre that frames it: the novel shows as little interest in the content of Crosbie's newspaper as Alexandrina feels in the content of her novel. Literalization again: what the newspaper "covers" isn't current events, but a character's body.

That absence can best be measured against the nineteenth century's most canonical variation on the quixotic theme. In *Madame Bovary*, physical gestures ("she turned the pages," "Emma greased her hands on the dust of reading-rooms," "delicately handling their fine satin bindings") serve to introduce the content of the books being read.

> So, when she was fifteen, Emma spent six months breathing the dust of old lending libraries. Later, with Walter Scott, she became enthralled by things historical and would dream of oaken chests, guardrooms, and minstrels. (Flaubert 34; the remainder of the paragraph describes the plot and characters of a historical novel)

> The girls used to read [gift books] in the dormitory. Handling their handsome satin bindings with great care, Emma . . . shivered as she blew the tissue paper off each engraving; it would lift up half folded, then gently fall back against the opposite page. There, beside the balustrade of a balcony, a young man in a short cloak would be clasping in his arms a young girl wearing a white dress . . . (35)

> She would even bring her book to the table and turn over the pages while Charles ate and talked to her . . . Paris, vaster than the ocean, shimmered before Emma's eyes in a rosy haze . . . The world of high diplomacy moved about on gleaming parquet floors, in drawing rooms paneled with mirrors, round oval tables covered by gold-fringed velvet cloths. (52)

Like virtual speech tags, these descriptions of books warn the reader that what follows should be attributed to another text. In Trollope, on the contrary, no paraphrase of textual content motivates or even follows the description of the hand or eye of the person holding the book.

We have a special word for persons when they're represented in fiction ("character"), but none for represented books. Yet both raise analogous questions. Is it legitimate to imagine an offstage life for either (for example, should we picture what news items Crosbie's newspaper contains)?

What's the relation between the use we make of the represented object and the use that we would make of its real-life referent? The fruit painted in a still life can't be eaten, but painted spines can sometimes be read. I say "sometimes" because, like any other object, books can be represented at varying degrees of resolution—not just in visual art, where, as Garrett Stewart has shown, print is conventionally recognizable but illegible, but also in words (G. Stewart, "The Mind's Sigh"; Butor 41–43).

If Flaubertian pastiche forms one extreme, the other end of the spectrum is anchored by Henry James's habit of withholding author and title but providing something like a descriptive bibliography: the color of the cover, the number of volumes, the size of the print. His references to "a small volume in blue paper" or "three books, one yellow and two pink," make the book as empty as a patent pill (*The Awkward Age* 934; *What Maisie Knew* 636).[5] Like Woolf describing Rachel Vinrace "stirring the red and yellow volumes contemptuously," James uses the visual to crowd out the verbal: a chromatic metonymy such as "yellowback" shares the dismissive form of a commercial term such as "penny dreadful" (Woolf, *The Voyage Out* 304). The narrator of "Greville Fane," too, brackets the content of his friend's writing when he measures her rate of production by the fact that "every few months, at my club, I saw three volumes, in a green, in crimson, in blue" (H. James, "Greville Fane" 233).

When the protagonist of *In the Cage* pulls out a novel "very greasy, in fine print and all about fine folks," the repetition of a single word in two opposed senses opens a gap between bibliographic form and mimetic content (H. James, *In the Cage* 119). The contrast between cheap typographical characters and rich fictional characters reverses the equally doubled logic of James's large-leaded volume about the petty bourgeoisie. Trollope goes even further, contenting himself sometimes with generic markers ("her novel"), sometimes with even more purely physical descriptions ("books and paper," "a paper or book"). These are phrases that an illiterate could have come up with.

If Trollope's narrator denies us access to the content being read, Thackeray's more often projects the invisibility of the inscribed text onto its own fictional readers. Thus at the Newcomes' breakfast table:

> "How interested you are in your papers," resumes the sprightly Rosey. "What can you find in those hard politics?" Both gentlemen are looking at their papers with all their might, and no doubt cannot see one single word which those brilliant and witty leading articles contain. (853)

By juxtaposing a husband's and father-in-law's concentration on "papers" with their insensibility to "words," the narrator changes the former from a metonymy for "newspapers" (and thus, transitively, for "news") to something more purely material: pieces of paper, regardless of the linguistic information that they "contain."

Bookmen

The newspaper takes on a different meaning in the home from what it bears in the public sphere that Benedict Anderson names under the metonymy of "the subway [and the] barbershop"—the very spaces where men can escape squabbles with their wives. The national community that he imagines depends on each reader's knowledge that distant strangers are reading different copies of the same newspaper—but also (although Anderson feels no need to spell this out) on the power of gender to provide a common denominator among readers who will never meet each other. Anderson's logic thus reverses Trollope's on every count: in one, not reading drives a wedge between opposite-sex intimates stuck in the same physical space; in the other, reading builds a bridge among physically distant same-sex strangers.[6]

The trope of the husband hiding behind the paper might appear to complement the strand of literary-critical feminism (not to mention feminist fiction) that sees reading as a peculiarly feminine source of interiority, individuality, or authenticity—a means, as we learned in the 1960s, to "find oneself." Barbara Sicherman's observation that in the nineteenth-century United States "reading provided space—physical, temporal, and psychological—that permitted women to exempt themselves from traditional gender expectations, whether imposed by formal society or by family obligation," is echoed a century later by Jan Radway's contrast between the patriarchal content of contemporary romance novels and the feminist force of the act of reading or even holding them. In the Cold War-era households that Radway describes, books mark women's personal time whether or not they're being read (Sicherman 202; Radway, *Reading the Romance* 213).

Books can screen mothers from children as easily as wives from husbands. The *Punch* cartoon entitled "How to Make a Chatelaine a Real Blessing to Mothers" finds its echo in a collection of librarians' anecdotes reporting that "at the library at Hull a young girl was heard to whisper to her sister: 'Don't get one of Miss Braddon's books. Ma will want to read it, and we shall have to wash up the supper things'" (Coutts 142). Sartre's memory of his mother vanishing into the pages of a book repeats Arthur Clennam's impression of his mother sitting "all day behind a bible—bound like her own construction of it in the hardest, barest, straightest boards"; Garrett Stewart recasts this trope in a psychoanalytic

HOW TO MAKE A CHATELAINE A REAL BLESSING TO MOTHERS.

Figure 2.1. "How to Make a Chatelaine a Real Blessing to Mothers," *Punch*, 24 February 1849, 78.

register, calling reading "the return of the repressed moment when your mother's voice first went silent to you" (Sartre 34; G. Stewart, "Painted Readers, Narrative Regress" 141; see also G. Stewart, *The Look of Reading* 105). If reading can distract from children, children can distract from reading. In an 1850 *Father's and Mother's Manual and Youth's Instructor* a story portrays a mother forgetting her novel as she "imprints upon [her child's] lips the kiss of love," but then returning to it until, even when her sick child cries out her name, "her ear caught the sound, but it made no impression upon her mind till it had several times been repeated . . . She was absorbed in a book; her very being seemed bound up in it." At one moment, the printed child takes the place of a book; at another, the "bound" mother becomes as deaf to the child's voice as an inanimate pile of paper (Park 141).

Even as the woman's book stands opposite the man's newspaper, the trope remains constant across media: a woman's sheet music can be counterposed to a man's newspaper as easily as can a codex, and complaints

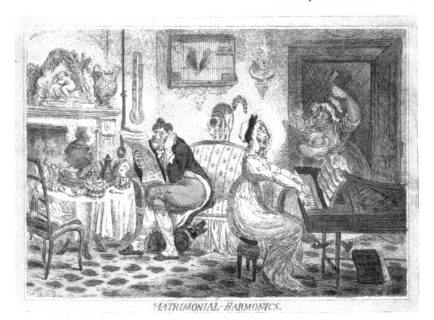

Figure 2.2. James Gillray, "Matrimonial Harmonics," 25 October 1805.
Prints and Photographs Online Catalog, March 2007, British Cartoon Prints
Collection, Library of Congress, 27 May 2011. http://www.loc.gov/pictures/
item/2007677406/.

about mothers engrossed in their smartphones update for the twenty-first
century a complaint that stretches back at least to the eighteenth.[7]

If the overinvested reader is prototypically female, the explanation
may be (paradoxically enough) that the demand for women to be especially attuned to those around them creates an especially strong demand
for escape. Yet the symmetrical structure of *The Small House at Allington*
suggests that two can play at this game. The husbands of New Woman
fiction who beat their wives for reading invert an older (but hardly dead)
tradition that counterposes an intellectual husband to a shrewish wife.
Women's hostility to books continues to be ridiculed in a mock-epic register as late as one *Macmillan's* contributor's lament that

> It is the custom of some ladies to make use of our works as weapons
> of offence in certain hostile emergencies, with which the peaceful arts
> should have nothing to do. A lady, who has differed in opinion with
> her lord and master, will not uncommonly retire behind a book and
> erect it into a sort of literary rampart. There is no making complaint of
> her want of attention to the matter under perusal *then* . . . Never was
> anything like her fixedness of attention. (C. A. Collins 162)

Figure 2.3. Weekly Caption Contest, *New Yorker*, 2006.
© Tom Cheney / The New Yorker Collection / www.cartoonbank.com.

This military language finds its musical counterpart in Jerrold's advice that "If [your husband] has a book, or would doze by the fire, immediately play the '*Battle of Prague*,' with all the cannon accompaniments" (D. W. Jerrold 194). And that remark itself makes explicit the theory behind Mrs. Caudle's twenty-eighth curtain lecture, entitled "In which Mr Caudle, in self-defence, takes a book," which climaxes when Mrs. Caudle notices what object her husband is using to screen himself from her tirade: "Why, what have you got there, Mr Caudle ? A book? What! If you ar'n't allowed to sleep you'll read? Well, now it is come to something! If that isn't insulting a wife to bring a book to bed, I don't know what wedlock is. But you sha'n't read, Caudle; no you sha'n't; not while I've strength to get up and put out a candle" (Jerrold and Keene 147). To the extent that it asserts the reader's separateness, "bringing a book to bed" insults the married state.

Symmetry in content, then, but hardly in tone. You've probably noticed that these scenes of women blocking their husbands' reading are played for laughs, while the reverse tends to be cast in a darker light—and not only because the former draw on a tradition of jokes about shrewish wives, the latter from gothic representations of male violence. The heroic narrative in which texts allow women to wrest their selfhood from male bullies finds its match in a more satirical tradition that sees the text as a shield from the demands of women who (as we'll see in chapter 7) destroy books to line pie plates or cut out dress patterns. Whether a wife or servant hardly matters: in stories like Ella D'Arcy's "Irremediable" or Gissing's "The Prize Lodger," the mutual avoidance of couples within a single social class gives way to an educated man's mésalliance with a vulgar girl who "hated books, and were he ever so ill-advised as to open one in her presence, immediately began to talk," or an ex-landlady who storms away when her husband opens a newspaper at the breakfast table (D'Arcy 114; Gissing, "The Prize Lodger" 152). Elias Canetti would update the theme with a housekeeper's (later wife's) hatred for her husband's library—a library that she sees, quite rightly, as a drain on the fortune that should be spent on her (Canetti).

Like televised sports in the twentieth century, newspapers (for the masses) and books (for the elite) provided men with a refuge from their wives. In *Vanity Fair*, the "study sacred to the master of the house" is focalized from the perspective of the wife outside its door: "George as a boy had been horsewhipped in this room many times; his mother sitting sick on the stair listening to cuts of the whip" (231). In Oliphant's *Kirsteen*, too, the paterfamilias "roused himself quickly with sharp impatience; though the doze was habitual he was full of resentment at any suspicion of it. He was reading in his room; this was the version of the matter which he expected to be recognized in the family . . . But he was not reading, though he pretended still to be buried in the paper" (40, 83).[8]

Books could stake out a claim not only to space within the household, but also to time and money. When Charles Darwin drew up a balance sheet to help him decide whether to marry, one of the entries in the "no" column concerned time and money for reading: "*Loss of time*—cannot read in the evenings—fatness and idleness—anxiety and responsibility— less money for books." At a time when the celibacy of Oxbridge fellows made the choice between reading and marrying quite concrete, women could be imagined at once as a drain on the time and money that could otherwise be spent on books, and also as a decorative item analogous to them: in the "marry" column, we find "Imagine living all one's day solitarily in smoky dirty London House.—Only picture to yourself a nice soft wife on a sofa with good fire, & books & music" (Darwin).

Anne Brontë's *The Tenant of Wildfell Hall* begins by pairing a husband's insistence on reading newspapers with his refusal to let his wife read novels: "he never reads anything but newspapers and sporting magazines; and when he sees me occupied with a book he won't let me rest till I close it . . . When he came in and found me quietly occupied with my book, too busy to lift my head on his entrance, he merely muttered an expression of suppressed disapprobation." Even the husband's reading, however, is quickly unmasked as a front: "The paper he set before him, and pretended to be deeply absorbed in its contents . . . When he had done yawning over his paper . . . he spent the remainder of the morning and the whole of the afternoon in fidgeting about . . . sometimes lounging on the sofa with a book that he could not force himself to read." Finally, more surprisingly, the narrator turns that accusation in upon herself. After the husband throws a book at his favorite dog, "I went on reading, or pretending to read, at least—I cannot say there was much communication between my eyes and my brain." By the time she confesses that "what the book was, that lay on the table before me, I cannot tell, for I never looked at it," the heroine becomes as indistinguishable from her husband as from his hypocritical lover who "employed herself in turning over the pages of the book, and, really or apparently, perusing its contents" (A. Brontë 164–69, 201, 43).[9] The dichotomy reappears in Sarah Grand's *The Heavenly Twins* (1893), where another sensualist husband tries to prevent his wife from reading, but her own "eye would traverse page after page without transferring a single record to her brain" (52). Even if we begin by associating mental acts with wives and eye service with husbands, the latter end up infecting the former: no stable ground from which to distinguish true reading from false.

How to explain the contradiction between these two tropes—one in which the most passionate and disinterested reading is attributed to women (a subset of the more general Victorian celebration of reading as a weapon of the weak, described in the next chapter), and another in which women are reduced either to blocking figures for men's reading or to philistines who value a book for its material properties—whether matching its binding to their dress and decor in the upper ranks, or disbinding its pages for dress patterns and pie wrapping in the lower? One hypothesis might be that women are associated with whichever activity is morally inferior—whether solipsistic overengagement with the text or social display of the book—and that the first, Bovaryesque scenario is reproduced even by feminist celebrations of women's authentic, individualistic reading. The culture's oscillation between feminizing the book and the text might also suggest, however, that women are imagined not as inferior to men but rather as higher variance. In relation to printed matter as much as to sex, that is, women gravitate toward either extreme: either the epitome of textual authenticity or the exemplar of bookish superficiality, either Madonna or whore.

Those value judgments may in turn reflect a socio-historical turning point in women's relation to reading. For most of British history, men's literacy rate outstripped women's; in the nineteenth century, however, the latter began to climb more steeply than the former, until around 1900 literacy was actually more diffused among women (Vincent *The Rise of Mass Literacy* 12–13; McKitterick, *The Cambridge History of the Book in Britain* 43; Colclough and Vincent 293–94). The feminization of literacy did not just depart sharply from historical precedent. It also made

Britain an anomaly on the international scene, since outside of a few rich countries, men were far more literate than women, as remains the case in developing countries today (Griswold 40). That pattern makes economic sense: investment in boys' education promises payoff in the form of high wages, while teaching girls to read withdraws their labor in the present. It's harder to explain either why this demographics changed in late nineteenth-century Britain, much less why cultural perceptions of reading anticipated that reversal by more than a century. Once Cervantes's hero reappeared in drag in *The Female Quixote*, women became associated less with illiteracy than with excessive reading—even if terms like "illiteracy" could be turned into metaphors to stigmatize the reading of the wrong books by the wrong persons. Today, women across Europe buy and borrow books more often than men, just as the very young and the very old read more than the middle-aged. Once a sign of economic power, reading is now the province of those whose time lacks market value.

Once the feminization of reading in the present-day West is recognized as an anomaly in both time and space, it becomes harder to explain by essentialist assumptions about women's greater capacity for empathy or imagination. Instead, the dependent variable seems to be status: associated with men when it's rare and therefore prestigious, literacy is feminized in societies (like ours) where ubiquity breeds contempt. Outside the West, reading is associated with mobility—both social and geographical; in modern societies, however, it becomes the refuge of those trapped in interior spaces: prisoners, children, housewives. As Charlotte Yonge observed in 1869, "There are so many hours of a girl's life when she must sit still, that a book is her natural resource" ("Children's Literature: Part III" 454).

MUTE CHAPERONES

A workplace as much as a leisure pastime, reading could also be used to define the relation *between* those two spheres, whether by filling the commute that separates them (you read on the subway to unwind before you get home) or by marking the telegraph girl's private time and space (official papers on the desk, lunchtime reading in the drawer). Among the "infinité de ces petites usages de convention" that one French contemporary credited the English with developing "pour se dispenser de parler," books moved fluidly across public and private spheres (Bulwer-Lytton, *England and the English* 23). The same objects that shield husbands from wives can screen commuters from strangers, parents from crying children, children from demanding parents, clubmen from one another—or even masters from servants, as when a conduct book warns servants not to take a master's "appearance" of reading at face value:

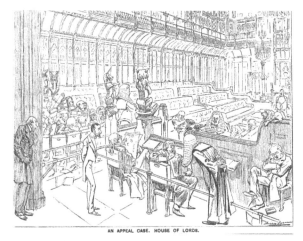

Figure 2.4. "An Appeal Case. House of Lords," *Punch*, 14 February 1891, 82.

> I know many ladies who have repented having spoken familiarly to
> their servants, finding that the girls have misunderstood their kindness,
> and sometimes have gone so far as to begin talking about their own
> affairs in the drawing-room. One lady who was thus annoyed, told me
> that she always took up a book and appeared to read, before she rang
> the bell, hoping thus to keep the girl from stopping to talk when she
> came up. (Motherly 16)

Yet even if marriage is not the only bond that books can buffer, its
privileged status nonetheless makes it an especially effective test case:
the strongest tie requires the sharpest solvent. A 2008 *New York Times*
article describing two Buddhist monks who replace "sexual touching" by
synchronized breathing and synchronized reading gains its shock value
from the traditional assumption that the book provides the last refuge
for bodies that are otherwise fused.[10] A fortiori, at the moment when separateness has very recently been disavowed—that is, the honeymoon—
the book's power springs into clearest relief. Mary Gladstone writes that
her mother "used to tell us, long afterwards, that it was something of a
shock to both sisters when, after marriage, any little waiting time, as the
railway station, which during their engagement would have been spent
in love-making, was now spent in reading—both husbands carrying the
inevitable little classics in their pockets. Out it would come and quickly
engross the owner" (Windscheffel 56). (In fairness to Gladstone, it should
be added that he read aloud to his wife not only during their courtship
and honeymoon, but also during their marriage [Windscheffel 65].) Alexandrina's hunch about the meaning of newlyweds' reading is confirmed

as late as 1904 in Sturgis's *Belchamber*, where a matchmaking mother pesters an eligible bachelor to recommend books for her daughter; the ploy works, but any hope of marital happiness disappears once we see the two reading on their wedding night. "Cissy sank deep in a big arm-chair, and appeared to be immersed in a novel she had brought with her. Sainty tried to read too, but his attention wandered; his eyes fell first on his companion, . . . the hands flashing with new rings that held the gaudy book-cover like a shield between her face and him" (Sturgis 188, 219, 20).[11] If oblivion to one's family can serve as a gauge for interest in a text, concentration on a book can just as well provide a yardstick for hatred of one's family.

One explanation for the prominence of conjugal reading in the nineteenth-century novel, then, is that it inverts the logic of courtship fiction. Unsurprisingly, Trollope's *non*fiction defines the relation of reading to marriage in mimetic terms: that is, his reviews, lectures, and autobiographical writings assume that we go to books in general, and novels in particular, in order to find out how to conduct a courtship.

> It is from them that girls learn what is expected from them, and what they are to expect when lovers come; and also from them that young men unconsciously learn what are, or should be, or may be, the charms of love,—though I fancy few young men will think so little of their natural instincts and powers as to believe that I am right in saying so. (Trollope, *An Autobiography* 220)

> There are countries in which it has been in accordance with the manners of the upper classes that the girl should be brought to marry the man . . . out of the convent—without having enjoyed any of that freedom of thought which the reading of novels will certainly produce; but we do not know that the marriages so made have been thought to be happier than our own. (Trollope, "Novel-Reading" 42)

But if Trollope's essays cast the novel as the genre of courtship, his own novels are more interested in what comes after. And a corollary to that shift from romance to realism is that his fiction upstages the texts that help their readers to reach marriage by the books that help their holders to bear it.

In that sense, the deployment of reading to mark a loveless marriage neatly inverts the age-old trope that makes dropping the book a preamble to courtship. (For Paolo and Francesca, the erotic charge comes not from reading, but from stopping.) That time line may help explain why the honeymoon forms such a crucial moment in Trollope's novels. If they define novel reading (or at least novel holding) as a postmarital activity, this isn't because of novels' sexual content (as in those other countries where girls can read fiction only once marriage has released them from the convent), or closer to home, as in yet another *Punch* cartoon, captioned "Emancipation." Trollope cares neither about what the choice of reading material suggests about a character's morals or tastes (as in, for example, in Eliot's contrast between Dorothea's high-minded appreciation of Pascal and Rosamond's middlebrow admiration for simpering verse), nor about the mimetic logic in which a character's reading of a novel about love either causes, echoes, or foreshadows her falling in love herself, nor even about some intertextual correspondence created by his placing in the hands of his characters a text named and known to his own readers.

When Emma reads opposite Charles at the table, Flaubert names names: the titles of periodicals and the authors of novels. For the space of two more pages, in fact, the narrator reproduces in free indirect discourse the content of the texts in front of her (52–53). In retrospect, the description of the breakfast table at which Emma is reading seems at best a lead-in to, at worst a pretext for, this uneasy mixture of pastiche and parody.

What interests Trollope, in contrast, isn't the relation between a person and a text so much as the relation, or lack thereof, that two persons can establish only in the presence of a printed third party. (Emma hates her husband because she reads romances, but Glencora reads romances because she hates her husband.) We usually think of the text as providing a connection to a writer who is both personally unknown to the reader and physically absent—even, in many cases, dead. In Trollope's inversion of that logic, the book offers disconnection from a known person, one who's all too physically present and all too intimately known.

In that sense, he one-ups the novel that set out to "adapt" *Madame Bovary* for English audiences, Mary Elizabeth Braddon's *The Doctor's Wife*, published in the same year as *The Small House at Allington*. During her honeymoon, Braddon's heroine shows less resemblance to Emma than to Alexandrina Crosbie. Where Emma reads without even registering Charles's proximity, Isabel (as befits a more straitlaced English bride) hesitates to read in her new husband's presence: "There were no books in the sitting-room of the family hotel; and even if there had been, the honeymoon week seemed to Isabel a ceremonial period. She felt as if she were on a visit, and was not free to read."[12] Like Alexandrina, Isabel wants to get out a book; like Alexandrina, she recognizes that desire as disrespectful to her husband. But there the similarity ends. Where Braddon makes the presence of another human being an impediment to reading, Trollope makes it the reason *to* read. Where Braddon contrasts self-indulgent reading with ceremonial visiting, Trollope parses reading itself as a ceremony.

That model fits neither with conduct-book homilies about reading aloud nor with novelistic celebrations of solitary reading: here, reading becomes most social when it's least sociable. Maud Churton Braby's

EMANCIPATION.

Young Bride of Three Hours' standing (just starting on her Wedding Trip).—"OH, EDWIN DEAR! HERE'S 'TOM JONES.' PAPA TOLD ME I WASN'T TO READ IT TILL I WAS MARRIED! THE DAY HAS COME . . . . AT LAST! BUY IT FOR ME, EDWIN DEAR."

Figure 2.5. "Emancipation," *Punch*, 5 December 1891, 270.

1909 conduct book *Modern Marriage and How to Bear It* decrees that when a man is at his club, "the wife can have a picnic dinner—always a joy to a woman—with a book propped up before her, can let herself go." In choosing a book as the marker of freedom from the husband's gaze, *Modern Marriage* sanitizes the fictional convention that made reading a symptom of marital breakdown. Yet its reasoning bears an equally uncanny resemblance to the 1857 conduct book advising women traveling alone that "civilities should be politely acknowledged; but as a general rule, a book is the safest resource for 'an unprotected female' "(quoted in K. Flint, *The Woman Reader* 100, 105). The oppressively intimate home mimics the excessively public railway carriage: in both cases, the opposite sex is what the book makes bearable.

Farther from home, Bill Bell's study of ships bound for Australia shows reading serving both as a social cement and as a guarantor of privacy. Sociable, because passengers read aloud, exchanged books, and produced and circulated manuscript newspapers; but also antisocial, as when Elizabeth Monaghan welcomed a storm because without the possibility of going on deck "I can be so much more alone, get a book and shut myself up in my cabin quite cosy." The book can be used to mark territory even in its owner's absence: as a newsletter for passengers instructs, "When the cushions at the after part of the saloon are arranged in a particular inviting manner and a book or glove is placed thereon, it may be surmised that the occupant of the couch is absent temporarily and that if another were to take possession it would be an intrusion" (136).

Public spaces like the ship and railway dramatize problems of privacy about whose domestic equivalents the novel has more to say than the conduct book. In fact, the tension between domestic and public reading etiquettes becomes one of the central themes of the professional press that springs up in the wake of the Public Libraries Act of 1850. The growing opposition between books marketed for collective reading at home and for individual use in public—on the one hand, the "Railway Libraries" founded in the 1840s and the Tauchnitz series of English-language books marketed to travelers on the Continent; on the other, series with names like "Parlour Library" and magazines called *Household Words* or *Family Paper*—simplified a reality in which members of the same family might read different books side by side in the parlor, while the same newspapers that commuters used to carve out privacy were sold by the cries of news-boys, read aloud, and passed around from hand to hand (Davies 49).

Even if the mid-Victorian novel exchanged guilt about solipsistic read-ing for cynicism about rhetorical reading, therefore, the presence of par-allel scenes in conduct books makes clear that this isn't the only genre or the only moment where cultural consumption looks like an avoidance tactic, any more than reading constitutes the only way to achieve that effect. On the one hand, the scene of a man fending off his wife with a paper would be cyclically redeployed in other genres and even other media. By the end of the century, the conjugal newspaper had figured so regularly in *Punch* that even a political cartoon could allude to the device in the confidence that readers would recognize it at a second remove. In the twentieth century, pre-Code Hollywood learned to cut between the husband unfurling his newspaper at breakfast to ward off his chattering wife and the wife, abandoned, staring blankly at a book in her solitary bed (as, for example, in Clarence Brown's 1936 *Wife vs. Secretary*). In the twenty-first, the railway novels that represented commuters hiding be-hind newspapers have given way to in-flight catalogs depicting husbands cocooned in an audiovisual equivalent to Anderson's "lair of the skull" (Anderson 35).[13]

On the other hand, reading is not the only way for novelistic charac-ters to carve out private space: Jeff Nunokawa argues, in the opposite direction, that George Eliot deploys "the sexual as the primal scene of so-cial withdrawal" ("Eros and Isolation" 839). Needlework can be rhymed with book holding as easily as newspaper reading with novel reading: in Trollope's *The Eustace Diamonds*, "Lord Fawn took up a book. Lady Fawn busied herself in her knitting" (66). Like a book, a piece of needle-work can be looked down to or up from; both can connote either oblivi-

MARRIED FOR MONEY.—THE HONEYMOON.

"Now then, Darling, put away your paper, and we'll have a nice long walk, and then come back to Tea in our own little Cottage, and be as happy a two little Birds!" said the Fair Bride— "Oh! Hang it!" mentally ejaculated the Captain.

Figure 2.6. "Married for Money—The Honeymoon," *Punch*, 1 January 1859.

ousness to others or awareness of others from whom one needs to hide. The mother in another novel enjoins her daughter not "to get into a habit of thinking, musing, and meditating, . . . sitting in a listless way with a book upon your knees which you are not reading, or a piece of embroidery between your fingers, which is continually being pulled out to correct false stitches" (Millington 395). Elsewhere, Trollope makes aggressive reading interchangeable with conspicuous sleeping: in *He Knew He Was Right,* for example, Colonel Osborne hides from his fellow traveler on the train by alternately "burying himself behind a newspaper" and pretending to sleep (219). When a Brontë character sleeps, she dreams; when a Trollope character sleeps, he shams.

In alternately flirting with and swerving from the question of what exactly it is that characters are reading, these fictions extend the traditional understanding of the novel itself as a placeholder or a blank. The reading *in* novels borrows its emptiness from the reading *of* novels. Put differently, novels project onto the newspaper their own task of reconciling what Coleridge called "indulgence of sloth and hatred of vacancy" (1:48–49).[14] Trollope himself (again, in a lecture rather than a work of fiction) complained of novel-readers' "listless, vague, half-sleepy interest over the doings of these unreal personages" (Anthony Trollope, "The Higher Education of Women" 85). Yet reading differs from sleeping or

THE HONEYMOON.

Wife (after a little "tiff"). "But you Love me, Dear"—*(sniff)*—"still?"
Husband ("Cross old thing!"). "Oh Lor', yes, the Stiller the better!"

Figure 2.7. "The Honeymoon," *Punch*, 17 May 1884, 230.

A PERFECT WRETCH.

Wife. "WHY, DEAR ME, WILLIAM; HOW TIME FLIES! I DECLARE WE HAVE BEEN MARRIED TEN YEARS TO-DAY!

Wretch. "HAVE WE, LOVE! I AM SURE I THOUGHT IT HAD BEEN A GREAT DEAL LONGER."

Figure 2.8. "A Perfect Wretch," *Punch*, 1 January 1851, 42.

"THE WANING OF THE HONEYMOON."

Right Hon. Arth-r B-lf-r (to himself). "WHAT! IS SHE TIRED OF ME ALREADY?"

Figure 2.9. "The Waning of the Honeymoon," *Punch*, 1 August 1896, 54.

sex—to state the obvious—in that it's also the activity on which those representations depend.

This isn't to say that the novels which embed reader-unresponse necessarily imagine provoking it themselves. At one extreme, Thackeray's descriptions of nonreading derive much of their shock value from breaking frame: thus a clubman "lies asleep upon one of the sofas. What is he reading? Hah! 'Pendennis,' No. VII.—hum, let us pass on" (*Sketches and Travels in London* 43). The narrator of *The Newcomes* instructs us to picture the page in front of us being read first by a woman in her husband's presence and then by a man trapped in a compartment with his wife:

> I think of a lovely reader laying down the page and looking over at her unconscious husband, asleep, perhaps, after dinner. Yes, Madam, a closet he hath; and you, who pry into everything, shall never have the key of it. I think of some honest Othello pausing over this very sentence in a railroad carriage, and stealthily gazing at Desdemona opposite to him, innocently administering sandwiches to their little boy—I am trying to turn off the sentence with a joke, you see. (151)

Such passages are as self-referential as any in *Don Quixote* or its imitators—with the difference that the commonality between character and addressee depends not on their shared reading (as it does in the passages I quoted earlier from *Madame Bovary*), but on the contrary in their shared failure to read. At the opposite extreme, Trollope avoids the second person in his descriptions of nonreading—an absence all the more striking given that his descriptions of marital unhappiness are elsewhere in *The Small House* punctuated by a refrain of "Who does not know . . . " His point is less that we must be paying as little attention to the description of Crosbie's marital battles as Crosbie is paying to the battles reported in the *Times*, than on the contrary that the dullness of the newspaper provides a foil to the interest of the novel.[15]

SELF DENIALS

Since *Don Quixote* at least, writers and critics alike have assumed that one defining feature of the novel lies in its investment in the act of reading. No other genre, this story goes, so inventively represents the act on which its own realization depends; none so ambivalently explores the pleasures and dangers of the absorption, the virtuality, and the selfhood (alias selfishness) that reading in general and fiction reading in particular exemplify. What, then, to make of novelistic representations of unread books—or, conversely, of the breakdown of narration at moments where we expect reading to happen?

One answer might be that such moments add up to an antiquixotic paradigm: a strand of realism that shares classic quixotism's obsession with the book, but that values bibliographic or social surfaces over linguistic or psychological depths. Where the novel from its beginnings has tended to imagine reading as heroically antisocial, Trollope makes it reductively other-directed. As pseudoreading replaces overreading, the old fear that fiction might prompt solipsism—as in a 1795 article in the *Sylph* that pictures mothers "crying for the imaginary distress of an heroine, while their children were crying for bread"[16]—finds its obverse in Trollope's understanding of silent reading as an interpersonal act. The *Sylph* assumes that novels bring readers closer to the consciousness of imagined characters or virtual authors only at the price of distancing them from others in their own world. But where that zero-sum logic imagines engagement as a limited resource—the more feeling we give textually mediated characters, the less is left over for our immediate surroundings— Trollope assumes the problem to be a shortage of excuses for *dis*attending. When Diderot described *Clarissa* as "a gospel brought onto earth to sunder husband from wife, father from son, brother from sister," he measured the novel's power against the strength of the social ties that it could override.[17] Trollope merely substitutes means for end.

Silent reading asserts not just one's own selfhood—as when Sarah Ellis warns that "the habit of silent and solitary reading has the inevitable effect, in a family, of opening different trains of thought and feelings, which tend rather to separate than to unite" (252)—but also, more aggressively, its right to be respected by others. Yet the more widely a book is "recognized" (and in *Kirsteen* Oliphant seems to mean both senses of that word) as a bid for what Goffman would later call "civil inattention," the ruder reading becomes. The impropriety of reading in the presence of one's spouse brings home the more general etiquette under which (as one guide rules in 1893) "a gentleman or lady may look over a book of engravings or a collection of photographs with propriety, but it is impolite to read in company" (Woodburn 219). In Elizabeth Sewell's Tractarian novel *Gertrude,* a girl asked to help her mother with some sewing immediately picks up a novel: "'Well, I will see about it presently,' replied Jane; and she went to fetch her book, and then, seating herself by the drawing-room window, forgot her mother's wishes" (7). (The mother who reads Miss Braddon instead of washing the supper dishes finds her match in a daughter who engages in leisure activities instead of contributing to the domestic economy.) Any Brontë reader will recognize the name and the window seat: the only difference lies in the value attached to solipsism by a didactic bildungsroman or an "antichristian composition."

What the bildungsroman codes as selfhood, didactic texts parse as selfishness. An 1894 conduct book trying to illustrate consideration for others can find no clearer example than that of a girl refraining from reading. "You are sitting, let us suppose, by a sleeping invalid, the third volume of your novel with its thrilling dénouement is on the mantel-piece just out of your reach. Your boots creak, or your dress rustles, you dare not stir; there you have to sit, perhaps in growing dusk, and you dare not light a candle. These are the kinds of little self-denials that really touch us" (quoted in K. Flint, *The Woman Reader* 93). Where we think of absorption as a virtue—to check Facebook is to succumb to laziness, to read a novel cover to cover is to find a stable self—Victorian conduct literature, and fiction, valued the willingness to be distracted. Another Elizabeth Sewell novel praises its heroine for "fetching her work directly she knew that it ought to be had; preparing for a walk as soon as the proper time arrived; *giving up an interesting book* whenever a superior duty claimed her attention"; the measure of her moral improvement is that "Margaret put aside her novel, though she had reached the most interesting part of the third volume" (*Margaret Percival* 1:240, 52; my emphasis).

In the midcentury tract *Susan Osgood's Prize* (of which my own copy is inscribed as a Sunday-school reward), the heroine's fascination with a copy of Edgeworth's *Simple Susan* given to her as a school prize instills two irreconcilable desires: to find herself in her namesake and to lose herself in the book. When her sister asks her to fill the kettle, "Susan was still busy over the picture, and was wondering which part of the story it described; she did not move at first. 'Now, Susan, dear,' said her grandmother, kindly, 'be brisk, I see you are not a "Simple Susan" yet.'" The desire to read exemplifies the desire to ignore others: Susan "had been put out when her father called her from her book, to weed his flower beds." Conversely, we know that the lessons of *Simple Susan* have been absorbed once Susan Osgood refrains from reading it: "The little book looked tempting on the table by Grannie, but Susan could not get to it yet. There was 'washing up' to be done" (Prosser 20, 29, 22). The logic is mechanical but irrefutable: if reading makes you a bad wife or mother or daughter (or, as we'll see in chapters 4 and 6, an even worse servant), then not reading must make you a good one.

Or, at least, not reading in the wrong place, at the wrong pace. Getting to the middle volume and stopping right there offers a bibliographic equivalent to the withdrawal method: just as sex becomes acceptable when interrupted, so reading is sanitized by discontinuity. If pacing one's reading by volume breaks implies self-restraint and even "self-denial," conversely the narrative momentum that overspills its material containers can be equated with selfishness and even aggression. The conduct book doesn't specify exactly what you risk doing to the invalid if you open a third volume the minute you close the second, but a discussion of page turning published two decades earlier provides a hint. In 1873, James Greenwood opened his article "Penny Awfuls" by invoking Harrison Ainsworth's Newgate novel *Jack Sheppard* (1839), whose highwayman hero was widely blamed for corrupting its readers. From the belief that texts *about* stealing cause boys to steal, however, Greenwood goes on to identify the stealing *of* books as the first step in a life of crime. Later in his article, a young thief testifies that his brother was first corrupted by reading *Tyburn Dick*—or rather, by *not* being able to read it. After gazing at a page spread displayed in the shop window, whose conclusion is cut off by the page break, both boys lie awake "wonderin' and wonderin' what was over *leaf*.'" The brother "wasn't a swearin' boy, take him altogether, but this time he did let out, he was so savage at not being able to turn over." Not content with word crimes, the brother steals a hammer to buy the desired number, the informant steals an inkstand to pay for the next, and a pattern is set (J. Greenwood 166).

Middle-class girls' reading might appear to pose a very different moral threat from working-class boys' literacy. What cuts across that divide between political and domestic registers, however, is a new awareness of the book as something more (and less) than a container. In both cases, the traditional model of imitative reading—in which a text *about* a crime (whether adultery or theft) is reproduced by its readers' real behavior—is upstaged by the possibility that the book itself might occasion antisocial behavior. Copies of *Jack Sheppard* lead boys not to imitate Jack Sheppard in stealing silver spoons, but to steal copies of *Jack Sheppard*. Appropriation of the text (as in Michel de Certeau's metaphor of "reading as poaching") gives way to appropriation of the book (Certeau).

More specifically, the linear form of the book becomes more threatening than its thematic content. For the traditional worry that the content of a text can corrupt its readers, Greenwood and the conduct book both substitute the fear that the material obstacles to narrative continuity—whether a page break or a volume break—can either test or overstrain the reader's self-control. Threatening an invalid's health or murdering a master: a multitude of sins can be blamed on the hunger to "turn over," or to see "what was over leaf." To surface on bibliographical cue is to deny oneself, while to be absorbed in the text is to kill the other.

DUMMY SPINES

Trollope's interest in faked reading makes it striking how devoid his fictional world remains of fake books: no dummy spines, no sofa-table albums. Nowhere does his moral code invoke the glib antimaterialism of Gaskell's contrast between the Thorntons' sofa table "with smartly-bound books arranged at regular intervals round the circumference of its polished surface, like spokes of a wheel," and the Hales' house where "books, not cared for on account of their binding solely, lay on one table, as if recently put down" (Jewsbury 13; Gaskell, *North and South* 79, 112). Nor should Eliot's quip about tulip petals be confused with the joke structuring a novel like Trollope's *Cousin Henry*, whose plot revolves around a will slipped between the pages of a volume of sermons that no one ever thinks to take from the shelf.[18] By representing the newspaper aggressively unfurled, and by turning "laid open" into a synonym for "unread," Trollope and Eliot reverse the trope of the closed book. Their language figures illegibility not in terms of outwardly visible signs like dust or uncut pages (remember Gaskell's description of "the great, large handsome Bible, all grand and golden, with its leaves adhering together from the bookbinder's press") but on the contrary in terms of the impossibility of distinguishing—even or especially on "closer view"—"reading" from "pretending to read" (*Mary Barton* 369).

An ethical tension, then: while the verbal content of novels forces readers to empathize with other minds, the material heft of the book allows them to block each other out. But also a formal problem: the unrepresentability of reading becomes a proxy for the incredibility of selfhood. The wedge that novels drive between the outside of books and the interiority of readers, or between material cover and verbal content, forces the genre to choose between describing the look of reading and its feel. I posited at the beginning of this chapter that any turn toward material media means a swerve away from both the text and the mind—as if the narrator needed to stake out a vantage point either inside the pages or outside the covers. In coding the handling of books as authentic and the reading of texts as a front, Trollope's comedies of manners upstage textually occasioned absorption by bibliographically assisted repulsion; but, more crucially, they abdicate any attempt to plumb psychological depths. Whenever the novel juxtaposes competing vocabularies in which to describe a printed object—whether it puns on "remembered her novel" or replaces "page" with "paper," or substitutes the opening of a book for the beginning of a text—it stages questions about the relation of the inner life to the object world.

In making behaviorism to solipsism what book is to text, novelists of manners also prefigure the challenge facing historians of reading: how to observe an activity against which the social defines itself. Chapter 4 will argue that the conventions for representing reading elaborated by competing subgenres of Victorian fiction have trickled down (through the intermediary of biography) into competing schools of twentieth-century historiography. It's to the second such subgenre that the next page turns.

CHAPTER 3

David Copperfield and the Absorbent Book

To ACKNOWLEDGE THAT the trope of repulsive reading cuts across classes, genders, and even media isn't to deny that the novel—traditionally the genre most invested in absorptive reading—lends that cliché a polemical edge. The comedies of manners that we've just seen enlisting the book in hostilities between husbands and wives mirror the scene of reading that opens the most canonical of Victorian bildungsromans. In these novels, the child's sense of self is jump-started not by reading, but by being hit with a book. (Or boxed on the ear with an encyclopedia, or poked in the ribs with a prayer book, or knocked off balance by having a book wrested out of his hands.)

This chapter asks what makes the novel associate printed matter with violence in general and interruption in particular. And what difference does it make that the act in which characters are interrupted consists not (contrary to what most readers remember) of reading, but rather of using an unread book as a material prompt or alibi for inwardness and abstraction? One answer, I'll suggest, is that the bait and switch that structures the midcentury bildungsroman sets us up to expect a novel about an agent *shaped by* books, only to reveal the protagonist instead as an object *compared to* books—metaphorically imprinted, bound, sold, and scanned.

READING AWAY

You'd think that bookishness began and ended in childhood. The child is small, lonely, sensitive, thoughtful, a strong imagination in a weak body; the child retreats to attics, lumber rooms, and window seats, fencing herself or himself behind a copy of Bewick's *Birds*, of *Robinson Crusoe*, of *The History of the Devil*, only to be jolted back into his or her surroundings by a poke with a prayer book or a volume thrown at her face or a box on the ear with a Latin Grammar. The child is baptized Jane or David, but will reappear under the names of Maggie Tulliver, Barbara Churchill, Aurora Leigh, Nell Adair, Beth Caldwell. By 1883, when Tant' Sannie catches Waldo reading Mill's *Political Economy* and "fling[s] the book at his head with much energy," the narrator of *The Story of an African Farm* can add that "books have been thrown at other heads before or since" (Schreiner 101). Without renouncing the Protestant tradition that celebrates texts as the weapon of the weak, these novels cast books, more literally, as weapons of the strong.

Like the Trollope novels that we saw in the previous chapter and the religious tracts to which we'll turn in Part II, the bildungsroman makes books a pawn in familial power struggles. Yet for the omnisciently narrated symmetry that pits the husband's unread newspaper against the wife's unread novel, first-person narrative (or more rarely, free indirect discourse) substitutes a less evenly matched battle between young and old—the former corresponding to the character through whose consciousness the narrative is focalized (nose in a book), the latter to characters viewed from the outside (book in hand). Where Trollope makes mutual avoidance an unlikely leveler, this more Manichaean subgenre projects the contrast between word and object at once onto a moral axis (a love for tattered pages signifies virtue; for morocco albums, vulgarity), a formal hierarchy (protagonists use the book as a mental prompt, minor characters as a manual prop), and a social structure (one of the many fantasies that the bildungsroman fulfills is that the reader's inner resources can overcome economic constraints).

Not that villains or minor characters or poor adults or rich children have no use for books. Even a dunce like John Reed knows that they can be owned and thrown: "I'll teach you to rummage my book-shelves: for they *are* mine." In return, all Jane Eyre can hurl is names—"'You are like the Roman Emperors!' I had read Goldsmith's *History of Rome*" (C. Brontë, *Jane Eyre* 17). Where Jane remembers an absent text, John holds an unread book. A recent plot summary that describes *Jane Eyre* beginning "with an assault upon a girl by a male text, or, more correctly, by a male *armed* with a text" confuses the text—whose contents John shows no sign of having read—with the brute materiality of the book (Frith 153).

In *David Copperfield*, too, the disembodied text faces off against a weaponized book. "I had a greedy relish for a few volumes of voyages and travels—I forget what, now," the narrator tells us,

> and for days and days I can remember to have gone about my region of our house, armed with the centre-piece out of an old set of boot-trees—the perfect realization of Captain Somebody, of the Royal British Navy, in danger of being beset by savages, and resolved to sell his life at a great price. The Captain never lost his dignity, from having his ears boxed with the Latin Grammar. I did. (60)

Dramatizing what Michael Fried calls "absorption," these scenes correlate the reader's attention to the text with his oblivion to the world around him (Fried, *Absorption and Theatricality*).[1]

Or at least the user's: for what the Murdstones do with the Grammar (or John Reed with Bewick's *Birds*) interrupts an act parasitic upon—but hardly synonymous with—"reading." In place of a book, David holds only a boot-tree: the book reduced to a prosthetic jackboot (to be thrown at those who would otherwise deserve kicking) matches the stick exalted to a prompt for the imagination.[2] Remember the analogy between novelist and shoemaker: "I can ride a boot-jack as I would a hobby-horse," boasts G. A. Sala in the pages of *Household Words*; don't "stretch your imagination with the boot-tree," warns *Judy, Or the London Serio–Comic Journal* (Sala; "Things It Is Better Not to Do"). But the boot-tree also forms a reductio ad absurdum of unread *books*' serviceability as a prompt for mental images—as when Edith Wharton recalls that in her childhood "if the book was in reach, I had only to walk the floor, turning the pages as I walked, to be swept off full sail on the sea of dreams. The fact that I could not read added to the completeness of the illusion, for from those mysterious blank pages I could evoke whatever my fancy chose" (Wharton, *Backward Glance* 33–34).

Unread books, or books read at some other time: nineteenth-century fiction intuited that what cognitive scientists today call "offline" processes (remembering, recombining, constructing coherence after the fact) could be more intense than the "online" act of reading itself.[3] Thus Madame Bovary's experience is most intense when she has ceased to see (let alone read) the page: "At length, her eyes growing tired, she would close them and see, in the darkness [a description of the world represented in the novel follows]" (Flaubert 52). Where Jane responds to a thrown book with the memory of having "read Goldsmith's *History of Rome*," conversely the box administered with a Latin Grammar punishes David not for reading but rather for remembering the already-read.

For Dickens as for Trollope, the moment of reading is as formally unrepresentable as thematically central. If visible, reading remains inauthentic; if meaningful, ineffable. To describe the experience is to debunk it. The materialism of a Reed or a Murdstone bears more resemblance to the external perspective of the Trollopian narrator than does the idealism of a Jane or a David. Where in *The Small House* the act of reading flies below the narrator's radar, in *David Copperfield* it soars above his descriptive powers. This vanishing point should be familiar to reception historians, dogged as we are by the paradox that the most engaged reading is often the most invisible: the more deeply a book marks its reader, the fewer the marks left on its pages (H. J. Jackson, *Marginalia*). In novels, too, those moments where reading is dismissed as a sham are hard to distinguish from those where reading is elevated to a secret.[4] The only difference is that the former involve children hiding from adults, the latter spouses hiding from spouses.

Books allow the child to withdraw into his mind and adults to drag him back into his body. When David's thoughts wander in church, "Miss Murdstone pokes me with her prayer-book"; when they wander at home, Mr. Murdstone (who sits "pretend[ing] to be reading a book") "takes the book, throws it at me or boxes my ears with it." The text remains in David's or Jane's head even when the book is no longer in their hands; conversely, the Murdstones and Reeds handle books instead of internalizing their contents. Mr. Murdstone "took a book sometimes, but never read it that I saw. He would open it and look at it as if he were reading, but would remain for a whole hour without turning the leaf" (57, 59, 128).

The manual gestures of the nonreader stand opposite the good reader's out-of-body experience. Helen Keller's assertion that "literature is my Utopia. Here I am not disenfranchised. No barrier of the senses shuts me out from the sweet, gracious discourse of my book-friends" invokes the consensus not only that books are the friends of the friendless, but that the ideal reader experiences through the mind rather than the body (Keller 117). Michel de Certeau speculated that for the past three centuries a "withdrawal of the [reader's] body, which is the condition of its autonomy, puts the text at a distance. It is the reader's *habeas corpus*" (175–76).

The throwing of individual volumes at the physically frail and socially disempowered will eventually snowball into the toppling of an entire bookcase that concludes *Howards End* (1910).[5] The interchangeability of the weak, poor, feminized clerk with the weak, poor, female or feminized stepchild (remember how often David Copperfield is referred to as "it") reveals how central gender is to the battle pitting a soul absorbed in a text against a body brandishing a book. In Sarah Grand's New Woman novel *The Beth Book*, women's retreat into a virtual realm is interrupted by books whose physicality shares the grossness of men's bodies. We fear the worst of Beth's father from the very first chapter when, exasperated by his pregnant wife, "he flung the book across the room." Later, Beth herself uses reading to ward off sex: "'Must I be embraced again?' she exclaimed one day, with quite comical dismay, on being interrupted in the middle of a book that was interesting her at the moment."[6] Like the Cold War–era housewives described by Radway, female characters use the right to read as a proxy for the right to withdraw from ministering to others' needs for sex (as in Beth's husband's embrace) or food (as in the *Sylph*'s "children crying for bread"). No surprise, then, that the materiality of a thrown book should provide a reminder of the materiality of a vulnerable body.

Victorian novels, that is, anticipate twentieth-century book historians' populist recuperation of reading as an act by which apparently passive social groups appropriate and subvert official dogmas. Michel de Certeau compares readers to "guerillas," describing reading as "the clandestine forms taken by the dispersed, tactical, and makeshift creativity of groups or individuals already caught in the nets of 'discipline' " (xv). A more local measure of the impact of this logic is that the humor of the opening scene of *Vanity Fair* hinges on readers' recognizing the trope of a book thrown at (rather than by) a dependent. Becky Sharp's audacity consists not just in attacking her benefactress, but more specifically in reversing the traditional direction of the book as it travels from richer and older assailant to poorer and younger victim. Yet even if Thackeray's governess plays John Reed to Miss Pinkerton's Jane, elsewhere he, too, imagines the weight of the book breaking the spell of the text: "Do I forget one night after prayers (when we under-boys were sent to bed) lingering at my cupboard to read one little half page more of my dear Walter Scott—and down came the monitor's dictionary upon my head!" (Thackeray, *Roundabout Papers* 334).[7] That violence literalizes the weight of the "heavier" schoolbooks whose content David Copperfield is punished for failing to memorize; if light reading gives solipsistic pleasure, heavy books signal painful contact (60). And where the text establishes the boundaries of the self, the book violates them. By century's end, the metonymic association of hitting a child reader with interrupting a child's reading is strong enough that one can become a metaphor for the other: the narrator of *A Little Princess* observes that "never did she find anything so difficult as to keep herself from losing her temper when she was suddenly disturbed while absorbed in a book. People who are fond of books know the feeling of irritation which sweeps over them at such a moment . . . 'It makes me feel as if someone had hit me,' Sara told Ermengarde" (Burnett 27). By making the text that fills minds coincide with the thing that thumps heads, these opening scenes dramatize two competing media theories. In one, the book's power comes from its material attributes (weight, heft, the volume of every volume); in the other, a sequence of words (no matter whether in the memory or on the page) crowds out any awareness of the physical world.

From the other side of the countless twentieth-century autobiographies that borrowed this trope from nineteenth-century fiction, it's hard to appreciate just how radically new its premises were. In the eighteenth century, references to the book as object or commodity can be found most often in the narrator's own voice—though "narrator" may be the wrong term, since such references tend to interrupt the narrative, breaking frame so violently that readers have no choice but to surface temporarily from their absorption. By the middle of the nineteenth, however, characters, not narrator, are now the ones who notice the look of the book. Where self-referential asides once broke readers' concentration, now a represented book is what interrupts characters' absorption. What once took place on the level of discourse now migrates to the level of story. Instead of an "editor's" paratextual jokes breaking into the narrative as in *Tristram Shandy*, now a book thrown by one character breaks into the text read by another. The violence of book throwing at the level of story replaces the violence of frame breaking at the level of discourse. Replaces, or at least supplements: for every time a novel reminds us of the sensory attributes of the object we're holding—and by extension, reminds us of our own eyes and our own hands—it shatters our concentration as violently as John Reed or Miss Murdstone breaches David's or Jane's. John aims the book to avoid breaking windows, but book throwing still ruptures the transparency of mimesis.

By 1850, paper falls under the same taboo as sex: "the page" (as Gissing put it in a different context) "scarce rustles as it turns." Scatological jokes and self-referential asides become equally recognizable as throwbacks to the eighteenth-century satirical tradition. When the narrator of *Vanity Fair* relates that "the curses to which the General gave a low utterance . . . were so deep, that I am sure no compositor in Messrs. Bradbury and Evans's establishment would venture to print them were they written down," the shock of the unnamed curse is conveyed through the breach of publishing decorum (285). Harking back to the era in which Fielding could entitle a chapter "Containing Five Pieces of Paper" (*Tom Jones* bk. 4, chap. 1), the eighteenth-century pastiche of *Esmond* and *The Virginians* gave Thackeray an alternative to a more modern domestic realism that was thematically close-minded and formally closure-driven, equally opposed to linguistic and to sexual digression and play. "Here it is—the summit, the end—the last page of the third volume": the fall from the metaphor of a "summit" to the literalism of a "page" and a "volume" registers the narrator's skepticism that happy endings can reflect anything more than the telltale compression of pages.

UNREAD BOOKS (*JANE EYRE* AND *THE MILL ON THE FLOSS*)

In the eighteenth century, a joke; by the nineteenth, a threat. The consciousness of the book's physicality that was witty in the mouth of the narrator becomes immoral in the minds of characters: to thematize materialism is also to stigmatize it. We might predict, then, that the bildungsroman quarantines any awareness of the book-object within the consciousness of its least sympathetic characters—that the text serves as a catalyst for sympathetic protagonists' daydreams, the book as a press for minor villains' flowers. Yet even the characters whom we remember as readers usually turn out to be doing something to the book that it would take a stretch to describe as "reading." What jump-starts the narrator's interiority isn't love of texts so much as hatred of books—whether fear of the books handled by the Murdstones or disgust at the books fingered by Uriah Heep. The child reader's out-of-body raptness finds its foil in Uriah "reading a great fat book, with such demonstrative attention, that his lank forefinger followed up every line as he read, and made clammy tracks along the page (or so I fully believed) like a snail" (Dickens, *David Copperfield* 222). A mind marked by the text recoils from a book marked by the body. The ink of marginalia looks no better than an animal's slime. Yet here as so often, clerical work threatens any distinction between Uriah and David—who, despite working at shorthand "like a cart-horse," finds himself "laboriously and methodically plod[ding] over the same tedious ground at a snail's pace" (505).

The corollary is that Brontë and Dickens hardly provide their own readers with role models. Between the girl retreating behind the covers of a book and the boy acting out a remembered text, reading dwindles to a vanishing point. In one case, inwardness is occasioned by the book, not the text; in the other, the text enters David's mind only once the book has left his hands. To call it "Jane's salvation to be a reader" or to describe her as an "avid and impressionable reader" is to misremember that from the very first page of the novel, Jane is staring at the pictures, not the text (L. Green, *Educating Women* 28; Brantlinger 115).[8] Jane "cares little" for the "letter-press" of Bewick's *Birds*, as little as Maggie Tulliver does for the text of the *History of the Devil*: the best she can say of the introductory pages is that "I could not pass [them] quite as a blank." The "quite" acknowledges a bibliographical version of the pathetic fallacy: the whiteness of the page seems to have rubbed off on the world that the pages represent, with its "death-white realms" and "forlorn regions of dreary space,—that reservoir of frost and snow." Like the encroaching margins of "Baxter's Procrustes" or the out-of-focus newspapers that Trollope represents, the book figures here as a negative space, "bleak" if not "blank."

The opening scene of *The Mill on the Floss* upstages content by heft: introduced "dreaming over her book," Maggie Tulliver soon "forget[s] all about her heavy book, which fell with a bang within the fender" (G. Eliot, *The Mill on the Floss* 18–19). Even as Eliot and Brontë both make the use of books a proxy for moral worth, then, the criterion is oddly negative: it's less that readers identify with characters who read, than that we distance ourselves from characters who recognize the book's material qualities. (We hate those who love the book-object as much as we love those who hate it.) Reading is one possible way of crowding out that awareness, certainly, but so are daydreams whose starting point lies no further into the book than the title page. No surprise that Mr. Tulliver buys the *History of the Devil* because "they was all bound alike—it's a good binding, you see—an' I thought they'd all be good books . . . They've all got the same covers, and I thought they were all o' one sample, as you may say. But it seems one mustn't judge by th'outside" (G. Eliot, *The Mill on the Floss* 251). What's more disturbing is that Maggie, too, "did not take the opportunity of opening her book": not content to deny that Maggie is reading, the narrator refuses even to go beyond the absence of the more minimal material bodily gesture that might or might not signify the mental act.

What catalyzes Maggie's "dreaming over her book" is not its contents but rather her looking away: "somehow, when she sat at the window with her book, her eyes would fix themselves blankly on the outdoor sunshine" (18, 299).[9] And that blank corresponds to a space that books should have occupied: "her eyes had immediately glanced from him to the place where the bookcase had hung; there was nothing now but the oblong unfaded space on the wall, and below it the small table with the Bible and a few other books" (251). Even when text does enter Maggie's consciousness, it stops at the title page.

> Then her brain would be busy with wild romances of a flight from home in search of something less sordid and dreary;—she would go to some great man—Walter Scott, perhaps, and tell him how wretched and how clever she was, and he would surely do something for her. But in the middle of her vision her father would perhaps enter the room for the evening, and, surprised that she sat still, without noticing him, would say complainingly, 'Come, am I to fetch my slippers myself?'" (300)

Scott appears as a secondary character in Maggie's fantasized autobiography, not as the author of a real book. For Maggie as for her father, the printed book functions less to contain a story than to occasion one—whether handwritten (in the case of the family Bible) or purely mental (in the case of Maggie's "vision"). Like marginalia unrelated to the content of the text being written in, daydreams can become untethered to the book being held.

More strikingly, the conjunction of Maggie's flight of fancy with her imagined "flight from home"—correlated in turn with Mr. Tulliver's more immediate consciousness of being ignored—rules out the zero-sum competition between attention to books and attentiveness to relatives that the novels discussed in the previous chapter took for granted. Here, Mr. Tulliver's realist slippers are displaced not by Scott's represented armor, but on the contrary by a daydreaming that crowds out the page and the world alike: instead of finishing *The Pirate*, Maggie "went on with it in my own head" (318).

The Mill on the Floss leaves no safe vantage point from which to condemn such lumping, for the character in whose consciousness readers become most deeply absorbed is herself no reader. On the contrary, Eliot places books in the hands of characters only to thwart our assumption that they will be read: unlike the gun whose appearance in the first act guarantees that it will be fired before the end of a play, *The Pirate*, whose first volume Maggie remembers having begun in the second volume of *The Mill on the Floss*, remains unfinished by the end of the third. Conversely, readers' noses are rubbed in the materiality of the book they themselves are holding: when Book II ends with the narrator announcing that "the golden gates of their childhood had for ever closed behind" Maggie and Tom, we can't help noticing that the covers of the volume are about to close behind us as well.

Chapter 5 will return to the question of what Maggie does with the books she fails to read. For now, consider the analogous problem in *Jane Eyre*. In the wake of John Reed's book throwing,

> Bessie had been down into the kitchen, and she brought up with her a tart on a certain brightly painted china plate . . . Vain favour! coming, like most other favours long deferred and often wished for, too late! I could not eat the tart; and the plumage of the bird, the tints of the flowers, seemed strangely faded: I put both plate and tart away.
>
> Bessie asked if I would have a book: the word *book* acted as a transient stimulus, and I begged her to fetch Gulliver's Travels from the library. This book *I had again and again* perused with delight . . . Yet, when this cherished volume was *now placed in my hand*—when I *turned over its leaves*, and sought in its marvellous pictures the charm I had, till now, never failed to find—all was eerie and dreary; the giants were gaunt goblins, the pigmies malevolent and fearful imps, Gulliver a most desolate wanderer in most dread and dangerous regions. I closed the book, which I dared no longer peruse, and put it on the table, beside the untasted tart. (28; my emphasis).

Unread book joins untouched tart: the content fails to live up to the promise of the container. Where remembered reading is invested with "delight," the book loses its magic once placed in the present or "placed in my hand." Manual gestures short-circuit mental operations: "I turned over its leaves" leads directly to "I closed the book." Like Captain Somebody's travels (or like Goldsmith's *History of Rome*), Gulliver's are read only offstage. "My own thoughts always swam between me and the page I *had usually* found fascinating": Jane is still a child, but it's never too early to start banishing reading to a prelapsarian past (27; my emphasis).

To the past, or to minor characters, for the absorbed reading that critics misattribute to Jane is in fact displaced onto Helen Burns: "absorbed, silent, abstracted from all round her by the companionship of a book, which she read by the dim glare of the embers" (65).[10] Where the opening scenes lead us to expect a protagonist's reading to be interrupted by minor characters, here, on the contrary, it's the heroine who shatters a secondary figure's textual "abstraction." The first time Helen pauses to turn a page, Jane interrupts her to take the book from her hands; when Helen begins to read again, "again I ventured to disturb her." Jane's only role in this scene is to interrupt Helen's reading—as if her cousinship with John Reed were as hard to disclaim as Maggie's with the Dodsons (60).

Bafflingly, Jane explains her urge to stop Helen from reading by the assertion that she herself likes to read: "I hardly know where I found the hardihood thus to open a conversation with a stranger; the step was contrary to my nature and habits: but I think her occupation touched a chord of sympathy somewhere; for I too liked reading, though of a frivolous and childish kind" (59). To be a reader is to claim an identity, not to perform an action. Or even to allow others to perform it unmolested: famously no fan of Austen's, Brontë may nonetheless be riffing on the scene in *Pride and Prejudice* where Miss Bingley, yawning over a prominently displayed book, interrupts Darcy's reading to remark, "How much sooner one tires of anything than of a book!" (37). Like Miss Bingley, Jane proclaims her love of reading at the very moment when she prevents others from engaging in it. Yet what begins as a cliché (of course a hypocrite will lack any authentic love of reading) becomes more unsettling when the action is transferred from a blocking figure to the heroine.

In the other direction, Brontë prefigures the Trollopian narrator's perception of the book as a buffer between men and women. At first, Jane recognizes her kinship with the Riverses by listening to them read aloud—from the outside of the window this time, in a mirror image of her silent perusal of Bewick's illustrations. Soon, however, she begins to describe their holding of books as an antisocial gesture: "St. John had a book in his hand—it was his unsocial custom to read at meals—." Later, St. John cuts short an awkward conversation on the excuse of asking her to tell him the location of a particular book:

> I showed him the volume on the shelf: he took it down, and withdrawing to his accustomed window recess, he began to read it.
> Now, I did not like this, reader. (438)[11]

By juxtaposing a direct address to the novel's reader with her dislike for another character's reading, the narrator makes visible the tension between reading as a bridge linking an author with an unknown audience, and reading as a barrier separating members of a single household.

St. John's "window recess" pays back Jane's long-ago window seat: St. John likes me, the implication goes, as little as I ever liked John Reed. Or, more crudely: when I read, it's interiority; when you read, it's hostility. Absorption can connote selfhood as easily as selfishness: Dickens and Brontë code as a psychological good what the conduct books quoted in the previous chapter cast as a moral evil. Even the sociable reading exemplified by Diana and Mary voicing Schiller—a use of books that, far from sundering characters sitting at the same table, draws in the outsider crouched on the wrong side of the glass—can never be more than a temporary recuperation of stories that will always be dragged back to their print origins. In her own window seat, Jane reflected that in Bewick "each picture told a story . . . as interesting as the tales Bessie sometimes narrated on winter evenings . . . taken from old fairy tales and older ballads; or (as at a later period I discovered) from the pages of *Pamela*, and *Henry, Earl of Moreland*" (15). The comparison reduces print to a way station between the visual and the oral. Moving from "letter-press" to picture to voice, the novel runs a technological time line in reverse—with the twist that Bessie's stories turn out themselves to be traceable back to print, coming full circle to the medium that Jane initially dismisses. Their bibliographical provenance is identified only offstage, in parentheses, in the future.

UNASSIGNED READING

In *David Copperfield*, a different set of punctuation marks elides editorial backstory. They occur immediately before the boot-tree incident:

> My father had left a small collection of books in a little room upstairs to which I had access (for it adjoined my own), and which nobody else in our house ever troubled. From that blessed little room, Roderick Random, Peregrine Pickle, Humphrey Clinker, Tom Jones, The Vicar of Wakefield, Don Quixote, Gil Blas and Robinson Crusoe came out, a glorious host, to keep me company. They kept alive my fancy, and my hope of something beyond that place and time—they, and the Arabian Nights, and the Tales of the Genii . . . I had a greedy relish for a few volumes of voyages and travels—I forget what, now—that were on those shelves; . . . When I think of it, the picture always rises in my mind of a summer evening, the boys at play in the churchyard, and I sitting on my bed reading as if for life. (60)

This passage must rank as one of Dickens's most memorable, and memorized. The final phrase continues to turn up Micawberishly today in venues as varied as a commencement speech, a movie review, a philosopher's memoir, and the acknowledgments to an academic monograph. Sometimes it's bowdlerized by the omission of the "as if"; sometimes intensifiers are added, as when an autodidact's memoir riffs that "I see myself in the far away time and cottage reading, as I may truly say in my case, for dear life" (Rose, *The Intellectual Life* 3); sometimes a single noun serves to conjure it up, as in Sartre's memory of posing with a book (before he even knew how to read) in the solitude of a lumber room (*cabinet de debarras*) (Sartre 36).[12] Even earlier, Amelia Edwards's otherwise *Jane Eyre*–inflected *Barbara's History* borrowed this passage almost verbatim, changing nothing but gender. After its young heroine wanders into a lumber room empty except for old boxes, "in one, the *smallest and least promising* of all, I found a dusty treasure. This treasure consisted of some three or four dozen wormeaten, faded volumes, tied up in lots of four or six, and overlaid with blotches of white mould." To the list of books that follows, the narrator adds: "Other books I had as well—*books better suited to my age and capacity; but these, being common property, were kept in the school-room*, and consisted for the most part of moral tales and travels, which, read more than once, grow stale and wearisome. Fortunate was it that I found this *second life* in my books; for I was a very lonely little girl, with a heart full of unbestowed affection, and a nature quickly swayed to smiles or tears. The personages of my fictitious world became *as real to me as those by whom I was surrounded in my daily life.* They linked me with humanity. *They were my friends, my instructors, my companions*" (Edwards; my emphasis).

The catchphrase's ubiquity suggests that "reading [as if] for life" provides convenient shorthand for a post-Romantic paradigm that makes reading the recourse of the poor, the lonely, the marginalized, the physically or socially powerless. The books inside the "smallest and least promising of all" boxes find their match in the smallest and least promising of children. Adults have little place in this model, except (at best) to remember their own past reading or (at worst) to punish children for reading in the present. *David Copperfield* celebrates the model of selfhood that the newspaper in Trollope's fiction so cruelly parodies: whether from the inside or the outside, the book is imagined as a shield against others who inhabit the same domestic space. Just as the books that constitute Edwards's metaphorical "friends" and "companions" crowd out human friends and companions, so the virtual "company" provided by literary characters isolates the reader from the human beings around him—from the "boys at play" as much as from "[every]body else in our house." Although David's "fancy" can take in the wildest counterfactuals, the one possibility to which his imagination never stretches is that his taste for reading might be shared.

Shared by relatives living in the same house, but also shared by the mass audience to which printed books are marketed. When Forster

quotes this passage, he points out that in translating his "autobiographical fragment" into the novel, Dickens made one silent change: "his omission of the name of a cheap series of novelists then in course of publication, by means of which his father had become happily the owner of so
large a lump of literary treasure in his small collection of books." The
metaphorical "price" for which the captain sells his life can be named,
but the price of the "cheap series" can't. Or even its title: David's "—I
forget what, now—" strips paratext away from text as aggressively as
Jane's "(as at a later period I discovered)" distances chapbooks from
storytelling. Both sets of punctuation marks quarantine the provenance
from which (and for how much) stories make their way into the child's
consciousness. Both children must wait until adulthood to discover that
their own imaginative investments are shared with tens of thousands of
other owners of mass-produced reprints.

Where Don Quixote learns that the one thing unmentionable in romances is the price that anyone paid for a night's lodging, by reading
bildungsromans we learn, closer to home, not to ask what books cost. To
inherit them is to preserve one's innocence. Even to steal them, like the
young narrator of Stendhal's *Vie de Henry Brulard*, is better than to buy.
The vulgar owning without reading epitomized by sofa-table books and
dummy spines finds its antithesis in reading without buying (the child
stumbling across a collection of books that are literally priceless), and
even reading without owning (remember the hero of *Ranthorpe* freeload-
ing at a bookstall). Twenty-first-century social-scientific surveys give the
lie to this logic: according to them, the number of books owned in a
household correlates more strongly with a child's academic success than
does the parents' educational level (Evans et al.). Reading aloud and library patronage are even weaker predictors: better to have books lying
around unread than to read books that don't belong to you.[13]

In our cultural imaginary, in contrast, reading undoes the book's commercial taint. A second bookstall scene in *Oliver Twist* shows engagement with the text making the marketplace disappear.

> [Mr. Brownlow] had taken up a book from the stall, and there he
> stood, *reading away*, as hard *as if* he were in his elbow-chair in his
> own study. It is very possible that he fancied himself there, indeed;
> for it was plain, from his *abstraction*, that *he saw not the bookstall*,
> nor the street, nor the boys; not, in short, anything but the book itself
> which he was *reading straight through*. (Dickens, *Oliver Twist* 74; my
> emphasis)

"Reading away" literalizes the sense in which reading *takes* Mr. Brownlow away from his surroundings; "reading straight through" literalizes
the sense in which the book becomes (if I can borrow the language that

David Copperfield later applies to a play) "like a shining transparency, through which I saw" (271). This time, "as if" reconciles the realities of the market with their banishment from the text. What Mr. Brownlow "fancies" is not the story he's reading, but rather the position of its reader: he imagines not that the fiction is true, but that he is standing somewhere other than the market.

Just as reading without buying is admired and buying without reading ridiculed, so a room dubbed the "study" but filled with guns and fishing-rods shows vulgarity as clearly as books found in a space not supposed to contain them (a garret, under the bedcovers, outdoors) bespeak good taste. Surprisingly, the scene of standing up at the bookstall makes the market another such space. In "George de Barnwell" (a parody of Bulwer-Lytton in *Punch's Prize Novels*), the young hero reads in the middle of a grocer's shop:

> Immersed in thought or study, and indifferent to the din around him, sat the boy. A careless guardian was he of the treasures confided to him. The crowd passed in Chepe: he never marked it. The sun shone on Chepe: he only asked that it should illumine the page he read . . . The customer might enter: but his book was all in all to him. (Thackeray, "George de Barnwell" 5)[14]

Here as in *Oliver Twist*, books' physical location within a shop highlights their textual power to block commerce out. And here as in Trollope, the book's power is established only negatively. Where Flaubert would have paraphrased the text being read, in *Oliver Twist* and "George de Barnwell" its place is taken by a description of the real surroundings from which the text grants oblivion.

The foundlings who give their names to David's titles—Tom Jones, Roderick Random, Humphrey Clinker—bring their stories to a close by discovering their father's identity. His own story, though, is set in motion by his discovering the father's books, themselves doubly orphaned—first by the death of their owner and then by the suppression of their publisher's name. And if you read this episode side by side with the autobiographical fragment, as Forster did, then it's hard not to notice that the commercial information suppressed in the case of books belonging to David's biological father resurfaces in the person of a father surrogate. Many chapters later, David's humiliation at the bottle warehouse is compounded by his being enlisted to dispose of Mr. Micawber's library to a tipsy book-dealer. On the one hand, a dead father who keeps books in the family; on the other, a living father figure (Micawber slotted into the place that John Dickens occupies in the memoir) who disperses them into the marketplace. The split between biological and fictive father maps onto the contrast between two bibliographical models: one in which texts magically (or at least Micawberishly) turn up; another in which books, lumped with spoons and other pieces of portable property that can be sold or pawned, change hands amidst embarrassment, declassment, and drink (Dickens, *David Copperfield* 160–61).

Common sense suggests that children get literacy, and books, from somewhere. But in Dickens's bibliographical riff on family romance, the reader emerges self-made: the father who bought the books killed, the publisher who reprinted them airbrushed out, the mother who taught reading banished to the past—"I can faintly remember learning the alphabet at her knee" (Dickens, *David Copperfield* 57)—and the only adults left standing likelier to wield a book than to read it. (That even the kindest adults fit this pattern suggests that age trumps morals: the first thing that David notices in the Peggotty household is a tea tray "kept from tumbling down, by a bible" [37].) That tension stands in parallel to the financial contradiction at the center of the novel's plot—the conflict between the individualist language in which David's professional rise is imagined, and the belated revelation that the loss of Betsey Trotwood's fortune was a fiction. As with money, so with books: as David Vincent argues, the self-made reader is as powerful a Victorian myth as the self-made millionaire (Vincent, *Literacy and Popular Culture* 259).

In *Jane Eyre*, an oral source can be named in full but printed origins only in retrospect; in *Copperfield*, too, the child remains as innocent of where stories come from as where babies do. As usual, Trollope provides a bathetic contrast to this Bounderbyesque self-mythologizing. Directly after the de rigueur reminiscence of his father knocking him down with "the great folio Bible which he always used," his *Autobiography* mentions a dozen rereadings of the only books found in the house, "the two first volumes of Cooper's novel, called 'The Prairie'", a relic—probably a dishonest relic—of some subscription to Hookham's library" (15). It's characteristic that what Trollope remembers is the commercial transaction—legitimate or not—by which the book reached his young self. In his novel *Ayala's Angel*, conversely, the narrator measures its young heroine's economic fall by the loss of access to new novels—cut off not by a puritanical guardian or wicked stepmother, but by the unavailability of the subscription fee. Before the crash, "that Mudie's unnumbered volumes should come into the house as they were wanted had been almost as much of a provision of nature as water, gas, and hot rolls for breakfast" (10). As the bathetic specificity of breakfast rolls replaces the bread to which texts are usually compared, the adults who traditionally block children's access to books give way to adults who bankroll it.

One early imitation of *Jane Eyre* even more explicitly calls the bildungsroman's bluff: where, it asks, do those persecuted heroes and heroines *get* the books that so richly furnish their imaginations? When the Rochester figure asks the governess whether she is fond of Shakespeare, her answer is not a literary critique but simply the confession that she has never been able to lay her hands on a volume:

> "Never read Shakspeare!" he repeated in an accent of surprise. "Had you assured me this morning you could read and enjoy that Greek poem I handed you down, I should have been less astonished."
>
> "But, sir, I have always been at school. And school girls have no opportunity of obtaining such works. At a school I was at in England, Miss Fenton's, there were some volumes of Shakspeare in the governess's private parlor, but I never saw any thing of them but their backs."
>
> "Have you no home—no parents?"
>
> "None."
>
> "Have you never read Byron?"
>
> "Oh no."
>
> "Nor any novels?"
>
> "No books of that kind."
>
> He looked at me with a half smile, standing with his back against a tree. "Your later years have been spent in France, I understood my sister to say; did you never get any French novels?"
>
> "Indeed no. Mademoiselle Barlieu would have been in fits at the bare thought. And since I left them I have been too fully occupied to read for recreation." (Wood 100)

The few Victorian novels that have entered our canon associate wealth with anti-intellectualism, poverty with a love for literature. In making books the refuge of the powerless, these bildungsromans forget that access to books requires a minimum of economic power: enough money to "obtain such works," along with (as the last line quoted from Wood acknowledges) enough money to afford the time in which to read them. The bildungsroman's association of reading with childhood, Wood reminds us, masks a reality in which adults had more access to books than children, and heads of households than dependents. Indeed, a second 1860s rewriting of *Jane Eyre*, Emma Worboise's novel *Thornycroft Hall*, associates reading instead with evil minor characters, in particular a cousin (the counterpart to the Reed children) "who keeps novels under her pillow and in a pocket," "who covers them with schoolbook covers." Where Jane calls John Reed a dunce, Worboise's orphan accuses her rich cousins of excessive reading: "Who took *The Secret Marriage* to *church*, and read it all through the sermon?" she taunts them (104). By casting books as objects that need to be bought or borrowed, Wood and Worboise question the bildungsroman's assumption that reading is free, in both senses.

Situating Shakespeare in the schoolmistress's parlor rather than in a remote attic, Wood reminds us that the books on which a schoolgirl is likeliest to get her hands are . . . schoolbooks. In contrast, a Dickens novel teeming with schoolmasters still manages to cast adults as blocking figures, rather than enforcers or even enablers of the child's alphabetization. More specifically, by pitting fond memories of pleasure reading against textbooks used as projectiles, *David Copperfield* represses the extent to which (as Catherine Robson has argued) literary memorization forms at once the result and the mechanism of adult violence against children (Robson). The record of nonfictional sources like diaries and schoolbooks (though not of retrospective autobiographies modeled generically after the bildungsroman) suggests that nineteenth-century children were likelier to be either punished for *not* having memorized assigned texts, or punished *by* being given lines to memorize, than to be beaten for internalizing forbidden pleasure reading. As Henry Tilney gently reminds Catherine Morland, "even you yourself, who do not altogether seem particularly friendly to very severe, very intense application, may perhaps be brought to acknowledge that it is very well worth-while to be tormented for two or three years of one's life, for the sake of being able to read all the rest of it. Consider—if reading had not been taught, Mrs. Radcliffe would have written in vain" (Austen, *Northanger Abbey* 80). Reading can "spell" work, in both senses: James's office worker fishes a novel out from under her paperwork to mark the lunch hour, but Thackeray's schoolboy slips a romance under his Latin grammar.

If textual "abstraction" becomes most visible when violently broken in upon by the book-object, so childhood absorption can be benchmarked most effectively against adult interruption. In an essay titled "On Fiction as an Educator" that appeared in *Blackwoods* at the end of 1870, the children's magazine editor Anne Mozley argues that every adult who looks back to "some particular book as an event in his inner history" knows that this book can never be among those supplied by teachers: "He will surely find that the book thus influential came to him by a sort of chance, through no act of authority or intention." On the contrary, even though a "snug and retired . . . window-seat" forms the best venue for children's reading, the book can work its effect only against a backdrop of hostile "observers."

> What shame in these tears—the shame that attends all strong emotions—as they are detected by unsympathising, quizzing observers . . . Who cannot contrast the weariness with which he now tosses the last novel aside, with the eager devices of his childhood to elude pursuit and discovery, to get out of earshot, to turn a deaf ear, when the delightful book is in his grasp which is to usher him into another world?

What ingenuity in hiding, behind hedges, in out-houses and garrets—nay, amongst the beams and rafters of the roof, to which neither nurse nor governess, nor mamma herself, has ever penetrated? (Mozley 195, 189)

As Charlotte Yonge put it a year earlier, "happy the child who was allowed to revel in [the true unadulterated fairy tale]—perhaps the happier if under protest" ("Children's Literature of the Last Century" 306). Augustus Hare remembers how "I used to pick the fragments" of the numbers of the *Pickwick Papers* that his grandmother read "out of the waste-paper basket, piece them together, and read them too" (135).

Coffee table against lumber room: children unearth and secrete books, adults display and deploy them. Governess's parlor against child's bedroom: when Conan Doyle remembers reading the Waverley novels "by surreptitious candle-ends in the dead of the night," he speculates that "the zest of crime added a new zest to the story" (Doyle 25). Auto-enlightenment commands more glamour than forced reading. Yet Samuel Richardson's boast that when he read "for Improvement of my Mind," "even my Candle was of my own purchasing, that I might not in the most trifling Instance make my Master a Sufferer," updates badly to an era in which the 'prentice-work of middle-class children has become, precisely, reading (*Letters* 229). Once schools assign "pleasure reading," the flashlight under the covers becomes an adjunct, rather than an interruption, to academic and ultimately economic success.

The rule is not just that, to be authentic, reading must be attributed to school-aged children but located outside of the classroom; it's also that absorptive reading ends at puberty, after which adults can remember past acts of reading more easily than undertake new ones. (Nothing left to do with books but throw them, display them, or unfurl them in your spouse's face.) Kate Flint's survey of Victorian autobiographies concludes that mentions of reading occur much more often before marriage than after (*The Woman Reader* 208). Even today, Matei Calinescu argues that reading about others' childhood reading "helps us to remember, and perhaps even to recreate," a kind of "quasi-hypnotic reading experience" no longer available in the present (96). In *David Copperfield*, too, the child's disinterested antisocial reading will give way to the adult narrator's profitable vocational writing.[15] In fact, oral storytelling replaces solitary reading at precisely the moment when David is sent away to school.[16]

If David's and Jane's relation to books sticks in English professors' minds as Crosbie's and Miss Hereford's doesn't, one explanation may be that the former numbs literary critics' discomfort with the instrumentality of our own reading—and, by extension, with our part in socializing younger readers. The frequency with which academics' memoirs end before they begin their academic careers suggests that those who read for a living would rather think about reading as if for life. One monograph on reader response published by a university press begins with the stricture that "I shall focus on reading that is done outside the school system, unguided if not completely uninfluenced by it" (Calinescu 92). When the NEA recently surveyed public participation in the arts, its questionnaire, too, excluded reading done within an institutional context: "With the exception of books required for work or school. Did [you] read any books during the LAST 12 MONTHS?"[17] Institutional or even commercial: like Mr. Brownlow or George de Barnwell, scholars today can combine fondness for traces of reading with hatred for traces of pricing. Alberto Manguel declares that "if a book is second-hand, I leave all its marking intact, the spoor of previous readers, fellow-travellers who have recorded their passage by means of scribbled comments, a name on the fly-leaf, a bus ticket to mark a certain page," but that "old or new, the only sign I always try to rid my books of (usually with little success) is the pricesticker that malignant booksellers attach to the backs" (*The Library at Night* 17).

Matthew Battles has explained modern intellectuals' ambivalence toward the research library by comparing the scholarly fantasy of stumbling on books whose significance you are the first to discern with "the rags-to-riches fantasies of the [American] penny dreadfuls—the dream of personal success unaided by unnamed others" (202). In this analysis, the reader's sense of self depends on his editing out those librarians or teachers—now often female—who provide the material conditions for his self-creation. At the same time as he blocks out others physically present, the reader also blots out the means by which the text has reached his hands. Dickens's occlusion of the publishing industry prefigures Mozley's exclusion of parents and teachers (not to mention children's magazine editors, like her). By reducing mediating figures to blocking figures, both turn a manufactured good into a found object.

This coyness about the book trade in fiction *about* children contrasts sharply with its prominence in fiction *for* them, as Katie Trumpener has shown in rich detail. Some early nineteenth-century children's books puff other volumes by the same publisher. Others show their characters coveting gorgeously bound books; Fenn's *Fables* begins, "You must have been good, else your ma-ma would not have bought a new book for you" (quoted in A. Richardson 133). Still others even name the bookseller in whose shop windows their young characters are staring.[18] Where fiction *about* children is idealist, fiction *for* children is materialist.

Such product placement takes on a darker hue in the Evangelical press, where one symptom of the child's fallen state is acquisitiveness: he or she wants to buy books, not to read them. In the prelapsarian model of childhood that the Romantics bequeathed to the bildungsroman, the child internalizes texts while adults wield books; in Evangelical literature addressed to children themselves, on the contrary, virtuous adults understand the text as something to be memorized, while greedy children fixate on the book as a thing to be owned. The preface to Watt's *Divine Hymns* (reprinted throughout the nineteenth century) advises parents to make their children memorize the hymns, turning "their very Duty into a Reward, by . . . promising them the Book it self, when they have learnt ten or twenty Songs out of it" (I. Watts ii).[19] In Mrs. Sherwood's tract "The Red Book," too, the father gives the eponymous object to his daughters on condition that they keep it on their dressing tables and read it every time they look in the mirror (M. Sherwood). Remember the Evangelical magazine that contrasts a young character who "puts books into his head" with those members of its own public whose books are "only on your shelves"; it speaks to the reach of this logic, across otherwise unbridgeable ideological divides, that its editor happens to be the clergyman on whom Brontë based Mr. Brocklehurst ("How to Read Tracts").

Today, once the reading child has become as iconic as the reading woman, the tension between a model in which children are pictured licking the page or stacking books into towers and another in which children are credited with a particularly pure and disinterested absorption in texts bears a striking resemblance to the tension between the belief that women are particularly rapt readers and the suspicion that women will arrange volumes by color, match bindings to their outfit, or otherwise reduce the book to a material thing. Both tensions reflect the historical shift that we saw in the previous chapter from a culture in which women and children had less access to books than adult males, to one in which the opportunity cost of reading (as measured by lost wages) is lower for women and children than for men. Women and children are now credited with as much disinterested inwardness as they were once blamed for vulgar superficiality (see also Armstrong).

Even as the Evangelical press associates children with the book-object and adults with its verbal contents—and the secular press, vice versa— both share two premises: that uses of the book change at puberty, and that whichever age group values the text over the book is morally superior. Both map age onto conceptual debates: should the book be understood as a material object or a mental prompt—situated on the shelf or in the mind, within or above the market? So far, so thematic. Only the bildungsroman, as the next section argues, aligns those debates in turn with a tension *between* diegetic and extradiegetic discussions of text— between (on the one hand) representations of *characters* writing and (on the other) analogies that the *narrator* draws to never-to-be-realized acts of inscription.

READ NONBOOKS

We've seen how often printed objects break into memories of, daydreams about, or role-playing based on absent texts. In the episode of Captain Somebody, that face-off occurs at the level of story; soon enough, however, it migrates to the level of discourse. Stuck in the bottle warehouse—where print takes only the degraded form of labels—David reverts to daydreams. Like his earlier identification with Captain Somebody, these take remembered but unnamed books as their launching pad. David's new fantasy of "going away somewhere, *like the hero in a story*, to seek my fortune" is introduced through a counterfactual: "but these were transient visions, daydreams I sat looking at sometimes, as if they were faintly painted or written on the wall of my room, and which, as they melted away, left the wall blank again" (Dickens, *David Copperfield* 132). Ultimately, of course, "whether I shall turn out to be the hero of my own life" depends on the narrator's writing—more precisely, on the printed pages that sit solidly before our eyes. Here, however, text takes the form, at best, of half-remembered reading (of the unspecified story whose unnamed hero inspires daydreams); at worst, of hypothetical writing ("*as if* they were faintly painted or written"). And even that counterfactual "melts away" when David's body is jerked back to the task of bottle labeling: here as in childhood, printed matter held in the hand interrupts stories stored in the mind.

What sets gentlemen apart is the ability not just to read books, but also to abstract them through metaphor. At school, David's metaphor of picking up "crumbs of knowledge" shifts to an even more explicit simile: "I could no more keep a secret [from Steerforth], than I could keep a cake or any other tangible possession" (95–96). Yet his downward mobility will soon be emblematized by a shift from figurative to literal bread: on his way to the warehouse, David will be embarrassed by "carrying my own bread (which I had brought from home in the morning), under my arm, wrapped in a piece of paper, like a book" (156). The middle-class schoolchild can compare the "knowledge" contained in books to loaves of bread; the boy whose work is to paste papers rather than to read them, in contrast, must redefine loaves as an analogue to the physical object sandwiched between paper covers. As bread compared to books replaces knowledge compared to cake, David descends from abstractions to things. (And not just any thing: let them eat cake.) David's biography projects onto the level of plot the bathetic thud of puns we saw in the introduction that substitute the Irishman's or servant's use of paper to wrap food for their betters' focus on the words it contains.

Like the works of cultural theory discussed in the introduction, too, *David Copperfield* can make the verb "to read" ubiquitous precisely because its meaning is rarely literal, and its object even more rarely a book.[20] "I believed that [Agnes] had read, or partly read, my thoughts"; and David's face allows Mrs. Steerforth to "look behind her, and read, plainly written, what she was not yet prepared to know" (788, 34). No need to invoke analogues in twentieth-century scholarship, however: in the other direction, Dickens is drawing on a long metaphorical tradition that remains alive in much mid-Victorian fiction. *North and South* juxtaposes the Thorntons' coffee-table books not just with the Hales' library, but also, even more sharply, with the metaphor of "reading" that the narrator invokes to describe the process through which middle-class and working-class characters interpret each other's expressions. On first meeting Nicholas and Bessy Higgins, Margaret asks for their address but realizes that her question "seemed all at once to take the shape of an impertinence on her part; she read this meaning too in the man's eyes." Nicholas returns the compliment: "Hoo thinks I might ha' spoken more civilly . . . I can read her bonny face like a book" (Gaskell, *North and South* 74). Later, when Margaret interprets Bessy's surprise at a dinner invitation as a sign of thinly veiled skepticism about her social status, "Bessy's cheeks flushed a little at her thought being thus easily read" (147). And finally, in order to convey Margaret's understanding of what the rioting workers must be thinking, the narrator tells us that "Margaret knew it all; she read it in Boucher's face" (176). The contrast between coffee-table book and battered text is subsumed here under a related chiasmus between closed books and faces that lay themselves open to interpretation. The more narrowly the book is reduced to its social or material attributes, the broader the metaphoric reach that the act of "reading" acquires.

Brontë presents readers with a slightly different choice. We've seen already that the action in which *Jane Eyre* enlists books rarely takes the form of reading; the converse is that what's "read" is rarely a book. Jane can "read [St. John's] heart plainly" (414), while Rochester's face is "no more legible than a crumpled, scratched page" (286); both men "read" her glance, her thoughts, and her face, "as if its features and lines were characters on a page" (154, 76, 396). Even a dunce's book throwing—apparently the furthest thing from reading—is explained as a result of the hatred that John Reed "read in my face" (16). (A dunce can read, just not books.) Reciprocally, from the moment where the "leaves of the book" win out over a "walk in the leafless shrubbery," books usurp the place of nonverbal objects (14): the boundary separating world from book dissolves even before we ourselves have crossed that threshold.[21]

If *Jane Eyre* hedges reading around with qualifications—remembered, predicted, delegated, or reduced to a figure of speech—*David Copperfield* does something similar to writing. Penmanship is introduced only as the analogue to a row of houses: David recognizes Traddles's landlord as Micawber thanks to the "character of faded gentility" that

> made [theirs] unlike all the other houses in the street—though they were all built on one monotonous pattern, and looked like the early copies of a blundering boy who was learning to make houses, and had not yet got out of his cramped brick-and-mortar pothooks. (373)

The street becomes a manuscript—the chirographic image replacing the print metaphor that had appeared in the earlier scene where David casts himself as a piece of paper, describing "Mr. Micawber impressing the name of streets, and the shapes of corner houses upon me, as we went along" (373). But this analogy goes further, reversing tenor with vehicle: the printed page represents a street, but the street in turn reproduces a line of writing.

When Dickens switches the literal with the figurative, the pivot often involves inscription: in *Our Mutual Friend*, for example, the scene of Bradley writing on the chalkboard is followed by Lavinia accusing her mother of looking as stiff "as if one's under-petticoat was a blackboard" (772, 876). In *Copperfield*, however, the pothooks point to a more systematic pattern of jokes about learning to write. When a hungover David tries to produce a letter that would justify the previous night's behavior to Agnes, the only analogue he can find for his mental struggle is a manual skill: "It took me such a long time to write an answer at all to my satisfaction, that I don't know what the ticket-porter can have thought, unless he thought I was learning to write" (341). The echo of the night before, when the building "looked to me as if it were learning to swim," provides a middle term linking the actually existing David to the hypothetical blundering boy: in one case the building's literal slope reflects human "learning"; in the other, the building itself is the one that metaphorically learns because literally leans. The "unless" here performs the same function as the "as if" in the description of the visions that David looks at "as if they were faintly painted or written on the wall": in both cases, writing is invoked for comparison purposes only. The counterfactual in chapter 25 borrows even greater force from its framing. Not only would it be absurd to confuse "writing" in the sense of formulating an apology with "writing" in the sense of inking a page, but where the narrator at least takes responsibility for the description of the Micawbers' street in propria persona, here the flight of fancy is handed off (like some unwieldy package) to the porter—that is, to a working-class figure as comic as the O'Connell of the *Dublin University Magazine* or the literalistic servants whom we'll encounter in chapter 7. Not, that is, to a downwardly mobile character who's gone from carrying knowledge to carrying bread.

To call this confusion "absurd" is hardly to dismiss it, for an analogous substitution structures *David Copperfield*'s critical reception. As *Copperfield* became the poster boy for late twentieth-century literary critics' deflationary mood, an older language of individual creativity and inspiration gave way to inquiries into professional authorship, intellectual property, and the material conditions governing the selling and buying of text. The tension between those competing definitions of "writing" was projected back into *David Copperfield*, as Alexander Welsh's psychoanalytically inflected analysis of David's copyist doubles was succeeded by Mary Poovey's analysis of the division of labor between David's invisible writing and Agnes's equally silent housework, itself followed by John Picker's and Jennifer Ruth's more topical discussions of Dickens's ambivalence toward professional authorship—as well as by Ivan Kreilkamp's argument that the process of learning shorthand comes to occupy the narrative space left empty by Dickens's refusal to tell us anything about David's authorial career (Welsh, *From Copyright to Copperfield* 116; Poovey, *Uneven Developments*; Picker 52–65; Ruth; Kreilkamp, "Speech on Paper"; Price, "Stenographic Masculinity"). In the context of this critical tradition, the porter's mistake begins to look like a figure for the arc of the novel itself, which introduces literal writing (the penmanship that's first joked about, then narrated) where we'd expect to find literary "writing" (the authorship that's taken seriously, but elided).

When Dora asks David to let her "see you write," he evades the request by exclaiming, "Why, what a sight for such bright eyes!" This is not the last time that writing will be described as a "sight": a few chapters later, David introduces Micawber's bill endorsing with the remark that "*to see* him at work on the stamps, with the relish *of an artist*, touching them *like pictures*, looking at them sideways, taking weighty notes of dates and amounts in his pocket-book, and contemplating them when finished, with a high sense of their precious value, was *a sight* indeed" (Dickens, *David Copperfield* 718; my emphasis). "To see him . . . was a sight": the tautology draws our attention to Micawber's idiosyncratic textual theory, which reduces letters to pure visuality. Micawber's sideways glance takes us back into the world of the pothooks to which his street is compared: a world in which even those markings that do look like alphabetical characters end up being looked at, not read. Writing disintegrates into two extremes: on the one hand, insufficiently tangible abstractions like David's memories and fantasies; on the other, excessively visual objects like Micawber's bills or women's account books, and excessively tactile objects like volumes used to poke, prod, or otherwise impress a child.

WRITTEN WORK

The *OED* defines "pothook" as "a curved or hooked stroke made with the pen, esp. as a component of an unfamiliar or unintelligible script or when learning to write." The example it cites comes from 1846: "*Swell's Night Guide* 128/1 Pothooks and hangers, short hand characters." Replacing product by process and mental operations by manual gestures, the pothook forbids us to reduce "write" to a metaphor for "think"— which makes it all the more striking that pothooks are introduced in *David Copperfield* as metaphor, only to be literalized chapters later. "As if I was learning to write," "like the early copies of a blundering boy who had not yet got out of pothooks": the plot will realize both counterfactuals when David comes to learn shorthand.

John Forster's echoes of *Copperfield* have taught us to read that episode backward—from Dickens's celebrity on the reading, lecture, and after-dinner-speech circuit, to novelist, to humorist, to journalist, to reporter, to clerk. Dickens's professional success can be measured by his progress from taking notes to having notes taken on him. Readers were quick to identify David's ten-shilling manual full of "marks like flies' legs" with Dickens's 1824 edition of Thomas Gurney's 1750 how-to book, *Brachygraphy, or, an Easy and Compendious System of Shorthand* (Steven Marcus). Reciprocally, identity politics ensured that *David Copperfield* was immediately transposed into shorthand—excerpted in a phonographic magazine, reprinted as a freestanding volume, and even mined for practice exercises in shorthand textbooks. This self-help literature both promised and represented upward mobility, whether for better or for worse: when R. H. Hutton asserted that "in some important intellectual, if not mechanical respects, Mr. Dickens did not cease to be a reporter even after he became an author," the social connotations of "mechanic" must have grated (Hutton 575). They prefigured our present-day reduction of "stenography" to a term of abuse—once again, for journalism that's mechanical rather than creative.

When Dickens bought Gurney's *Brachygraphy*, he could just as easily have chosen Tachygraphy, Zeitography, Zeiglography, Semigraphy, or Semography. For centuries, an alphabet soup of mutually unintelligible notation systems had vied for the loyalty of court recorders, parliamentary reporters, diarists like Pepys, theater-goers making unauthorized transcripts, and clergymen plagiarizing each other's sermons. Published in the same year as *Pickwick Papers*, Isaac Pitman's *Stenographic Soundhand* began as one among many such systems, spread virally through a counterculture of early adopters: spirit-rappers, teetotalers, vegetarians, pacifists, antivivisectionists, antitobacconists. Like the open-source movement a century and a half later, Pitmanism was idealistic, distributed, and male. The First International Congress and Jubilee of Phonography was transcribed by "an army of phonographers . . . not at all concerned with the economic rewards of shorthand, important as these are, but only with the service—personal, social—even professional—which one Pitmanite can render another in any part of the world" (Cope 130). The delegate who termed shorthand a "bond of brotherhood"[22] corroborated one textbook's accusation that longhand spelling invested English, "a strong and masculine language," with "a garb altogether unfitted for it."[23]

Within Dickens's lifetime, however, the American Civil War began to leach men from the workforce; the typewriter, first commercially distributed in the year of his death, would later be joined by the phonograph to create demand for white-collar, and then pink-collar, labor. A magazine promoting the competing system launched in 1887 by Gregg opened with a column called "Our Ladies' Chatterbox" that presented its alphabet as user-friendly and specifically female-friendly: "Dear Girls, You will be glad to hear we are to have our corner in the magazine all to ourselves . . . After all, we have heard lately about ladies and light-line and its being the ladies' accomplishment" (Butler 149). Gregg was to Pitman as Windows to Linux: a technique that could be universalized from male nerds to female clerks only once stripped of the ideological baggage that had originally impelled its spread. An identitarian ethos gave way to a utilitarian skill: by 1901, the *Phonetic Journal* was complaining that "the great majority of young girls study simply for the proficiency which will enable them to enter business" ("Midland District Conference of the National Federation of Shorthand Writers' Associations"). Even a pacifist like Pitman acknowledged that his system needed "an army of advocates" (Pitman 7). A magic lantern slide advertising Pitman's method captioned a portrait with the remark that "many knights have won their knighthood with their swords; Isaac Pitman won his with his pen" (*A Lantern Lecture on Isaac Pitman*, slide 4). One journalist praised "the drill of the pen" by analogy with the parade ground ("Excerpt from *Hereford Times*").

No surprise that Dickens should be enlisted in this rearguard action—both as role model and as content provider. Potted anecdotes of his life became shorthand for a golden age when stenography was still a prelude to authorship, not to marriage. *Half Hours with Popular Authors, printed in the advanced stage of Pitman's shorthand* (1927) excerpts Dickens's speech to his "brethren" at the Press Club in 1865 (who proceeded, of course, to record it in shorthand). Within the speech, Dickens already shows nostalgia for a youth spent transcribing shorthand notes "on the palm of my hand, by the light of a dark lantern, in a post chaise and four, galloping through a wild country, all through the dead of night, at the then surprising rate of fifteen miles an hour." The coach's gallop across "wild country" magnifies the hand's race along the page: nothing less like David's image of the stenographer going in circles ("my imbecile pencil staggering about the paper as if it were in a fit"), let alone Uriah's finger making "tracks along the page like a snail" (Dickens, *David Copperfield* 504, 222). As the office shifts to the wilds and the day job to "dead of night," words per minute conspire with miles per hour to prefigure the speed with which Dickens will rise from a deskbound clerk to an uncommercial traveler on the international lecture circuit.[24]

Ivan Kreilkamp has argued that the bravado of this speech (of which he considers only the longhand version) echoes Dickens's earlier attempt to align shorthand with masculinity in *David Copperfield*.[25] That biographical parallel makes it all the more puzzling, however, that neither *Gleanings from Popular Authors* (1888) nor the *Popular German Reader* (1897) chooses to reprint the most obvious passage from *Copperfield*—that is, the extended description of David learning stenography. In its place, both anthologies quote the scene of a woman copying—in longhand.

> Dora told me, shortly afterwards, that she was going to be a wonderful housekeeper. Accordingly, she polished the tablets, pointed the pencil, bought an immense account-book, carefully stitched up with a needle and thread all the leaves of the Cookery Book which Jip had torn . . . Her own little right-hand middle finger got steeped to the very bone in ink; and I think that was the only decided result obtained. Sometimes, of an evening, when I was at home and at work—for I wrote a good deal now, and was beginning in a small way to be known as a writer—I would lay down my pen, and watch my child-wife trying to be good . . . She would take up a pen, and begin to write, and find a hair in it. Then she would take up another pen, and begin to write, and find that it spluttered. Then she would take up another pen, and begin to write, and say in a low voice, 'Oh, it's a talking pen, and will disturb Doady!' . . . Or, if she were in a very sedate and serious state of mind, she would sit down with the tablets, and a little basket of bills and other documents, which looked more like curl-papers than anything else . . .
>
> I occasionally made a pretence of wanting a page or two of manuscript copied. Then Dora was in her glory. The preparations she made for this great work, the aprons she put on, the bibs she borrowed from the kitchen to keep off the ink, the time she took, the innumerable stoppages she made to have a laugh with Jip as if he understood it all, her conviction that her work was incomplete unless she signed her name at the end, and the way in which she would bring it to me, like a school-copy, . . . are touching recollections to me. (596)

Mary Poovey has shown that in the novel itself, Dora's domestic labor fills the space that David's literary works might be expected to occupy

(*Uneven Developments* 89–125). In the Pitman reprints, however, what Dora crowds out is David's vocational copying. Instead of replacing the aesthetic by the utilitarian, that is, the *Gleanings* present us with an even (and gender-neutral) trade: the ledger for the steno notebook, the cookbook for the "approved scheme of the noble art and mystery of stenography (which cost me ten and sixpence)" (Dickens, *David Copperfield* 504). Unlike the father's books, a cookbook or a how-to manual can be priced.

Why excerpt a scene about Dora instead of about David? Pitman's eagerness to cultivate a male audience may make this choice appear even more perverse. His other 1888 abridgments projected their own gender politics backward, sometimes in a form recognizable only to that subsection of their readership who remembered the full-length original (as when he reprints Dobbin's schoolyard fight but excises Miss Pinkerton's Dixionary), sometimes under a cruder—and more self-contained—thematic guise. The same series reprints the scene from Frederick Marryat's *Mr Midshipman Easy* in which Mrs. Easy tries to coax her son past "B" in the spelling book, he retaliates by pouring a boiling tea urn into Mr. Easy's lap, and Mr. Easy packs him off to a school run by a master with a caseful of canes displayed on the wall like billiard cues. The cane that's mightier than the pen could have looked continuous with Pitman's interest in Dickens's interest in the embodied nature of writing, except that the gender roles are neatly reversed. The Marryat excerpt assigned men the task of realigning mind with matter; in contrast, the Dickens quotation takes David's "writing" to refer to abstract literary composition, delegating to Dora the more literal writing "smeared" by association with the material world and the body. Whether her body, David's, or even Jip's hardly matters: just as the noisy pen usurps the articulateness that Dora herself never quite achieves, so the hairy pen recalls Dora's habit of distracting David from *his* writing by trying to curl his hair. Her own curlpapers, recycled from the account book, literalize the equivalent ledger that Clara Copperfield ruined by putting "curly tails to my sevens and nines" (Dickens, *David Copperfield* 17). On his end, David turns the house into an office, whether by spending his own evenings "at home and at work" or by nagging Dora to align housekeeping with bookkeeping;[26] on hers, Dora turns the study into a kitchen, juxtaposing pothooks with pots, borrowing aprons and bibs, replacing reading by stitching, and "steeping" her finger like a cup of tea, like a dog's inked paw or like Uriah's hand.[27] In Phiz's illustration, even the bookshelf contains a jar marked PICKLES. The spines of the books between which it's jammed are blank: in this topsy-turvy household, food can be read while books only provide raw material for curlers.

By the time Pitman reprints the scene, the handover of clerical work from husband to wife—the progression from David as stenographer to

Figure 3.1. "Our Housekeeping," David Copperfield, 1850.
Phiz [H. K. Browne], "Our Housekeeping," *The Personal History of David Copperfield* by Charles Dickens (London: Bradbury and Evans, 1850).

David as author whose works are copied, or "copied," by Dora—has come to look like ontogeny anticipating phylogeny: the novel's plot writ large in the replacement of male clerks by women. In the pages of *Gleanings*, Dora's writing takes over the logic that David's has outgrown. Yet a century of pink-collar work makes it hard to notice that *Copperfield* is crowded with male characters who copy: not only the young David, but Uriah, Traddles, mad Mr. Dick, and bad Jack Maldon.[28] Far from inertly reflecting any existing vocational practice, David's marriage anticipates the higher-level parallels—woman is to man as book to text—that will enable a new division of clerical labor to emerge a generation later. To those of us who stand on the other side of those turn-of-the-century developments, the opening battles between children reading texts and adults handling books look like the groundwork for the gendered division of labor that emerges after puberty. The difference is that the latter reinforces the social hierarchy that the other overturned. In the first case, powerful adults are associated with brute materiality, powerless children with airy abstraction; in the second, "writing" can refer metaphorically to men's production of ideas, or literally and comically to women's reproduction of pages.

TALES OUT OF SCHOOL

Competing definitions of writing rest on not just by whom it's done, but where: study or kitchen, kitchen or law office, law office or school, school or kitchen. ("The way in which she would bring it to me, like a school-copy . . . are touching recollections to me.") Dickens identifies that problem from the outset of his novelistic career, when Squeers organizes a spelling lesson around "c-l-e-a-n, clean, verb active, to make bright, to scour" (*Nicholas Nickleby* 8). Once the drunken cook's "written character, as large as a proclamation," comes to echo the equally oversized, and equally unreliable, sandwich board that earlier labeled David "a dangerous character" (*David Copperfield* 81), school begins to align itself not (as Mary Poovey suggests) with the home created by Agnes's self-effacing housework, but on the contrary with the domestic chaos created by Dora's all-too-visible labors (*Uneven Developments* 101). The space in which the sandwich board does its work complicates that analogy further. "The playground was a bare gravelled yard, open to all the back of the house and the offices; and I knew that servants read it, and the butcher read it, and the baker read it" (*David Copperfield* 82). At midcentury, the word "office" was still in transition from a term for the kitchen and servants' workrooms to a term for the space occupied by male clerks like Uriah. The problem with Mr. Creakle's school is marked, even before he ever appears, by the relocation of reading from white-collar spaces (including the schoolroom but also the "office" in its newer sense) to the space associated with trade and with manual labor. Trade (in the person of the butcher and the baker) prefigures Creakle's belief that education's goal is to put money in an ex-hop-grower's pocket; manual labor (in the form of the servants), his confusion of teaching with laying on hands.

The taxonomy that the novel elaborates doesn't just distinguish real writing surfaces from imagined or counterfactual or metaphorical ones. Its metaphors also distinguish different *kinds* of writing surfaces, corresponding to different technologies of inscription. At one end of the novel, David's back is reduced to a blank slate by the arithmetic lessons that drive him to rub chalk into his skin; at the other, "the palms of my hands, hot plates of metal" imply a body waiting to be engraved, not inscribed (Dickens, *David Copperfield* 339). More literally, Creakle removes the sign reading "HE BITES" in order to make good his threat that "you won't rub out the marks that I shall give you": like Micawber drawing "the ruler from his breast (apparently as a defensive weapon)," Creakle confuses bodies with pages (689).[29] But he's not the only one to equate handwriting with bringing up by hand: the whisper that "impressed" David with Creakle's cruelty echoes Murdstone flogging David "with an impressive look" and Steerforth's graffiti "cut very deep and very often" as closely as it will be echoed in turn by the donkey's owner "leaving some deep impressions of his nailed boots in the flower-beds" (84, 90–91, 199). Even before the cane makes David the butt of an oral image—a biter bit by what Creakle calls the "tooth" of his cane—Mr. Creakle's long-awaited entrance into the novel is announced by a print metaphor: "A profound *impression* was made upon me, I remember, by the roar of voices in the schoolroom suddenly becoming hushed as death when Mr. Creakle entered after breakfast, and stood in the doorway looking round upon us like a giant *in a story-book*" (90; my emphasis).

Where children's bodies became writing surfaces, their minds were compared to writing surfaces. A longer philosophical tradition made multiple media available as possible analogues for the mind: Locke's tabula rasa itself comes from Plato's comparison of the soul to a *wax* tablet, while Rousseau's comparison of education to "engraving" imagines the mind as a metal sheet, not a paper one (A. Richardson 131). The metaphorical slates and real boards that punctuate David's childhood forbid us to take for granted the nature of the surface to which his mind is compared and his body assimilated. Within the novel itself, writing media extend beyond paper to cloth (David's clothes labeled "in indelible marking-ink" as well as the sailor's waistcoat with "Skylark" spelled out across its chest), brass (the Micawbers' doorplate), mugs, walls (real or imagined), and even desks (marked by graffiti) (Dickens, *David Copperfield* 206, 154, 10). This array reflects Dickens's long-standing interest in the historically changeable—but also socially specific—range of alternative writing surfaces. In a discussion of clerks in *The Uncommercial Traveller*, blotting pads are described as "the legitimate modern successor of the old forest-tree: whereon these young knights (having no attainable forest nearer than Epping) engrave the names of their mistresses" (Dickens, *Uncommercial Traveller* 337).

Where critics in the age of the ballpoint fixated on the metaphor of inscription, nineteenth-century writers drew finer distinctions within that metaphor: was the body being compared to paper or parchment, and were the marks on it assimilated to writing or printing? A generation before Creakle's school, the Lancastrian system had replaced the traditional manuscript signs pinned to dunces' foreheads with preprinted punishment labels bearing captions such as "Idle," "Talking," "Playing," and "Dirty boy" (Rickards and Twyman 104). And the child's scarred skin might more aptly be compared to parchment than to paper: as a clerk like Dickens would well have known, parchment itself *is* skin. In Gaskell's contribution to a Dickensian round-robin, a memory of parchment may be buried in a mother's almost cannibalistic lament that "London is as bad as a hot day in August for spoiling good flesh, for [her son] were a good-looking lad when he went up, and now, look at him, with his skin gone into lines and flourishes, just like the first page on a copybook!" (Reynolds 89).

Like a piece of parchment, too, David's skin bears witness to his history. Years after he notices that in Mr. Creakle's classroom "My head is as heavy as so much lead," Dora can still remark, "'Oh, what ugly wrinkles in my bad boy's forehead!'"; "and still being on my knee, she traced them with her pencil; putting it to her rosy lips to make the mark blacker, and working at my forehead with a quaint little mockery of being industrious" (Dickens, *David Copperfield* 92, 586). (Melville would mechanize the metaphor in the voice of the narrator of "The Tartarus of Maids" watching two women working in a paper mill: "I looked upon the first girl's brow, and saw it was young and fair; I looked upon the second girl's brow, and saw it was ruled and wrinkled" [325].) Where Agnes "reads" David's "thoughts," Dora can only write his face; where Agnes's depth psychology helps David develop, Dora can age him only visually (Dickens, *David Copperfield* 788). Yet Agnes's own face, too, is marked by an "indelible look" (24). The pencil's power to reduce David to a flat character prevents the novel from confining the violence of inscription safely to Creakle's schoolroom or even Heep's showdown. In a novel where rulers are more often used to mark schoolboys' skin than to line schoolboys' paper, even the most loving relationships equate impression with aggression. Here again, however, Dickens's metaphors draw fine distinctions among writing surfaces: the novel's media ecology reduces David to a manuscript only after assimilating him to something more like a published book, bound between boards labeled not with his name but rather with "HE BITES."

In the context of David's stenographic career, the analogy between skin and slate may point back to Cassian, the patron saint of stenographers, martyred by being stabbed to death with his pupils' styli. With two differences: in *David Copperfield*, the metonymic relation between scarring and writing gives way to a metaphorical one, and the man attacked by boys gives way to a boy attacked by men. The only case in which a child is the attacker—Steerforth scarring Rosa with a hammer—occurs before the novel begins; all David sees, belatedly, is the scar "lengthening out to its full extent, like a mark in invisible ink brought to the fire . . . I saw it start forth like the old writing on the wall" (278). When David evoked the visions rising in his mind "as if they were faintly painted or written on the wall," a solid but counterfactual inscription provided the metaphorical counterpart to real but intangible images. Here, in contrast, an imaginary wall competes with an implied sheet of paper as terms of comparison for very real skin.

The violence of inscription may help explain David's oddly bifurcated experience at school. Forbidden (and pleasurable) oral performances of remembered texts differ as starkly from required (and painful) physical punishment for forgotten lessons as—well, as night and day. What redeems David from the tale the sandwich board tells about him by day is his own ability to tell tales by night—not to an encyclopedic public of butchers, bakers, and candlestick-makers, but rather to a clandestine coterie of schoolboys centered on his self-styled aristocratic patron, Steerforth.

Transforming him from the wearer of a written sign to the reciter of printed books, David's storytelling prefigures his later career not only in its "hard work" but in its schedule: both the telling of stories and the learning of shorthand depend on early rising (94).[30]

The novel is structured, then, by a double progression: first, a fall from women's speech to men's writing (usually on David's hapless body); second, a rise from David's own feminized speech (whether at school or in the bottle warehouse) to the writing that he himself comes to perform on arriving at manhood. From "the gentleness of my mother's voice" as she reads aloud and prompts his lessons in stage whispers, David is exiled to a classroom whose "quiet" is underlined by chirographic overload: the "scraps of old copybooks" littering the floor, the graffiti carved into the desks, and the "beautifully written" sign that David himself is forced to display (80–82).

Like Captain Somebody's collision with the Latin Grammar, David's nighttime storytelling sessions translate absent books into mental images that are externalized in turn, whether in the form of playacting at home or recitation at school. And both performances derive from the same lumber room: like *Jane Eyre* banishing the act of reading to its backstory, *David Copperfield* represents only the state of having read. In the presence of a "litter" of paper, ink, and other school supplies, but the absence of printed storybooks, oral transmission turns David himself into a walking reprint series.

Where David's domestic education pitted the abstract pleasures of acting out remembered texts against the physical pain of being beaten or boxed with a book, here a mental storehouse of stories is contrasted with the schoolchild's bodily suffering. What the pedagogy of pain shares with the pedagogy of pleasure is that both equate learning to write and read with coming to *be* written and read. The only difference is whether what's marked is the child's mind, or his skin. In the Renaissance, according to Walter Ong, pain was integral to the learning of Latin—and vice versa.[31] At the same time as he translates that symbiosis into a modern vernacular idiom, Dickens also adds a corollary: the pleasures of orality provide a foil for the pain of inscription, or more precisely of being inscribed. It makes sense, in that context, that Steerforth should try to change writing instruments into vocal aids—counteracting David's hoarseness, for example, with cowslip wine "drawn off by himself in a phial, and administered to me through a piece of quill in the cork" (Dickens, *David Copperfield* 94). Instead of ink, wine; only a hundred pages later, once David begins to sell off the Micawbers' books to the tipsy bookseller, does the oral storyteller too disinterested to be paid in any currency stronger than home brew morph into the businessman who reduces books to money that can be exchanged in turn for drink. Ontogeny recapitulating phylogeny: David's life can be parsed either as a fall from a gift economy to the market, or as a Whiggish progression from the face-to-face audiences figured here by Steerforth's pseudoaristocratic "patronage" to the anonymity of a modern, professionalized public sphere.

In laying the sign on David's own back open to the "back of the house," which lacks the expected division from the middle-class front, the novel embodies the underside of Dickens's own ambition to be read by a representative cross-section of English society. The servants and tradesmen who read David's sandwich board form a dystopian mirror image of the figures whom Victorian critics conventionally used to embody the classlessness of novel reading. For Bagehot, Dickens was the one writer "whose works are read so generally throughout the whole house, who can give pleasure to the servants as well as to the mistress, to the children as well as to the master" (Bagehot 459); according to *Fraser's* review of *David Copperfield*, its author "has done more for the promotion of peace and goodwill between man and man, class and class, nation and nation, than all the congresses under the sun . . . [His novels] introduce the peasantry to the peerage, the grinder at the mill to the millionaire who owns the grist" ("Charles Dickens and David Copperfield" 700).

What do those claims about the power of reading to transcend class do to the story of David's own fall out of the middle class and ascent back into it? For one thing, they complicate the child's downward mobility, because storytelling is the one act that spans the divide between a before and an after, school and warehouse, which the novel otherwise so strenuously keeps apart. Where the return of the middle-class schoolmates who can "read" David is preceded by his being "read" by the servants, the baker, and the butcher, conversely David's storytelling to Steerforth prefigures his storytelling to the other boys in the warehouse. And just as the sign erases the distinction between front and back of the house, so the only time his workmates venture to call him by his first name is when they become "confidential" under the influence of his storytelling (Dickens, *David Copperfield* 159). The novel leaves open whether to explain that fact by the theory that stories debase their teller or on the contrary that they cut across class lines. What is clear, however, is that storytelling in the warehouse looks back on the one hand to storytelling in the dormitory, and on the other to self-display in the schoolyard. The narrator's assertion at the bottle warehouse that "how much I suffered it is, as I have said already, utterly beyond my power to tell" mirrors his statement that "what I suffered from that placard, nobody can imagine"—so closely, in fact, as to make a reader wonder whether that "saying already" refers to the schoolyard rather than the workplace (57, 81).

Helen Small has argued that the mid-Victorians' hope that novel reading in general (and reading Dickens in particular) could suspend or transcend class divisions culminated in the public readings of the end of Dickens's life, where the transformation of texts into speech became inseparable from the dream of uniting the nation in the act of listening (Small). In contrast, the sandwich board figures the mass reading of visual signs as a nightmare. Trollope's celebration of a society whose cohesion is measured by the fact that novels are "read right and left, above stairs and below," is reversed in the democratically "open" schoolyard where the backstairs outdo the front stairs in reading and objectifying the future author (Anthony Trollope, *An Autobiography* 220). In the warehouse window as in the bare schoolyard, shame can best be vehicled through writing. To be labeled is to be exposed—not just for David, but for the Micawbers, whose social decline is emblematized by the fact that "the centre of the street door was perfectly covered with a great brass-plate, on which was engraved 'Mrs. Micawber's Boarding Establishment for Young Ladies'" (Dickens, *David Copperfield* 154).

Yet the novel manages to recuperate even the worst humiliations: an upward trajectory leads from the school where David wears a label to the bottle warehouse where he pastes labels. The bottle warehouse may not represent the opposite of literary production, then, so much as its analogue: the progression from being labeled to labeling foreshadows the shift from a body that's written upon to a mind that writes.[32] As the plot turns a child who is acted upon into an adult who acts, its trope shifts from metaphor (a child who resembles a book, as the next chapter argues in more detail) to metonymy (an adult who makes one). As a result, *David Copperfield* turns only belatedly into a proto-Smilesian account of self-help—of salvation by books such as Gurney's *Brachygraphy*, unaided by human agents such as Mr. Spenlow. Its first debt is to an older genre that, far from celebrating self-help, associates selfhood with helplessness and passivity—more specifically, that locates consciousness not in a person marked by books, but in a book marked by readers. That genre forms the subject of the next chapter.

CHAPTER 4

It-Narrative and the Book as Agent

UNTIL NOW, THE subjects of my sentences have been human agents. Whether decoding words or throwing volumes, whether facing a page or hiding behind a paper, these persons do something with (and to) books. What they don't read, they still use. So far, so conventional. No matter how energetically book historians distance themselves from the aesthetic, we remain no less attached than literary historians to narratives centered on agents: the author, the editor, the reader, or (even more literally) the literary agent. Such scholarly accounts mirror the structure of their sources, whether authors' biographies, company histories, or readers' memoirs. They also recapitulate a more diffuse tradition—both religious (specifically Augustinian) and literary (specifically Wordsworthian)—that relies on the encounter with a book to account for the development of a self.

Even when book historians choose examples that happen to fall outside the literary canon, the language in which they describe their own scholarly practices remains parasitic on those novels and memoirs that thematize reading. The previous chapter argued that one subset of that tradition, the bildungsroman, has both generated and limited the stories scholars tell about reading. Where else, then, might we look for models that make the book narratable? This chapter contends that the most productive overlap between recent book-historical scholarship and the longer tradition of bibliographically themed life writing lies not in their common interest in human subjects, but rather in their shared attention to the circulation of things.

Analytical bibliographers have taught us that books accrue meaning not just at the moment of manufacture, but through their subsequent uses: buying and selling, lending and borrowing, preserving and destroying. A history of the book that took that whole range of transactions as building blocks (rather than focusing on the fraction of the book's life cycle that it spends in the hands of readers) could usefully borrow its formal conventions from the "it-narrative": a fictional autobiography in which a thing traces its travels among a series of richer and poorer owners.

THE BOOK AS VICTIM, THE BOOK AS SUBJECT

When late twentieth-century critics rediscovered the it-narrative, they were thinking of a late eighteenth-century genre. By 1800, its babbling banknotes, canting coins, prosing pocket watches, and soliloquizing snuffboxes seem to have talked themselves out. It might not be too far-fetched to explain the recent vogue of it-narrative as a stick with which to beat the Victorians. On the one hand, the it-narrative's obsession with two-dimensional objects—whether metal or paper—prompted literary critics beginning with Deidre Lynch to reexamine the metaphor that dismissed early fictional characters as "flat" and therefore valueless (*The Economy of Character*). On the other, the metaphor of the "rise of the novel" deflated. The eighteenth-century fiction that had once led upward to later realism now bled outward to contemporaneous nonfictional genres: the advertisement, the economic treatise, the slave autobiography, the letter of credit. Synchronic juxtaposition replaced diachronic succession. To the extent that eighteenth-century it-narrative could be made to prefigure anything at all, its telos was no longer nineteenth-century fiction but twenty-first-century "thing theory."

In the decades hopscotched over by those critics, however, it-narratives never stopped being read, or even written—with two differences. First, the guineas, rupees, and banknotes whose histories, adventures, and lives had formed the genre's stock-in-trade were now replaced by talking books; and instead of addressing middle-class adults, it-narratives now went down-market to those too young, or too poor, to choose the books they owned. Both shifts register in *Middlemarch*, a novel whose own plot is structured, it-narrative-like, by the circulation of "a bit of ink and paper." When Lydgate is driven to a dusty encyclopedia entry on anatomy—until which "he had no more thought of representing to himself how his blood circulated than how paper served instead of gold" (143)—it's only because he's already tired of *Chrysal, or the Adventures of a Guinea* (1760–65). Ontogeny recapitulates phylogeny: the child draws his reading material from the infancy of the novel, the adult from modern medical research. The displacement of the inanimate by the human over the course of Lydgate's life mirrors the novel's shift from focalization through things (in the it-narrative) to focalization through human characters (in the strand of "subject narrative" that culminates in Eliot's novel for grown-up people). Yet Lydgate's own fantasies continue to feature nonhuman heroines: "the primitive tissue was still his fair unknown" (272).

Likewise, it-narratives continued to be not only recirculated and reprinted, but even composed from scratch. All that changed was their audience. Although some had always trickled down to children, eighteenth-century it-narratives were adult—occasionally in the strong sense of the word, as the narrator of *Middlemarch* acknowledges in describing *Chrysal* as "neither milk for babes, nor any chalky mixture meant to pass for milk" (143). Crébillon's *Le sopha* (1742) and Diderot's *Les bijoux indiscrets* (1748) prefigure Prince Charles's latter-day it-fantasy of being reincarnated as a tampon. But around the same time that the handpress was dethroned, the genre became G-rated: witness *The History of a Religious Tract Supposed to Be Related by Itself* (1806); *The History of an Old Pocket Bible* (1812); *Adventures of a Bible: Or, the Advantages of Early Piety* (1825); *The History of a Pocket Prayer Book, Written by Itself* (1839); *The Story of a Pocket Bible* (1859); *The Story of a Red Velvet Bible* (1862); *Handed-On: Or, the Story of a Hymn Book* (1893). Moving from one kind of printed paper to another, the it-narrative shadowed two competing disciplines: first numismatics, then bibliography.

As good books replaced bad coins, officious thing-exposition upstaged confidential thing-confession.[1] But religious books weren't the only ones that talked, nor Evangelical publishers the only ones who gave them voice. As late as 1873, a copy of *Robinson Crusoe* could narrate Annie Carey's heavily illustrated account of papermaking and binding, *The History of a Book*, commissioned by the same firm that had published her earlier *Autobiography of a Lump of Coal; A Grain of Salt; A Drop of Water; A Bit of Old Iron; and A Piece of Old Flint* (1870).

Two innovations united both strands of it-narrative religious and scientific. One was that in neither case did the end user correspond to the buyer. Whether gifts presented to children by their parents, or tracts thrust upon poor adults by philanthropists, both reached readers through a more than purely commercial transaction. The other was that both shared a new kind of protagonist. Where banknotes had once exemplified circulation across class lines, paper now changed hands in the form of bibles, hymnbooks, prayer books, tracts.

To readers familiar with the classic phase of the it-narrative, this turn may come as a surprise. But the talking book doesn't come out of nowhere. Eighteenth-century object narratives already allude to the life cycle of books, beginning with *Chrysal* itself. There, the narrator of the preface happens on a fragment of "some regular work" in the paper wrapping the butter served to him by a poor family; he goes on to seek out more of the same manuscript by going to the chandler's shop that they patronize, "as if for some snuff, which, as I expected, was given me on a piece of the same paper" (Johnstone x–xi). As Christina Lupton has argued, the classic it-narratives are "in the first place, the life story of a pile of paper, and only in the second, the story of the objects represented there" ("The Knowing Book" 412). In *Adventures of a Black Coat* a manuscript is used as a potholder, and in *Adventures of a Banknote* an author can't even afford enough coal to dispose of his rejected verses—although his neighbor later burns them to revive him from a concussion. Christopher Flint has therefore interpreted it-narratives as an allegory of authorship, arguing that "the speaking object figures the author's position in print culture" (212). If you read backward from the nineteenth century, however, eighteenth-century it-narratives begin to look less invested in a figurative representation of the author than in a literal representation of the book. To the extent that early it-narratives frame themselves as found objects, they already emphasize the circulation of paper, changing hands as it passes from manufacture to sale to resale to disposal.

All that changes is where that emphasis occurs. In the first generation of it-narratives, the book-object is mentioned around the edges: in prefaces, introductions, and other paratexts. Only around 1800 does its representation migrate from frame narrative to plot. In that sense, the development of the it-narrative mirrors the shift that I've argued characterizes the novel as a whole in the nineteenth century. If the bookish self-referentiality that the eighteenth-century novel situates in the voice of an "editor" gets replaced by the nineteenth-century novel's more thematic interest in *characters'* uses of books—and if, in the process, bibliographical materialism migrates from beginnings and endings to middles, or from paratext to text—similarly the it-narrative goes from joking about wastepaper in its front matter or frame narratives, to taking papermaking and printing as the subject of its plots. In the process, the novel's bookishness—its allusions to the material forms that it takes and the social transactions that it occasions—goes from exemplifying the reader's labor to instancing the buyer's passivity.

Book, Prisoner, Slave

Where eighteenth-century it-narratives taught readers the rules governing cash and credit in a commercial society, the *Stories*, *Histories*, and *Adventures* that straggle in after 1800 take on a narrower topic: how one very particular kind of consumer good—books—should be bought, sold, given, borrowed, and disposed of. More specifically, it-narratives commissioned by religious publishers struggle to reconcile the competing imperatives of a person's relation to his books (imagined as less alienable than other belongings) and a person's relation to other persons (vehicled, in the world that tracts both represent and inhabit, by the exchange of printed matter).

After 1800, as secular it-narratives shifted their focus to manufacture, religious publishers kept alive the genre's traditional emphasis on circulation. What mechanical technologies are to one, social relations are to the other. Talking tracts allot as little space to their own conception as any human narrator does: only a secular volume like the 1873 *History of a Book* could end, Tristram Shandy–like, at the moment when its narrator first goes on sale. In asking how books are shared, and only secondarily how books are made, it-narratives put out by Evangelical publishers anticipate the 1847 pamphlet that urged the Religious Tract Society to stick to its professed aim of "circulation, leaving production to individuals" (Fyfe, "Commerce and Philanthropy" 176).

The distribution of printed matter forms the central problem of the "Appendix, containing Anecdotes calculated to shew the utility of distributing religious tracts" that is tacked on to the *History of a Religious Tract Supposed to Be Related by Itself*—itself the inaugural volume in a series of tracts that appear under the title of "The Cottage Library of Christian Knowledge." "Should the Story of the Little Red Velvet Bible have the effect of arousing any," another it-narrative declares, "to the conviction that the noblest work in which a Christian man or woman can be employed, is that of circulating the Bible amongst all classes of the community, both at home and abroad, it will not have been written in vain" (Horsburgh 95).

Founded in 1799 for the purpose of moving books across classes and continents, the Religious Tract Society called for books to be "used, worn out, and worn to pieces."[2] Handling trumps hoarding; books should be transferred, not treasured. As speaking books replace speaking coins, the Enlightenment faith in exchange gives way to its Christian equivalent— the hidden hand to the parable of the sower. The narrator of one American *History of a Bible* tells us nothing about the moment of its making, choosing instead to begin its life story at the moment when a buyer "liberates" it from being a "close prisoner" in a bookseller's shop. Elated when praised for "the elegance of my dress" (that is, binding), the bible soon discovers that its new owners "would not permit me to say one word": instead, these "jailors" force it "to sit upon a chair in the corner of the room." One owner, worried that the bible will turn her son into an "enthusiast," "in the heat of passion locked me into my old cell, where I remained in close confinement"; others "joined in scandalizing my character; and I was again confined to my old cell"—that is, to a bookcase where, to make matters worse, it has to rub shoulders with vicious companions. To be read, in this metaphor, is to be "discharged from prison"; even the glass front of the bookcase becomes a "prison door" whose locking and unlocking determines the narrator's fortunes (*History of a Bible* 1, 5, 6, 7).[3]

Although it reappears, for example, in *The Story of a Pocket Bible*'s description of its narrator "shut up" in the "prison-house of a strong chest," the carceral metaphor is hardly unique to the Evangelical press (Sargent, 1st ser., 2). The same image is recognizable in the graveyard of metaphors that is Emerson's 1858 essay titled "Books" in the *Atlantic*: "In a library we are surrounded by many hundreds of dear friends, but they are imprisoned by an enchanter in these paper and leathern boxes; and though they know us, and have been waiting two, ten, or twenty centuries for us,—some of them,—and are eager to give us a sign, and unbosom themselves, it is the law of their limbo that they must not speak until spoken to" (344). The episodic structure of it-narrative dovetails equally well with the Evangelical imperative to circulate as with the secular logic that privileges use above display and reading above collecting.

That convergence makes it doubly puzzling, however, that not every bible seeks out readers as wholeheartedly as does the narrator of the *History*. A tug-of-war between circulation and stasis structures the *Story of a Pocket Bible* serialized in the RTS magazine *Sunday at Home*.[4] Given to a child on his birthday, willed to a spendthrift, unloaded onto a bookseller, windowshopped by a laborer, bought for the sake of its binding, inherited by a profligate, passed along to his sick sister, stolen by her servant, resold to a "romanist merchant," stumbled on by a houseguest, stomped on by a priest, cast into prison along with its owner, auctioned to a tradesman for use as wrapping paper, donated by a pious bookseller to a godless Chartist, shoved under a floorboard by the Chartist's Catholic wife, trampled (a second time) by striking workmen, traded for a dram, its peripatetic narrator ends up, like a foundling in romance, rescued from a used bookshop by its original owner's children (Sargent). By this point, the book is not secondhand, but nineteenth-hand: a picaresque narrator must always move on.

Like the title of *Handed-On: Or, the Story of a Hymn Book*, the plot of *The Story of a Pocket Bible* is driven by perpetual motion. Yet every step the narrator takes has violence and theft for its mechanisms, death and bankruptcy for its catalysts. This may help explain why the Pocket Bible worries less about being consigned to solitary confinement than about falling into the wrong hands. Because the *History of a Bible*'s "close confinement" is matched by the Pocket Bible's equally long stint on a "shelf of banishment," we might expect a sigh of relief when, after weeks gathering dust on a licentious gentleman's bookshelf, the narrator finds a duster approaching, closely followed by a hand. Instead, it shrinks from the servant's advances:

> After looking around in her trepidation, and wiping her hands with her apron, she mounted on a chair, and reached me from my shelf of banishment . . . It was not right; none have so emphatically condemned the slightest approach to unfaithfulness in servants as I . . . Had this poor girl, therefore, been better acquainted with me, she would scarcely have ventured to seek an interview at that time . . . Those stolen min-

utes of communication were improper; and on one occasion, I had it in my power to hint as much to her. I was recounting the duties of the various classes of persons to whom I had messages to deliver, and among other things I had something to say to servants. "Servants," I said, "obey in all things your masters . . . " Hannah blushed deeply when this reproof reached her . . . She hastily replaced me; and after that, though she cast many longing looks towards me, she did not take me again from my place of repose. (Sargent 38)

Nothing less Augustinian: instead of being converted through reading, the maid is converted to not reading. Where one conversion begins with the phrase "tolle lege," the other ends with the decision not to "take" (or read) the book. At least, not the master's copy on the master's time: the language of stolen interviews and stolen minutes makes it hard to remember that the volume itself never leaves the room. Here as in *Susan Osgood's Prize*, the lesson of reading is not to read—as if the quixotic logic of conduct literature could be extended from the dangers of reading romances to the dangers of reading anything at all. The bible provides a reductio ad absurdum of that logic—proof that the meaning of reading depends on its circumstances more than its content.

As we'll see in more detail in the final chapter, a long literary tradition privileges mental operations over manual ones, associating handling with either the female servants who dust books, the tradesmen who tear them apart to wrap groceries, or the even more vulgar nouveaux riches who display them on sofa tables. In this case, the cheapness of the tract in which the anecdote is contained makes it all the more striking that the maid who turns her mind to the pages is urged instead to turn her duster to the binding. By the end of this paragraph, the talking bible almost recoils from being read: what was introduced as "my shelf of banishment" is now dubbed "my place of repose."

The past two chapters have posited printed matter as a tool to ward off or replace human relationships: the thrown book, the unfurled newspaper, the text that allows the child to lose any consciousness of the adults around him. The Evangelical press turns that relation around: far from competing with human associations, the book enables them. It-narratives hover uneasily between these perspectives— the book as block, the book as bridge. Their form remains torn between the demands of subject and object, relations among readers and relations with books. The history that, as we say today, "personalizes" belongings—a bond that the book trade enshrined earlier in the phrase "association copy"—tugs against the human relationships that can be forged only in the process of alienating possessions: lender and borrower, buyer and seller, carrier and pickpocket, Sunday-school teacher and pupil. The stop-and-go rhythm of

THE

SUNDAY AT HOME:

A Family Magazine for Sabbath Reading.

A NEW PAGE IN MY HISTORY.

THE STORY OF A POCKET BIBLE.
PART IV.

fulness, and toil on, from day to day, and all in vain, without that blessing which maketh rich, and adds no sorrow to success. I could also

Figure 4.1. "A New Page in My History," *The Story of a Pocket Bible*, 1st series, *Sunday at Home*, 22 March 1855, 177.

it-narrative registers this double bind. At moments when characters treat their book as a "friend" or "companion" too dear to sell—metaphors that project the anthropomorphic logic of the genre onto the world represented within it, as if the characters somehow knew that they were inhabiting a genre that grants books voice—the plot stands still. Only when characters reduce their book to a commodity valued for no more than it can fetch at the pawnshop do they come into contact with other characters; only then can narrative give way to dialogue.

Books therefore balk not only at being stolen, but even at being legitimately sold. When the hero of "The History of an Old Pocket Bible" (published in the Methodist *Cottage Magazine; or, Plain Christian's Library* in 1812) is exchanged for a ribband, its only consolation is to remind itself that "Judas sold his divine master for thirty pieces of silver" (129). The narrator of the *History of a Religious Tract Supposed to Be Related by Itself* secularizes the metaphor, complaining that "very soon after I came into the world I was sewn up into a book, and sent to a certain Depository in Stationers' Court, where I was exposed to sale with as little remorse as cattle in Smithfield, or Negroes in a slave market" (1). Like slaves, it-narrators find themselves lumped together with inarticulate objects at the moment of changing hands. The Old Pocket Bible, for example, protests being included in a list of stolen goods consisting of "a silver ink-stand, a lady's pocket-book, a small tea-caddy, and myself" ("The History of an Old Pocket Bible" 241). The narrator of *The Story of a Pocket Bible* complains more explicitly about the "state of slavery" that makes it liable to being, as it were, sold down the river to a godless stranger (Sargent, 1st ser., 19). The problem is not simply that books can be sold: it's more specifically that they can be resold, rather than faithfully handed down to the descendants of its first and last buyer.

Owners prove as ambivalent about alienating their books as books are about being alienated. In fact, the former's motives often remain opaque to the latter: when the narrator of *The History of a Religious Tract* is abandoned on a table, for example, it remains unsure "whether from neglect, or from the hope of my benefitting some other person" (2). The poor man who figures in the 1825 *Adventures of a Bible* agonizes before parting with the narrator, explaining that "it appears to me a kind of sin to sell my bible" (18). At the level of the text, bibles demand rereading (you wouldn't return one to the circulating library once you'd finished it). At the level of the object, too, devotional books exemplify sentimental value: handwritten births and deaths reduce not only their resalability but even their suitability as gifts. Made to be reread on a cyclical schedule, neither disposable like a novel nor supersedable like an almanac, devotional books would ideally change hands only once in a lifetime.

A second explanation for this fear of circulation is that where coins (the heroes of the classic it-narrative) are meant to pass through richer and poorer hands, books are marketed to specific social classes—even, or especially, the Bible. What Sargent's narrator forbids Hannah to read isn't the Bible so much as a bible—a book, not a text. "Hannah did not rest satisfied," it adds, "until she had procured the services of another of my own family—the very counterpart of myself, indeed, except in mere externals" (Sargent 38). Books are personified by being subjected to the same sumptuary laws as persons: like a human body, a bible requires clothing befitting its station. One Victorian journalist sneers that "books in handsome binding kept locked under plateglass in showy dwarf book-cases are as important to stylish establishments as servants in livery, who sit with folded arms, are to stylish equipages" (Coutts 120). Reciprocally, a bible—as much as an article of clothing—requires a socially appropriate owner. The question of what texts are appropriate for different social classes to read is upstaged by the question of what bindings are appropriate for them to own or even (as in the servant's case) to touch without owning.

The Story of a Pocket Bible was itself multiply reprinted, both in magazine and volume form. One American edition, subtitled A Book for all Classes of Readers, abridges the narrative "for the purpose of so reducing it in size and price as to adapt it to the wants of Sunday schools, though nothing essential to its main purpose has been left out" (Sargent 1). Once again, the text can circulate only at the price of mutating. The universality of the bible's own content paradoxically depends on the variability of its "externals"—lettering, binding, paper, size. But the fact that this humbler "counterpart" never gets to tell its story short-circuits the self-referential conceit with which The Story of a Pocket Bible began. Self-referentiality breaks down just as dramatically in tracts that invite readers to "wear [them] to pieces" but show unremarkably few traces of use: across the several dozen copies of the genre that I've been able to examine, wax and smoke stains are surprisingly absent, let alone intentional marks like underlining or turned-down corners.[5]

The failure of Hannah's own bible to narrate reminds us that these books about books are, more specifically, cheap tracts about expensive bibles. The Story of a Red Velvet Bible is covered with paper, not velvet; the Pocket Bible that considers itself too good for a servant to read narrates nothing better than a tract serialized in a penny weekly—a publication whose implied reader shares Hannah's class position, even if its implied buyer is likelier to resemble her master. (A run of this periodical is found in one of the few surviving servants' libraries, at Cragside in Northumberland [Stimpson 7].) Even more paradoxically, the aspirational book vehicles a text that slums. Composed by middle-class (or adult) writers mimicking the voice of uneducated (or not yet educated) readers, the language of tracts talks down; the material forms that it represents are both more durable and more upmarket than those it inhabits. Bibliographical "externals" short-circuit linguistic content: the form of the object undermines the circularity on which the text's feeble humor depends.

The History of a Religious Tract—a tract about a tract—forms one exception to this rule. Another, which turns the conventions of Evangelical it-narrative to secular and even commercial ends, is "The Life and Adventures of a Number of Godey's Lady's Book. Addressed Particularly to Borrowers, Having Been Taken Down in Short-Hand from a Narration Made by Itself, When the Unfortunate Creature Was in a Dilapidated State, from the Treatment Received at the Hands of Cruel Oppressors."[6] Here, the autobiography of a number of *Godey's* appears in—a number of *Godey's*; what's more, the conditions under which *Godey's* is bought and borrowed form its subject. "You must not suppose that I was always the wretched, coverless, soiled, dog's-eared and torn object you see," the narrative begins. "I was an intellectual individual. I knew it; I surveyed my own cover with a proper degree of pride, a little abated, however, by the reflection that I could be bought and sold for twenty-five cents." Unlike the Pocket Bible and the Religious Tract, however, the Number of *Godey's* is not complaining about the fact of being put up for sale. On the contrary, its ambition is to fetch an even higher price: "I felt I was worth, at least, a dollar; and to dispose of me for less was a poor reward for all my wit and wisdom" (425). The question here is not (as in the Smithfield metaphor) whether books should be conceptualized as something more or less than commodities, but—more technically—whether the transaction through which they change hands should consist of buying, renting, or borrowing. "When in the course of human events," the narrator declaims, "it becomes necessary for people to borrow boot-jacks, salt, or cucumbers, let boot-jacks, salt, or cucumbers be loaned. But let indignant subscribers to the 'Lady's Book' declare their independence of borrowers" (427).

That commercial message drives the complaints that structure its "Life and Adventures": "One visit would lose me a leaf, another a plate . . . My music got enamoured of a piano at my fifth stopping-place, and shamefully deserted me forever. The great gap you see on one of my pages was occasioned by the scissors of a young lady, who clipped out a beautiful poem, by Mrs. Neal, for her scrap-book . . . One careful housewife, to complete my degradation, after she had read my contents, used me as a duster" (426). Where the Pocket Bible's reluctance to circulate contradicts its Evangelical mission, the equivalent structure makes perfect sense in a magazine story designed to boost sales. After parting mournfully from an honest schoolmaster, "never to look on his face again," the Number of *Godey's* complains that the thoughtless girls whom he teaches demand to have his copy passed along to them, on pain of being fired by their fathers. A dialogue between two of his pupils drives the point home:

"What! Injure a man because he hesitated to suffer you to use his property as freely as though it were your own?"

"But, it is only a book."

"Very true, and that yonder is only a bonnet. How would you like to have that passing from head to head, when it came from the milliner, being borrowed in turn by all the girls of the village."

"Yes, but a bonnet is a necessity."

"And the 'Lady's Book' becomes a necessity very soon: it is mental food." (426)

Women read, but men buy: even the goody-goody who refuses to join the twenty-eight women who club together for a single subscription cites the authority of her father, who thinks so highly of *Godey's* that he has given her permission to take out her own subscription. When the schoolmaster laments that "the copy I have to pay for" is extorted by girls who "spend yearly, on folly, more money than would suffice to support me," he endorses the very equation between person and thing on which this blackmail depends: the "yearly . . . money" could just as well be the narrator's name for the annual cost of supporting (that is, subscribing to) itself. The magazine competes with the schoolmaster as an instructor of youth.

The structure of it-narrative is used to reframe a debate about the rights of persons (in this case, publishers) as one about the rights of objects (in this case, magazines). Where the Religious Tract is insulted at being bought, the Number of *Godey's* is wounded instead by a bibliographic promiscuity that forces it to serve the pleasures of more successive subscribers than it can stand. Beginning its life "dressed by the binder," the Number ends up with "my cover taken entirely off, leaving me in a distressing state of nudity." If clothing is both personal and subject to an annual fashion cycle—who would wear a secondhand bonnet?—so are magazines. To pass "from hand to hand" is to be sullied. Like the Religious Tract, the Quire of Paper, the Old Pocket Bible, and the Book, the Number of *Godey's* feels most vulnerable at those moments where it's endowed with skin that can be scarred and limbs that can be amputated:

> The brown mark on one of my corners came from the hot ashes of a cigar. Every step that I took was marked with fresh indignity and additional mutilation . . . Here I am, prematurely old, and ready to fall to pieces from continued ill treatment. At one house I would find the face of one of my plates, smudged with candy from a child's fingers; at another, the eyes of a lady in the fashion plate were ornamented with an enormous pair of spectacles. (426)[7]

Whether the narrator's wounds are metaphorical (inflicted on the bodies it resembles) or metonymic (inflicted on the bodies it represents), the topic that moves books to speech is the wear and tear to which they have been subjected. "The History of an Old Pocket Bible" begins, "I am at present in a most tattered condition. One of my covers has long since been missing, and the other hangs only by a single thread. A great part of my leaves are torn out, and the remainder are so doubled down and spoiled as scarcely to be legible. Indeed I daily expect to be cast into the flames" (89). The narrator of *The History of a Book* introduces its description of the printing process by punning that "I was yet to undergo a great deal of sharp usage" (Carey 141). The eponymous Quire of Paper translates that bruised and battered condition more specifically into a loss of voice.

> What horrors I endured when after being borne through several dark apartments, I saw before our eyes a dreadful machine, whirling round with terrible velocity, and roaring with so loud and tremendous a voice for prey that every ear was deafened, and every sound lost near it! Think what my situation must have been, when I discovered that *I* was the kind of *food* this monster craved for, and amongst the number of its devoted victims. All language were weak to describe to you the terror and anguish I felt when I was thrown between its gaping and voracious jaws. ("Adventures of a Quire of Paper" 449)

The physical violence of being mangled is doubled by the psychic violence of being silenced. Yet the indignity of a literal voice's being drowned out by a machine's metaphorical "roar" is undermined by the fact that that voice belongs to an equally inanimate object: if the narrator's "language [is] weak," what else could we expect from a sheet of paper?

Like the repentant sinners that it depicts, the genre becomes self-conscious only on its deathbed. One of its final instances, *Handed-On: Or, the Story of a Hymn Book* (SPCK, 1893), transposes the it-convention from a formal to a thematic register, shifting away from the first person but recapitulating every other cliché of the genre. Here again, the plot charts the colonial and domestic travels of a volume whose flyleaf is inscribed: "This is a wandering hymnbook. The finder will please mark some line or verse and then pass it on" (30). Here again we find a debate over the value of circulation, beginning with the eponymous book's being spotted by the side of road:

> I shall pick it up, and see what it is, and whether there is any name inside the cover.
>
> Don't, Ethel! How can you tell who touched it last, or from what house of illness it may come? said the fashionably dressed lady, who always declared that she was in constant terror of infection from one cause or another, and that she never ventured into the quarters of the city where the poor congregated. (9)

And here again the daughter can imagine the hymnbook as a person only by dint of imagining it as a victim: "Poor little hymn-book!" she said softly. "Oh, mother, you need not be frightened, for it is . . . a fellow-countryman of ours" (9). In fact, the maid embarrassed to offer her mistress "a dirty old book like this" is proven wrong. Once again, Christianity changes wear and tear from faults to virtues. Another servant (the former nursemaid of the young man whose death precipitates his hymnbook's travels) balks at passing along the book as its flyleaf requests: "If I did what I liked best I'd ask to have it buried with me" (71, 25). For her as for the young man's mother, the hymnbook is both metaphor and metonymy for its dead owner: to pass the book along is to accept the sacrifice of the son.

Handed-On does recuperate the form of the it-narrative, but only as an unrealizable fantasy: its third-person account of a "wandering hymnbook" frames the axiom that "no book can tell the story of its ups and downs in the world, or describe how and why it began to pass from hand to hand" (6). The it-narrative reappears only vestigially, in the person of a character who "wished that a book could tell its own true story—or rather the story of its various owners" (46). In *The Story of a Pocket Bible*, that sentiment had been placed in the mouth of the book itself: it's the bible who, on the last page of its autobiography, overhears its owner say, "O, if this Bible could speak, what a history it would have to tell!" By 1893, not only can a book no longer tell its story, but that story is itself imagined as a placeholder for human stories. Yet the book itself continues to have a story, and its narratability continues to anthropomorphize it: another character asks, "Where is the book that does not or will not possess a story? . . . A book is for me somewhat like a man or woman; I dream and speculate as to the scenes where it has been" (*Handed-On* 76). Characters' consciousness replaces the narrative itself as the place where books are imagined to possess a self.

BOOK, VETERAN, INVALID, PROSTITUTE

One explanation for the it-narrative's turning bookish at the end of the handpress era, then, is that at a moment when mechanization is making books more closely identical to one another, it-narrators must struggle harder to individuate themselves. When the copy of *Crusoe* refers to "one of my brothers in better plight than myself," it makes each copy part of a family, or at least a litter—not, in any case, identical twins ("Adventures of a Robinson Crusoe" 191). And the Number of *Godey's* reports that "after having been dressed by the binder, myself with eighty thousand nine hundred and ninety-nine companions were carried over to the publication office" ("The Life and Adventures of a Number of Godey's Lady's

Book" 425) before going on to describe being cut out and penciled on: only at the moment of transmission do mass-produced objects acquire a unique life story. As Walter Benjamin reminds us, "*Habent sua fata libelli*: . . . not only books but also copies of books have their fates" (61).

A catch-22: too little handling appears as dangerous as too much, being a wallflower as bad as being raped. Lamb lends the book a "voice" only when it bears the mark of dirty thumbs: "How beautiful to the genuine lover of reading are the sullied leaves, and worn-out appearance, nay, the very odour (beyond Russia), if we would not forget kind feelings in fastidiousness, of an old 'Circulating Library' Tom Jones, or Vicar of Wakefield! How they *speak of the thousand thumbs*, that have turned over their pages with delight!" ("Detached Thoughts"; my emphasis). The bible that confesses that its "personal appearance had begun to assume that of a veteran in my Master's service" (even though, it adds sarcastically, "the parental warning given to my young owner, to take care of me, was so far unnecessary that there was no fear of my becoming further worn or soiled by his frequent usage, whether good or bad") draws on Lamb's reference to his books, a moment later, as "battered veterans" (Sargent, 2nd ser., 33–34) . Angus Reach writes, in the same vein, that "the books may sometimes be a little greasy, to be sure, the paper stained and thumbed, and the leaves dog-eared. But what of that? We respect a stained dog-eared book. It is a *veteran* who has seen service—not a mere gilt ornament to an unread library" (Reach 248; my emphasis).

Tract-distributors themselves scanned for signs of use more eagerly than any reception historian can. One fictional colporteur testifies that in entering a former Catholic's house "I found the Bible he had purchased from me lying on the table; it bore marks of frequent usage, for it was quite worn out." In another house a convert "drew out a New Testament in 12mo., which was all in tatters, so much had it been used" (*Fifty-Sixth Report of the British and Foreign Bible Society* 31, 10, 14). In a novel about a Protestant colporteur in France, meanwhile, the hero

> observed with joy that the book betrayed tokens of constant reading. Every page appeared to have been frequently, though carefully turned; and various favourite passages were marked, one by a ribbon, another by a dried flower or leaf, another by a bit of thread or tiny scrap of paper; while some pages were doubled in half, others turned down at the corner above, others at the corner below. There could be no doubt of its being the vade mecum of a Bible Christian. (Manning 203)

And a missionary from the Institution for the Evangelization of Gypsies testifies that "in one of the former families there was a Testament, which had been presented by the Committee in Southampton, and bearing date

1830. It bore marks of frequent usage. Many single leaves were turned down as marks for certain passages" ("Institution for the Evangelization of Gypsies" 170). The hero of one 1859 novel even falls in love the moment he notices that the heroine's bible is "not one of those velvet things with gilt crosses that ladies delight in, but plain-bound, with slightly soiled edges, as if with continual use" (D. M. Craik 59).

Keeping a book too clean is as bad as letting it get too dirty. One London bible-woman reports: "Called on a man in C—— street. His answer was, 'No, missus, I do not want a Bible. I have one in my box, and it is one hundred and two years old.' I replied, 'I should like to see it.' He took it out, and I was obliged to say, 'It looks as if every page condemned its several owners.' 'How so, missus?' 'It has always been kept in the box, and not a leaf is soiled'" (Ranyard 95). The story has a happy ending, as the man eventually agrees to subscribe for a large-print bible. Instead of a person judging books, books sit in judgment over persons. When a character in *Ministering Children* confesses that "a locked-up Bible is a bad witness against me," the evidence of the book trumps the testimony of its owner: empty protestations of faith can be faked, but the wear and tear on the page doesn't lie (Charlesworth, *Ministering Children* 72).

If the happiest women have no history, the same could be said of the happiest books (G. Eliot, *The Mill on the Floss* 400). Conversely, it-narrators must suffer, because the only voice with which prosopopoeia can invest them is the passive. For the narrator of the *History of a Religious Tract Supposed to Be Related by Itself*, the very basis on which books can be assimilated to speaking subjects is their vulnerability. "Much indeed do I resemble man," it begins, "not only in the vast variety of my members, but in the delicacy of my constitution. As human 'Life contains a thousand springs, and dies if one be gone,' so the loss or misplacing of a word sadly disorders me, and the fraction of a page is death" (1).[8] The narrator of the 1873 *History of a Book*, too, matures by being humbled: "Not for long was I allowed to remain in this inflated state of mind with regard to my probable size as a book. My sheets were taken and . . . passed between two iron rollers . . . This 'rolling-machine' compresses the sheets so very much more than the old 'hammering' did [that] the result, to my mind, was to make me 'feel small'" (Carey 137).

In the *Adventures of a Bible* (1813), a characteristically Christian reversal makes that same smallness a source of power: "I was indeed but six inches in height," acknowledges the narrator; "but with this I was by no means discontented, as I thought that, probably, I should be more frequently brought into use, than if I had been of a larger size; and I knew that, small as I was, I could teach and do as much as the largest" (Boston Society for the Religious and Moral Improvement of Seamen 6). No less than the undersized and underfunded protagonist of a Dickens or Brontë novel, it-narrators exemplify the hidden powers of the physically and socially insignificant. A Christian theme, but also a political subtext: a battered paper-covered volume demands as much respect for its contents as the spotless leather-bound twin from which it's been separated at birth. A book's a book for a' that.

Like abused children, too, the book evokes our empathy by reporting mistreatment. In secular accounts of book production, the narrator gets pressed, trimmed, and hammered; in religious accounts of book circulation, the narrator gets pawned, stolen, torn, kicked, and trampled on. If books' accounts of their own martyrdom borrow from the conventions of missionary autobiography, they also draw on a long tradition in which Christ's body was compared to a book (Kearney, *The Incarnate Text* 14). The word made flesh—but as in Kafka's *Penal Colony*, flesh most legible when martyred.

Paradoxically, then, those moments when the book usurps its human owner's agency occur not when it's most powerful, but when it's most abused. Caroline Wilder Fellowes's "A volume of Dante," for example, opens, "I lie unread, alone. None heedeth me. / Day after day the cobwebs are unswept / From my dim covers" (White, *Book-Song* 53). In fact, this logic exceeds the it-narrative proper: even in a human-narrated tract, Mrs. Sherwood can warn that "Bibles are now so abundant in England, that the rich supply, we fear, rather tends to cause a contempt for the gift than a spirit of thankfulness; but let it be remembered, that every Bible which has lain neglected on the dusty shelf may, some time or other, rise in judgment against its careless possessor" (M. Sherwood 315).[9] Here as in *Ministering Children*, it's precisely the fact of being silenced that gives the bible standing to complain. Across the ocean, the *Life of William Grimes, the Runaway Slave* (1825) concludes: "if it were not for the stripes on my back which were made while I was a slave, I would in my will, leave my skin a legacy to the government, desiring that it might be taken off and made into parchment and then bind the Constitution of glorious happy and free America. Let the skin of an American slave bind the charter of American liberty" (quoted in Fabian 87–88). And if a slave's skin could be imagined as a book's binding, a book's cover could be imagined as a slave's skin: the most heavily subsidized bibles bore a "charity brand"—a stamp referred to, once again, in the language of slavery (Howsam 122).

"If you would know how a man treats his wife and his children, see how he treats his books": the aphorism often attributed to Emerson (probably apocryphal) endows the book with personhood at the price of vulnerability. An 1882 article titled "The Library" in the *Gentleman's Magazine* that begins with a hackneyed quotation from *Areopagitica* ("as good almost kill a man as kill a good book") goes on to make clear that books bear less resemblance to men than to women: "He would think that he richly deserved the six months' hard labour which London magistrates deal to brutal husbands who kick and jump upon their wives, could he bring himself to double up the backs of his books" (Watkins 101).

Such analogies pair the book's capacity to feel pain with its inability to protect itself. Or herself, for the most minimal grammatical marker of the genre—the "it" contradistinguished from those more conventional narrators who must be resolved into a "he" or a "she"—is contradicted both by the sexual metaphors that convey the book's vulnerability, and by the metaphors of dress that spark the book's reflections on its own vanity. (In *Romola*, too, selling off your father-in-law's library brands you capable of monetizing his daughter.) This isn't to say that Emerson (or whoever came up with the aphorism) is advocating a chivalrous refusal to lay hands on the book: unlike the seduction narratives invoked by *Godey's*, these analogies figure the book as legitimately married to her (male) owner. By that logic, a book that's been used to pieces would reflect as badly on its master as does a book that's respected to the point of not being used at all.

Book, Child, Narrator: *David Copperfield*, Again

Surprisingly, then, the book speaks most loudly when its words go unread. Or maybe not so surprising. A voice that emerges from a body small enough to be overlooked; a narrator that eloquently analyzes its sensations but can't talk itself out of a beating; a narrator whose physical pain is compounded by the humiliation of being silenced; a narrator, more fundamentally, whose subjectivity is never acknowledged by other characters: if the it-narrative's combination of strong focalization with represented weakness sounds at once counterintuitive and familiar, the reason may be that the same contradiction vertebrates a better-known genre, the bildungsroman. Each genre endows its narrators with consciousness while stripping them of power; each contrasts the narrator's fluency with other characters' refusal to recognize its standing to speak.

It would be stating the obvious to acknowledge that bildungsroman and it-narrative stand at opposite poles of Victorian fiction: one on the rise, another on the wane; one centered on subjects, the other on objects. More crucially for our purposes, the it-narrative reveals what the bildungsroman conceals: the backstory by which books reach their readers. Nothing could be further from the fantasy of the self-made reader and the self-distributing text that we saw in the previous chapter than the it-narrative's understanding of books as vectors for human relationships. Nothing further from the bildungsroman's representation of books as found objects—stripped of price tag, Micawberishly turning up in uninhabited garrets, finding their way into children's hands through an agency as invisible as that which supplies gas and running water. In endowing books with a life story, the it-narrative restores what the bildungsroman suppresses. Or more precisely, what the bildungsroman confines to human subjects: David develops a *Personal History* (as the full title of the novel has it) at the expense of his books' being stripped of one.

Those contrasts would seem to undermine the logic of much recent research on it-narrative, which mines the genre for intertexts to more familiar works of (human-narrated) fiction. Thus Jonathan Lamb and Lynn Festa have described slave autobiography as giving voice to a piece of property, and Paul Collins has reclassified *Black Beauty* as at once an it-narrative whose object happens to be animate and a slave narrative whose hero happens to be equine.[10] These readings take as their benchmark the earlier, numismatically themed phase of it-narrative. The bildungsroman borrows something far more specific from later it-narratives: in both cases, I want to suggest now, questions about agency are routed through a particular category of object, the book.

The child metaphor doesn't just construct the book as a particular kind of subject; more specifically, it also constructs the book as a particular kind of narrator. A sensitive yet powerless protagonist who forms the object rather than the subject of action, a camera eye whose plight serves as an index to the morals of those with whom it comes into contact: if it-narratives draw at once on slave autobiography (as in *The History of a Religious Tract Supposed to Be Related by Itself*) and seduction plots (as in the "longing looks," "blushes," and "stolen moments of communication" in the *History of a Bible*, or the *Godey's Lady's Book*'s descent from pride in its fine clothing to humiliation at the violation of its body), they would go on to be drawn upon by the bildungsroman and the narrative of animal rights. Whether the seduced women upstream of it-narrative or the child narrators downstream, the "wife and children" of Emerson's analogy mirror the book's ability to sift the virtuous who recognize its subjectivity from the vicious to whom it remains a tool. The it-narrative functions at once as a user's manual (treat your books as well as you treat yourself) and a litmus test (beware the suitor in whose home you notice spread-eagled books or whimpering dogs).

Lynn Hunt and Joseph Slaughter have argued that vulnerability to suffering defines the limits of the human. "In the eighteenth century," Hunt contends, "readers of novels learned to extend their purview of empathy. In reading, they empathized across traditional social borders between nobles and commoners, masters and servants, men and women, perhaps even adults and children" (Hunt 40; Slaughter). But also, thanks to it-narratives, between animate and inanimate beings. If, as Slaughter suggests, our modern conception of "human rights" comes from the novelistic enterprise of imagining the pain of unprotected beings like women, children, and slaves, that list could also include books—which, as we've seen, become anthropomorphized at precisely those moments when they are assimilated to a "prisoner," "captive," "slave," "wife," "child," or even war invalid.

Outside the boundaries of the it-narrative, judicial punishment has traditionally anthropomorphized books: a volume put on trial and burned by the public hangman looks more human than a book that's being read. Heine's prediction that book-burners will come to burn people transposes the pseudo-Emersonian aphorism into a judicial register. Predictably, Ray Bradbury spins that comparison out in a plot that ends with persons "becoming" books in the hope that bodies will prove harder to burn than pages; a draft of *Fahrenheit 451* endows books with body parts at the moment of destruction, making a character declare that "we just gave them the bullet behind the ear" (quoted in Seed 238).

Remember that the books which distract Julien Sorel and Hugh Trevor from their surroundings get drowned and burned: where the bildungsroman links book with person through metonymy (a book is knocked out of a child's hands before the child himself is knocked down), the object narrative and the legal regime of censorship both link books with persons through metaphor (a book becomes most humanized when manhandled in the it-narrative, or burned by the public hangman in real life). If, as we saw in the previous chapter, the bildungsroman makes those characters who judge a book by its cover identical to those who objectify human beings, Heine's phrasing reverses that logic: for him, the same metaphor that anthropomorphizes books can also objectify humans.

By extension, the same it-narratives that elevate books to narrators can reduce narrating subjects to grammatical objects. The Lockean metaphor in which "a profound *impression* was made upon" David Copperfield is literalized in the account that the narrator of *The History of a Book* gives of its equally painful entrance into consciousness.[11] "No words can express the secret agony of my soul," David tells us, when he begins his work at the bottle warehouse; "I mingled my tears with the water in which I was washing the bottles; and sobbed as if there were a flaw in my own breast, and it were in danger of bursting." The sandwich board, too, turns "me" into "it": "the servants read it, and the butcher read it, and the baker read it." Here as in the earlier scene that reduces him to a parcel—stowed on top of the scale in the luggage office "as if I were weighed, bought, delivered, and paid for"—what assimilates David to an object is their common state of being written upon (Dickens, *David Copperfield* 152, 82, 76). To be labeled with words is to be made wordless—for David as much as for the narrator of the *History of a Bible*, "confined in the family prison, called the library," where books "had their names written upon their foreheads, but none of them were allowed to speak" (1).

For David to imagine himself traveling is also to imagine himself being sold or read. When he runs away to Dover, too, "I began to picture to myself, as a scrap of newspaper intelligence, my being found dead in a day or two, under some hedge; and I trudged on miserably, though as fast as I could, until I happened to pass a little shop, where it was written up that ladies' and gentlemen's wardrobes were bought, and that the best price was given for rags, bones, and kitchen-stuff" (Dickens, *David Copperfield* 174). The scrap of newspaper that David "pictures" proves no more durable than the visions that arise in his mind "as if they were faintly painted or written." In the first case, labels on actual bottles make writing fade away from a hypothetical wall; in the second, an imagined scrap of newsprint gives way to the raw material of paper manufacturing. The installments in which Goroo dribbles out payment for David's jacket would remind us of serial publishing even if his real name weren't Charley, but the place that he occupies in the printing industry lies far upstream of that. Any reader in 1850 would have known that the page in front of her was made from rags, collected in shops like Goroo's to be turned into pulp and then paper. When we speak of a newspaper as a "rag" (or of fiction as "pulp"), metonymy converges with metaphor: paper comes from old clothes; paper resembles old clothes. *Copperfield* confronts the two figures of speech, pitting a pictured scrap of newspaper against a dealer in real scraps of clothing.

That these metaphors occur during travel suggests that what it-narrator and bildungsroman protagonist have in common is not just weakness, but more specifically vulnerability to being moved. We've seen that anthropomorphic metaphors and metaphors of debasement (slave, animal) both cluster most thickly at those moments when it-narrators change hands. That clinging to an original owner can be compared in turn to the bildungsroman protagonist's resistance to moving from country to city, or even (as Franco Moretti has argued) to relinquishing his childhood sweetheart. The only difference is that one circles back to a first love, the other to a first owner (Moretti, *The Way of the World*).

Persons *were* in fact least in control of their own fates when on the move. John Newton reports that on slave ships like those that he captained before turning abolitionist, "the slaves lie in two rows, one above the other, on each side of the ship, close to each other, like books upon a shelf" (248). Dickens's reduction of a runaway child to raw material for paper echoes his borrowing of that metaphor to describe an American canal boat where going below, I found suspended on either side of the cabin, three long tiers of hanging bookshelves, designed apparently for volumes of the small octavo size. Looking with greater attention at these contrivances (wondering to find such literary preparations in such a place), I descried on each shelf a sort of microscopic sheet and blanket; then I began dimly to comprehend that the passengers were the library, and that they were to be arranged, edge-wise, on these shelves, till morning. (Dickens, *American Notes and Reprinted Pieces* 87)

Just as it-narratives anthropomorphize the book at its moments of greatest vulnerability, Dickens bibliomorphizes persons at moments when other characters are treating them no better than objects.[12] Betsey Trotwood's insistence on referring to David as "it," like his own confession to being "cherished as a kind of plaything," projects onto characters Dickens's own tendency to confuse persons with things. But in the context of one of his few book-length fictions to be narrated in the first person, that confusion begins to look more specifically like an incursion of the it-narrative into the bildungsroman.

Even within the object narrative, this sense of the self as marked rather than marking migrates from things to persons: when the *Story of a Hymn-Book* (an 1881 production of the Wesleyan Conference Office) introduces a smoking, drinking gambler with "a yellow parchment-like skin, bearing significant brown stains about the mouth," the handling "signified" by the marks of ink or drink on a page becomes the model for the traces left on the body by a human character's life story (Yeames 80).

Formally, it-narrative and bildungsroman look like polar opposites: one centered on a human being through whose hands texts pass, the other on a book that passes through characters' hands. Thematically, both give voice to the voiceless: in focalizing so much of his fiction through child characters, Dickens extends the eighteenth-century humanitarian project implicit in a genre that made snuffboxes speak. From coins to lapdogs to adult slaves to free children, the it-narrative steadily extends the borders of subjectivity. Yet to assimilate books to persons is not to grant them agency, if only because the voice with which they're endowed is so rarely the active.[13] *David Copperfield*'s grammatical opening gambit ("I am born") constructs an analogous paradox, in which the expression of self becomes indistinguishable from the lack of power. Not only do both genres endow their protagonists with voice while stripping them of agency: more specifically, both frame that contradiction in bibliographic language. "How much I suffered," "what I suffered": David is reduced to a helpless object by being bound between boards. Or more generally, by being labeled, inscribed, read, and sold—sold, in Goroo's shop at least, in installments that turn out to be more valuable for their rag content than for the human subjectivity that they contain. When drunk, David feels "as if my outer covering of skin were a hard board"; the metaphor doesn't just plunge the prosperous young man back into the childhood nightmare of the sandwich board, but also equates the debasement of a human being with the process of being assimilated to a book (339). If David Copperfield is bound between labeled boards, so is *David Copperfield*; in fact, the *Fraser's* review reminds us that both sets of boards are degraded by being scanned by the butcher and the baker.

These formal borrowings from the it-narrative may help explain George Levine's argument that "it is one of the curious facts about the most virtuous heroes and heroines of nineteenth-century English realist fiction that they are inefficacious, inactive people" (33).[14] If novelistic heroes appear curiously weak, one reason may be that they're modeled after protagonists who are not even recognized as human. The only difference is that the bildungsroman situates this suffering consciousness in childhood, and the it-narrative in old age: one focalized through a character who has not yet reached the social invulnerability of adulthood, the other through an object that becomes aware of its identity in the process of realizing its own mortality. Between maturation and dissolution, the narrator just born and the narrator about to die, the two genres stake out complementary portions of the life cycle. For it-narrators, to be born is already to be assaulted; in contrast, David can age only by being assimilated to a book, when Dora pencils lines onto his forehead. Yet that contrast looks more like similarity if we emphasize instead both narrators' hatred of traveling.

Think back to the prayer book and Latin Grammar that appeared earlier in *David Copperfield*: the teachers who mark David's skin *like* a book take a cue from adults who already marked it *by* a book. In both cases, the book's encounter with the body upstages the text's with the mind. Yet as the metonymic logic of the boxed ears and the poked side gives way to the metaphoric logic of the inked forehead and the rubbed-out welts, something more fundamental changes. Once reduced to an object (valued for its outside rather than for the words that it contains), the book is now exalted to a subject—or, at least, to something to which a human being can be compared.

If the tattooed arrow pointing in the same direction as the sailor's bare arm mirrors Dora's finger steeped in ink to provide a baser counterpart to Agnes pointing upward, both also make the body parts that we expect to *perform* writing, *undergo* writing (Dickens, *David Copperfield* 728). As we saw in the previous chapter, David's adult career as a writer is preceded by years spent as a surface to be printed upon like a page of a book, ruled like a piece of blank paper, and scarred like parchment. As upward mobility and personal development conspire to turn a child who is written upon into an adult who writes, the novel's key trope shifts from metaphor (a child who resembles a book) to metonymy (an adult who makes one).

FROM *THE HISTORY OF A BOOK* TO "THE HISTORY OF THE BOOK"

I suggested in the previous chapter that the swerve away from David's composition to minor characters' copying prefigures the intellectual-historical shift from a literary criticism interested in authorial genius to one interested in more mundane inscription practices, such as accounting and shorthand. *David Copperfield* in particular, and the Victorian novel in general, have served as magnets for what could be called "thematic materialism"—whether they read questions of intellectual property into the bottle warehouse, or of branding into the sandwich board on David's back. The danger is a typological interpretation that reduces the text to an allegory of its own manufacture—or, if you move from supply to demand, its own distribution. In this kind of reading, a history of the material conditions of production and consumption can either redundantly corroborate some formal or thematic explication of the text itself, or irrelevantly contradict it.

One way out of that impasse would be to trace different models of textuality back to their *generic* origins—to determine in what contexts, both formal and material, competing understandings of the author, the reader, and the book take shape. The it-narrative would then correspond to a bibliographical project that traces the transmission of textual content and of printed objects; the bildungsroman to those methods that chart the development of an individual's literary sensibility (whether under the name of reception history or of close reading). No matter whether that individual is the critic in propria persona or is a historical figure like Carlo Ginzburg's Menocchio or Darnton's Ranson or John Brewer's Anna Larpent—readers whose diaries, letters, or inquisitorial files have deposited a record of response. In those scholarly traditions no less than in the bildungsroman, the guiding thread remains a human subject's formation through a series of texts. Analytical bibliography and it-narrative substitute an object's accretion of meaning via a series of uses. One is structured by the growth of a child's mind (caused or at least catalyzed by books), the other by the aging of a book (worn out as it passes from hand to hand). Competing Victorian fictional subgenres thus anticipate twentieth-century scholars' oscillation between tracking persons and tracing objects. One could be aligned with Robert Darnton's "communications circuit," structured around human agents such as bookseller, author, smuggler, and binder; the other with Thomas Adams and Nicolas

Barker's alternative paradigm, which reduces humans to stations along the path that the book follows from commissioning, to composition, manufacture, circulation, and disposal (Adams and Barker 12, 15). The early twenty-first-century vogue for what is called "the history of books and reading" has blinded us to the possibility that those two histories are distinct and even competing projects. In that context, you could trace reception theory back to the post-Romantic psychologizing of the bildungsroman (*David Copperfield* and *Jane Eyre*), the history of the book back to the interest in the outsides of books exemplified by Trollope and Thackeray—or indeed analytical bibliography to the it-narrative.

More narrowly, it-narrative could offer a precedent for the internalist account of reading audiences that Roger Chartier has called for. Chartier urges book historians to look, not at the reading habits of a group defined by "a priori social oppositions," but rather at "the social areas in which each corpus of texts and each genre of printed matter circulates" (Chartier, *The Order of Books* 7).[15] This method follows the order of it-narrative in that instead of starting from a person and asking what books he owned, it starts from a book and asks into whose possession it came. In this model, the book would exemplify Arjun Appadurai's argument that while "from a *theoretical* point of view human actors encode things with significance, from a *methodological* point of view it is the things-in-motion that illuminate their human and social context" (5).

In shifting their gaze from the meeting of minds to the manipulation of things, early twenty-first-century book historians not only draw on the bibliographical tradition but mimic the turn of late twentieth-century historians of science toward the object world—as in Sherry Turkle's interest in prostheses or Bruno Latour's case study of a driverless train (Turkle; Latour, *Aramis*).[16] What Turkle says of objects in general holds particularly true of books: "Behind the reticence to examine objects as centerpieces of emotional life was perhaps the sense that one was studying materialism, disparaged as excess, or collecting, disparaged as hobbyism, or fetishism, disparaged as perversion . . . So highly valued was canonical abstract thinking, that even when concrete approaches were recognized, they were often relegated to the status of inferior ways of knowing, or as steps on the road to abstract thinking" (Turkle 6).

The it-narrative, I've suggested, bequeaths a powerful set of conventions to the bildungsroman—even if a formal gimmick in the first becomes an ethical commitment in the second. Another place to look for the legacy of the genre, however, is in *non*fiction discussions of the book, both in their own time and after. The habit of slotting books into roles normally occupied by human beings soon spilled out from it-narratives into the religious press at large. Like the Number of *Godey's* describing its "visits," RTS magazines such as *The Visitor* (1833), *The Weekly Visitor* (1836), and *The Monthly Messenger* (1844) figure themselves as persons paying social calls, not as objects being bought and sold.[17] Termed "silent messengers," "silent monitors," or "silent preachers," tracts are also characterized as "preachers which penetrate where no voice of man could reach"—even at the risk that such speaking objects can too easily be confused with the "dumb idols" that they are designed to displace (Jones 43, 238, 594, 174, 360, 32). Even in the secular press, in fact, Charles Knight's *The struggles of a book against excessive taxation* (1850) casts the book as an agent whose desire is to circulate as widely as possible.

Such metaphors don't only draw on the conventions of the it-narrative: they reach back to a much older and less localized tradition of comparing books to friends, which stretches from Petrarch right through to Groucho Marx's remark that "Outside of a dog, a book is man's best friend; inside of a dog, it's too dark to read" (Emerson and Lubbock 35). Here as in the jokes we saw in the introduction, books become the pivot between a figure of speech and its bathetic literalization. But Marx also draws on an equally old metaphor of book as animal (think of the religious tract "exposed to sale with as little remorse as cattle in Smithfield") that in turn invokes the metonymic derivation of books from skins. The tenth-century anthology of Anglo-Saxon poetry known as the Exeter Book poses a question in the first person (essentially, who am I?), to which the answer appears to be "a book," but Bruce Holsinger argues that the answer can also be construed as "an animal," the animal whose slaughtered corpse provides the parchment (622).[18] Here as in the it-narrative, suffering is what gives voice to the book-object: the "I" becomes a book only once it's been killed and flayed. More specifically, the marks of wear and tear are what remind handlers of the book's animal origins: as one scholar notes, "handling a parchment page even hundreds of years after it has been rendered into a writing surface, the reader is often well aware of its history as flesh: one can see hair follicles, tiny veins, discolorations where the living skin carried scars or blemishes" (Kearney, *The Incarnate Text* 7).

As we saw in relation to sumptuary codes, anthropomorphism hinges on a sartorial metaphor that makes the printed sheets to the body what the (cloth or leather) binding is to clothing—a metaphor taken to its logical conclusion in the Sikh tradition, whose holy book is covered with light cloth in summer and heavy cloth in winter (G. S. Mann). In an 1851 description of a railway bookstall, "here and there crouched some old friends, who looked very strange indeed in the midst of such questionable society—like well-dressed gentlemen compelled to take part in the general doings of Rag-fair" ("The Literature of the Rail" 7). Earlier, Hester Piozzi had praised "our Leather-coated Friends upon the Shelves; who give good Advice, and yet are never arrogant or assuming" (H. J. Jackson, *Romantic Readers* 121). Dickens draws at once on the trope of binding as clothing and on the logic that anthropomorphizes battered or abused books when he prefaces the Cheap Edition of 1847 with the hope that the book will

> become, in his new guise, a permanent inmate of many English homes, where, in his old shape, he was only known as a guest, or hardly known at all: *to be well thumbed and soiled in a plain suit* that will bear a great deal. (Dickens, *Prospectus for the Cheap Edition of the Works of Mr. Charles Dickens*; my emphasis)

Or, reciprocally, humans' clothing can be assimilated to books' binding, when the same journalist who compares dog-earers to wife-beaters adds that "he is no true lover of books who suffers his volumes to remain in yellow paper and blue boards. Would he like to see his wife, the very apple of his eye, go about a dowdy?" (Watkins 101). To compare binding to clothing is to endow books with a human body, but not necessarily a human soul.

It's a mark at once of respect for the Bible and dismissal of social inferiors that missionaries often make the book the grammatical subject and the reader the object, rather than the other way around. "Moreover, the book goes where missionaries and other workers do not; where under present circumstances, for lack of numbers or for other reasons, they cannot go. Be it that the book remains unread, that it is used by the women for putting their silks in . . . " (Watkins121). In choosing not to say "be it that no one reads the book," the writer avoids naming the natives. More playfully, the tract writer Legh Richmond writes to his daughter: "I wish very much to know how you are behaving since I saw you; what character will your pen and your needle give of you when I ask them? And what will your book say? Your playthings, perhaps, will whisper that you have been very fond of them" (Pugh 55). The trope that we've seen already of the book rising to testify against its owner sharpens here into the fantasy that conversation with books might altogether replace conversation with the daughter.

Tracts are imagined not only talking but walking: in China in 1814, "such is the political state of the country at present, that we are not permitted to enter it, and publish, by the living voice, the glad tidings of salvation. Tracts may, however, penetrate silently even to the chamber of the emperor. They easily put on a Chinese coat, and may *walk without fear through the length and breadth of land*" (Jones 474; emphasis mine). The young hero of another RTS publication, coming across a bible in his new home, "recognised [it] in a moment by its binding; and a feeling of joy came over him, as if an old friend had met him" (Millington 52).

The promise of the *Girl's Own Paper* (published by the RTS from 1880 onward) to be "to its readers a guardian, instructor, companion, and friend" (report of the Committee, 1880, quoted in S. G. Green 128) borrows doubly from the logic of the it-narrative: in anthropomorphizing the book and in relocating agency from persons to objects. More specifically, reviewers trying to describe the material attributes of books find a ready-made vocabulary in the it-narrative—or, at least, this seems like the most plausible explanation for the pronouns in a 1846 review of a volume newly reissued in a tenpenny binding, where the *Baptist Magazine* remarks archly that now "he appears in clothing which will facilitate his reception into good company, and conduce to his preservation from the casualties of the way" (quoted in Fyfe, *Science and Salvation* 159). In fact, one midcentury *History of the British and Foreign Bible Society* describes a Protestant Bible hidden in a cradle rocked by a young girl (Zemka 112).

To make books narratable was—and is—to make them human. The metaphor of books as "witnesses" rising in judgment against their owners reappears in secular guise in the 2009 printing of the standard textbook on bibliography: "analytical bibliography considers books," it declares, "as witnesses to the processes that brought them into being" (Williams and Abbott 10). The metaphor of the book as a living being who ages and dies persists, too, in scholarly titles like "The Biography of a Book" (the first chapter of Robert Darnton's *The Business of Enlightenment*) and in Cathy Davidson's play on italics in "The Life and Times of *Charlotte Temple*: The Biography of a Book."[19] The master trope of book history has always been personification. Elizabeth Eisenstein's *Printing Press as an Agent of Change*, Jerome McGann's "socialization of texts," and Paul Duguid's *Social Life of Information* anthropomorphize books as thoroughly as any it-narrative does. All three draw on the long tradition of users ventriloquizing books, like the inscription in one early modern copy of *The Treasury of Amadis of France* that reads "Dale Havers oweth me / he is my veri tenet [owner] / and I this booke confesse to be / quicunque me invenit [whoever finds me]" (Sherman 17). Outside of the object narrative's covers, no book speaks for itself. But the speech produced in (and by) its pages may provide one model for the stories that get told about books and their users.

It's also worth asking, however, what the personification of the book elides. When tract-distributors imagine volumes walking around China or penetrating where persons can't, they conveniently forget that books need someone to carry them: that only in the most metaphorical sense do ideas have legs. This isn't to say that the agent of their transmission is necessarily a person endowed with legs and hands. To Evangelicals, the force that drove books around the world was divine rather than human; and as we'll see in the final chapter of this book, even a thinker as defiantly secular as Mayhew could substitute literacy for Protestantism as the motor of a providential narrative in which pages are drawn almost magnetically into readers' field of vision. Information may not want (as Stewart Brand once claimed) to be free, but some ideological force—whether God or progress—is usually imagined as causing the written word to spread. Nineteenth-century Evangelicals thus anticipate the providentialism of twentieth-century cybertheorists: "in the claim that information will circulate freely once liberated from the book," Paul Duguid points out, "information is personified and endowed with the desire to transmit itself" ("Material Matters" 74). Between the self-propelling book of it-narrative and the providentially propelled book of master narratives (whether Christian or techno-utopian) lies a vast middle ground in which human agents take responsibility for forcing printed matter into the right hands. It's to those agents that the next chapter turns.

PART II

Bookish Transactions

CHAPTER 5

The Book as Burden: Junk Mail and Religious Tracts

ANTHROPOCENTRISM MAKES ORPHANS' hunger for books more recognizable than pocket bibles' quest for owners. Yet as the traditional dearth of text (and paper) gave way to a scarcity of attention (and shelves), books struggled harder to reach the proper audience. Old problems of production gave way to new problems of distribution: the modern genres to which this chapter turns—religious tracts and junk mail—were impelled less by new manufacturing technologies than by new social networks through which printed matter could be exchanged, donated, requested, accepted, or, increasingly, rejected.

Natalie Davis's foundational essay subtitled *Books as Gifts in Sixteenth-Century France* took the gift as the paradigmatic interpersonal transaction. Today, we continue to assume that (as Lewis Hyde puts it) "in a market society . . . getting rather than giving is the mark of a substantial person" (xiii). I want to suggest, in contrast, that in the nineteenth century the book came to feel like a burden. Printed matter was foisted upon, more often than given to; receiving a book now connoted powerlessness as often as privilege. That change doesn't just respond to the increasing abundance of books themselves; too much information and too much paper were compounded by too much contact with fellow handlers. Every reader is cumbered by an excess of books, and every book by an excess of readers—each overwhelmed in turn by the consciousness that others have touched the same book that he or she is now holding, and thereby gain some hold over him or her.

The Victorians pioneered institutions—whether secular (the post) or religious (the tract society)—that allowed printed matter to be distributed at the expense of someone other than its end user. By disjoining owning from choosing, those transactions challenged Enlightenment assumptions about the relation between reading and identity. Where the secular press trusted print to lift individuals out of their social origin, the niche marketing pioneered by Evangelical publishers and commercial advertisers alike vested it instead with the power to mark age, gender, and class. If the content of tracts interpellated new audiences by matching characters' demographic to readers', so did the different material forms that each text took—reprinted on different paper, sold at different price points, distributed in gross and in detail. Even a secular novel as expensive as *The Mill on the Floss*—rented, not given away; proscribed, not prescribed—could present its characters torn between forming their identity by imitation of fictional persons about whom they read, or by communion and competition with fellow readers, along with fellow owners, fellow handlers, and fellow inscribers of books, whether living or dead.

PAPER DUTIES

Maria Jane Jewsbury described hers as "an age of books! Of book making! Book reading! Book reviewing! And book forgetting" (quoted in Newlyn 3). "The difficulty of finding something to read in an age when half the world is engaged in writing books for the other half to read is not one of quantity," noted one journalist in 1893, "so that the question, 'What shall I read?' inevitably suggests the parallel query, 'What shall I not read?'" ("A Few Words About Reading" 226). In *The Choice of Books*, Frederic Harrison, too, presented the problem in negative terms: "the most useful help to reading is to know what we should not read, what we can keep out from that small cleared spot in the overgrown jungle of 'information.'" Even the jungle soon gives way to the rubbish heap: "I often think that we forget the other side to this glorious view of literature—the misuse of books, the debilitating waste of brain in aimless, promiscuous, vapid reading, or even, it may be, in the poisonous inhalation of mere literary garbage and bad men's worst thoughts" (1, 3). For decades, alarmists continued to reprint a chart showing that as many books were published in 1868 alone as in the first half of the eighteenth century (Ackland). Reprints such as this themselves compounded the problem.

Where autodidacts' autobiographies represent a hunger for books, middle-class commentators more often equated reading with indigestion. As George Craik remarked in his classic compendium of exemplary biographies, *The Pursuit of Knowledge under Difficulties*, "If one mind be in danger of starving for want of books, another may be surfeited by too many" (22). A decade later, an article in *Victoria Magazine* presented those problems as mirror images: "Of the underfed, in these days of education of women, education of ploughboys, education of curates, we are sure to hear enough, but of the sufferings of their scarcely less pitiable antipodes, whose complaint is overfeeding, we are not so likely to be well informed" (Butterworth 500). An 1869 article found "as curious cases of moral delirium, dyspepsia, and decay from the abuse of mental stimulants as there are records of physical injury from gluttony" ("Excessive Reading"). Like food, books had to be carried, stored, preserved, maintained: Ruskin compared knowledge to "at best, the pilgrim's burden or the soldier's panoply, often a weariness to them both" (*The Works of John Ruskin* 66).

Nothing new about any of these images: the caricaturists who represent Brougham thrusting the *Penny Magazine* down readers' throats with a broomstick draw on analogies between teaching and force-feeding that stretch back at least to Rabelais and Montaigne, by way of Jeremy Collier's axiom that "a man may as well expect to grow stronger by always eating, as wiser by always reading."[1] Newer technologies did replenish the stock of metaphors: references to Brougham's "steam-intellect society" riff more topically on steam-printing, introduced by the *Times* during the Napoleonic wars. Such figures of speech shouldn't obscure the fact that the speed and cost of the printing process mattered less than the cost of raw materials. The reduction of paper taxes in 1836 and their abolition in 1861, together with the advent of machine-made paper in the early decades of the century and of esparto grass and wood pulp in its second half, cheapened paper almost as dramatically as digital storage has cheapened in our lifetime. Paper production went from 2,500 tons in 1715 to 75,000 tons in 1851; measured per person, it shot up from 2.5 pounds per year in 1800 to 8.5 in 1860 (Welsh, *George Eliot and Blackmail* 38). Between 1841 and 1911, similarly, the number of persons employed by the paper, printing, and publishing trades increased sixfold (Vincent, *The Rise of Mass Literacy* 82). Where once paper was the scarce and valuable resource, now time looked in shorter supply—along with shelf space and room in the wastepaper basket.

As a result, the long-standing search for technologies to make literary production more efficient shifted to the consumption side. Francis Jeffrey replaced speed-writing by speed-reading when he speculated that "if we continue to write and rhyme at the present rate for 200 years longer, there must be a new art of short-hand reading invented—or all reading will be given up in despair" (472). In 1893, Herbert Maxwell contrasted "the number of books that a single bookworm" could consume (9,000, in his generous estimate) with the number produced (20,000 annually added, in a phrase that once again invoked furniture, "to the shelves of the British Museum")—without even counting "the vastly greater mass of journalistic literature which consumes part of everyone's time and attention" (1047). It's true that periodicals multiplied more rapidly than books: Simon Eliot calculates from tax returns that in the first half of the nineteenth century the production of stamped periodicals increased fivefold (*Some Patterns and Trends in British Publishing* 78–86). One reason was the abolition of the newspaper tax in 1855. The end of "taxes on knowledge" opened the floodgates not (just) for knowledge but for ephemera.[2] Periodicals were blamed for crowding out more durable works: competition between the two was taken for granted in comparisons like the calculation that "at the end of the year [the *Times*] is comprised in a book larger than all the classics and all the standard histories of the world put together" (L. Stephen, "Journalism" 60). In 1825, Charles Lamb filled a few column inches of the *Times* by comparing books to Sisyphus's boulder: "No reading can keep pace with the writing of this age, but we pant and toil after it as fast as we can. I smiled to see an honest lad, who ought to be at trap ball, labouring up hill against this giant load" (C. Lamb, "Readers Against the Grain," 238).

Newspapers and magazines had only themselves to blame, since their own pages were padded out with hackneyed statistics about the number of newspapers and magazines. Indeed, the *Pall Mall Gazette* for 12 January 1886 proposed marking Victoria's Jubilee by a year in which "the literary soil should be allowed to lie fallow," with an embargo on the production of new literature—except, of course, for newspapers (quoted in Mays 189). Printed attacks on printed matter always risk self-referentiality, if not quixotism. "We find the 'Quarterly Review' anathematising circulating libraries with great force," notes an 1871 article titled "Circulating Libraries," but "this is very hard on libraries now-a-days, especially as no inconsiderable number of the 'Quarterly Review' is taken in by Mr Mudie" (Friswell 519). Conduct books remarked that trashy reading took time away from outdoor activity, but neglected to count the hours eaten up by their own perusal.

W. H. Wills estimated in 1850 that the daily papers produced in 1848 added up to "1,446,150,000 square feet of printed surface"; a decade later, another journalist observed that "there are persons who will count up the number of acres which a single number of the *Times* would cover if all the copies were spread out flat, or illustrate the number of copies by telling us how long the same weight of coal would serve an ordinary household . . . Every morning, it is said, a mass of print, containing as much matter as a thick octavo volume, is laid on our breakfast-tables" (Wills 238; L. Stephen, "Journalism" 60). The comparison with coal emphasized the material aspect of paper but also lent it a factitious ephemerality: newspapers may count as consumables, but, unlike coal, they don't consume themselves with use.

As a result, readers had to cope not just with new material being added to the old, but also with the survival of existing books and the reprinting of existing texts. The *Quarterly*'s review of the seventh edition of the *Encyclopedia Britannica* (1842) complains of "the imperishable nature of books, the cheapness with which they are now produced, and the rapidity and extent of their production."

Unfortunately for authors there is no epidemic among books, to thin their ranks, and render necessary a new supply; and the fire-proof inventions of the present day extinguish the hopes that were sometimes realised from the timber boards of our books and the wooden carpentry of our libraries. There is, therefore, no law of mortality by which the number of books is regulated like that of animals. ("The Encyclopedia Britannica" 71)

In a review that also critiques the *Britannica*'s article on fire, the comparison of library collections to animal populations inverts a Promethean logic: far from symbolizing enlightenment, fire appears here as a destroyer of cumbersome knowledge. Libraries expand as geometrically as populations (books beget books), but books lack animals' fixed life span.[3] In fact, the *Quarterly* itself confesses to contributing to the problem, to which the *Britannica*'s miniaturizing strategies adumbrate a solution:

We have before us now an octavo volume, containing about 1150 pages of double columns, and printed on paper so thin that the thickness of the volume (though not beaten) is only two inches, and in so small a type that the quantity of matter that it contains is equal to above TWELVE NUMBERS of this Review, supposed to be all printed in its ordinary type . . . A bookcase might thus contain a large library, and a moderate one might be packed in a traveller's portmanteau. Books now forwarded by tardy conveyances might be sent by post . . . These processes too might be aided by a stenographic representation of the terminations of many of our long words, and even by a contraction of the words themselves; and in the spirit of these changes authors might be led to think more closely, and to express their thoughts in the shortest and fewest words. ("The Encyclopedia Britannica" 72)

Reprints form the problem, but also the solution. In fantasies of epidemics and fires, the library becomes an image at once for comprehensiveness and for permanence. After comparing the air to "one vast library, on whose pages are forever written all that man has ever said or woman whispered," Charles Babbage added that "no motion impressed by natural causes, or by human agency, is ever obliterated" (37). The backlist could crush readers: as Jack Goody observed in a different context, "literate society, merely by having no system of elimination, no 'structural amnesia,' prevents the individual from participating fully in the total cultural tradition" (58).

Worries about disposal may help explain why Victorian discussions of the book so often take on a ghoulish tinge. Tomes as tombs: Mudie's Circulating Library justified "selection" (i.e., censorship) on the grounds that "no library could provide space for all the books that might be written, and as bad and stupid novels soon die and are worthless after death—no vault could be found capacious enough to give them decent burial" (Mudie 451). "Burial" was no figure of speech: short on shelf space and forced to stock multiple copies of novels whose backlist value dropped quickly, Mudie's resorted to

> a charnel-house in this establishment, where literature is, as it were, reduced to its old bones. Thousands of volumes thus read to death are pitched together in one place. But would they not do for the butter-man? was our natural query. Too dirty for that. Not for old trunks? Much too greasy for that. What were they good for, then? For manure! Thus, when worn out as food for the mind, they are put to the service of producing food for our bodies!" (Wynter 278)

Burial wasn't, of course, always so literal. Gladstone's essay on library design reluctantly envisages movable shelves described by analogy to a "book-cemetery," "what I will not scruple to call interment": "The word I have used is dreadful; but also dreadful is the thing. To have our dear old friends stowed away in catacombs, or like wine-bottles in bins: the simile is surely lawful . . . [but] it can hardly be contemplated without a shudder at a process so repulsive applied to the best beloved among inanimate objects" (386–87).

The metaphor of companionship that should be familiar from the previous chapter sounded comforting when Gladstone first deployed its most hackneyed form: "In a room well filled with [books], no one has felt or can feel solitary. Second to none, as friends to the individual." But once those friends are imagined to be dead, the tone shifts; the body to which the book is compared becomes a wine bottle, the person to which the text is compared a corpse. Once the body replaces the soul as the vehicle of the metaphor, mortal book upstages immortal text—or rather, a book whose inconveniently bulky body remains after the soul has departed. If "the binding of a book is the dress, with which it walks out into the world" and "the paper, type, and ink are the body, in which its soul is domiciled," then old books deserve to be treated with as much reverence as dead bodies (385).

Or as little, for a fourth analogy is less respectful. "The artist needed for the constructions required [i.e., the movable shelving] will not be so much a librarian as a warehouseman" (396). To compare a librarian with a manual laborer is also to question what sets the book apart from other, nontextual objects. And as chapter 7 will show, it's at the end of the book's life that its physical bulk becomes most visible. Once advances in papermaking drove down the cost of production, disposal became a problem for genres as various as religious tracts and blue books. One MP complained: "[I] object to having tons of papers, which are never opened, sent to my lodgings . . . [I can] not exchange them for books, for that would be selling them; [I can] not burn them, for that would be voted a nuisance" (1865, quoted in Frankel 308). Paradoxically, the re-sale value of blue books' raw material made matters worse: Edwin Chad-wick campaigned to substitute octavo for folio as the standard format for official papers on the grounds that larger formats—especially useful as wrapping—encouraged waste and overproduction (Frankel 314). Two decades later, a Stationery Office committee would debate how to dispose of unsold stocks of Record Office publications, in language that uncan-nily echoed contemporaneous arguments about cremation in the *Lancet*: like human bodies, books needed to be disposed of (McKitterick, "Orga-nizing Knowledge in Print" 557).

"[I can] not exchange them for books, for that would be selling them": one measure of bibliographical overload was the number of books that stood outside the market. Early Victorian innovations in production tech-nology look insignificant when compared to the revolution in distribu-tion systems—in particular, new networks for funneling printed matter (whether religious tracts or advertising circulars or political pamphlets) into the hands of a captive audience.

JUNK MAIL

Liberal intellectuals celebrated the cheapening of postage from 1839 on-ward and the establishment of prepayment in the following year. But the March of Mind devolved into a parade of paper. Even the reformer Charles Knight acknowledged that triumphalist statistics about the rise of the newspaper press need to be adjusted for the fact that "price-currents, catalogues, and circulars" could legally be mailed as newspapers (*The Old Printer and the Modern Press* 291).[4] The same postal reforms that prompted the invention of the postcard spawned the advertising circular, the chain letter, and the postal scam.[5]

Reformers' high-minded abstractions about the diffusion of knowl-edge were soon countered by conservatives' reminders that the post was being used to convey not just ideas, but things: "specimens of vegetable seeds; cuttings of trees from Professor Henslow's shrubberies; . . . new manures, books of patterns, . . . pills, patent medicines, . . . and turtle" (*Administration of the Post Office* 103); "a great-coat, a bundle of baby-linen, and a pianoforte" (Hill and Hill 1:241). Specimens of vegetable seeds could be understood as a metaphor for text via the biblical parable of the sower, eventually to be secularized in the metaphor of "broadcast-ing" (Matthew 13).[6] As Louis James points out, the 1823 religious tract that begins with the image of a rider who "every now and then pulled from his coat-pocket a bundle of tracts and scattered two or three in the road" would have been understood by any reader as a reference to sowing (135). Yet the same seeds that provided a figure for letters could also function as their antonym—as a form of scientific or agricultural knowledge that competed with the written word. On the other side of the Atlantic, the narrator of Melville's "The Tartarus of Maids" engages in "the seedsman's business (so extensively and broadcast, that at length my seeds were distributed through all the Eastern and Northern States, and even fell into the far soil of Missouri and the Carolinas)"; the papers that he sends through the post are envelopes "made of yellowish paper, folded square" (321–22).

Between parcels and letters, books occupied an ill-defined middle ground. The book post established in 1848 at sixpence per pound allowed unmarked volumes to be sent more cheaply than any other object of equivalent weight; only later, however, was it extended to encompass secondhand books, newspapers and circulars (in 1856), and patterns (in 1863) (Lewins 167). Railway companies' objections were answered with the high-minded claim that "the exceptions taken in the case of the book-post were only to books and printed matter intimately connected with . . . the diffusion of knowledge" (Lewins 233). Knowledge, not paper: the division of labor separating post from rail treated books and newspapers *as if* they lacked material dimension. Pianofortes could travel by rail, sheet music by post; initialed handkerchiefs by rail, printed rag pulp by post. The postal system thus institutionalized the exceptionalist logic of those etiquette guides that excluded books from the advice against unmarried ladies' accepting gifts from gentlemen: "Gentlemen, as a rule, do not offer ladies presents . . . Should the conversation, however, turn upon some new book or musical composition, which the lady has not seen, the gentleman may, with perfect propriety, say, 'I wish that you could see such or such a work and, if you will permit, I should be pleased to send you a copy'" (Cooke 123).[7]

Yet the abstractions of information overload were doubled by the more literal weight of parcels—a complaint that appears throughout the correspondence of even, or especially, those social reformers who devoted their time to reading and writing letters in support of widening access to the postal system. Harriet Martineau was one of the most prominent apologists for postal reform ("we are all putting up our letter boxes on our hall doors with great glee, anticipating the hearing from brothers and sisters—a line or two almost every day"), but her own collected correspondence teems with apologies for troubling her correspondents and complaints about the logistics of receiving, storing, and disposing of unsolicited paper (Hill and Hill 1:390). To the architect of postal prepayment, Martineau writes, "Dear Mr Hill, I write not to trouble you for an answer, about which I always feel most scrupulous, but to thank you for sending me ['The State and Prospects of Penny Postage']" (Hill and Hill

2:14). When Martineau writes to the queen, she "particularly request[s] that no sort of acknowledgment—no notice of my letters whatever should be thought of"; another letter insists, "I may add that I desire no reply" (Martineau, *Selected Letters* 80). Martineau's etiquette code anticipates the postal system's change from payment on delivery to prepayment by the sender. Both take pains to shift the cost of correspondence (whether measured in pounds or in hours) from reader to writer.

Until 1840, when prepayment replaced payment on delivery, those two costs coincide: time is money. At the beginning of her career, in fact, Martineau very literally paid the price for her fame by being sent an unmanageable mass of fan mail, hate mail, and prank mail. "I dreaded the arrival of a thirteenpenny letter, in those days of dear postage," she writes in the *Autobiography* (Martineau, *Autobiography* 109).

> In the preface to 'Society in America,' I invited correction as to any errors in (not opinion, but) matters of fact. After this, I could not, of course, decline receiving letters from America. Several arrived, charged double, treble, even quadruple postage. These consisted mainly of envelopes, made heavy by all manner of devices, with a slip of newspaper in the middle, containing prose paragraphs, or copies of verses, full of insults. (370–71)

And Martineau measures the popularity of the *Illustrations of Political Economy* (1832) less by the size of their own print run than by the volume of fan mail and hate mail that they bring upon her. The first sign of their success, according to her *Autobiography*, is the postmaster's warning "that I must send for my own share of the mail, for it could not be carried without a barrow—an announcement which, spreading in the town, caused me to be stared at in the streets" (136). The embarrassing wheelbarrow emblematizes a postal regime that makes recipients pay for letters that they never requested.

To advocates of prepayment, then, the mechanics of cash on delivery become a metaphor for the burden of reading. When Hill declares that "I really am at a loss to discover any case in which it is desirable for one person to write to another at the expense of the latter," he measures in money the problem that Martineau frames in terms of time: how correspondence can be carried on without imposing on a recipient who never asked for it. "Imposition" in both senses: Martineau was herself the target of a postal hoax, and Hill complains that the existing system encourages letters

> such as ought not to be sent unpaid, as letters soliciting orders, subscriptions, &c.; or such as ought not to be sent at all, as those written by vindictive people for the purpose of putting the receivers to the

expense of postage . . . [Under Prepayment those] letters would undoubtedly be suppressed: but this, so far from being an objection, is no inconsiderable recommendation to the proposed plan. I would deprive the thoughtless, the impertinent, and the malicious, of a means of annoying others, which is now but too often resorted to; and no one, I presume, would regret the small amount of revenue which would be sacrificed in obtaining so desirable a result. (Hill 97–98)[8]

The result, however, was the opposite of what Martineau predicted. Far from suppressing solicitations, postal prepayment created the material conditions for the proliferation of unsolicited mail that we know today. Postal prepayment had assimilated letters to books: in theory at least, the reader was expected to pay for access to either. In practice, however, by the nineteenth century neither books nor letters were goods whose cost was borne in some simple way by the end user. Prepayment made perfect sense in a culture where paper had become a more abundant commodity than access. One is reminded of some early twenty-first-century blogger reveling in the lack of wireless on airplanes when Martineau gloats during an ocean journey that "I have enjoyed few things more in life than the certainty of being out of the way of the post" (Martineau, *Autobiography* 332).

None of this means that scarcity ceased to exist: that wastepaper continued to command resale value (as we'll see in chapter 7) makes clear that books and even broadsheets remained a scarce resource for the poorest Londoners. Yet at the same time that the costermongers depicted by Henry Mayhew were hard put to find enough paper to wrap their wares, middle-class readers struggled to find enough wheelbarrows and incinerators.

GIFT BOOK, WASTEPAPER

Twenty-first-century intellectuals inherit an eighteenth-century understanding of literacy as a precondition for psychological interiority and political self-determination—along with a nineteenth-century infrastructure that thrusts printed paper into our letter slots, our faces, our hands, our fields of vision, and even the bedside tables of our hotel rooms. ("No Menus Please" is the New Yorker's declaration that his home is his castle.) In theory, a self formed by print; in practice, a mass assaulted by printed matter. Rather as prophets of the paperless office soon realized that the personal computer had increased the rate of paper consumption, so the strand of mid-twentieth-century sci-fi that predicted a postliterate future looks increasingly quaint in an era of print and pixel pollution.

In the political sphere, campaigns against censorship make it hard to remember that even those twentieth-century regimes most closely associated with biblioclasm spent more energy distributing books than burning them. Some estimates call Mao's *Little Red Book* the most-circulated book of its time; *Mein Kampf*, too, is often described as a best seller, thanks to the ten million copies in circulation by 1945 (almost one per German household). It might be more accurate to coin a neologism like "best giver," for those copies were donated, not sold; from 1936 onward a copy was handed out at every wedding (Fritzsche 796). (Reciprocally, the US Office of Information described its propaganda newspaper airdropped over enemy territory as "the German newspaper with the world's largest circulation" [Rickards and Twyman 11].) When Macaulay estimated in 1823 that "there is scarcely one Englishman in ten who has not belonged to some association for distributing books, or for prosecuting them," the symmetry of his syntax already lumped positive with negative controls on reading (T. B. Macaulay, "On the Royal Society of Literature" 20). If *A Clockwork Orange* turns out to be more prescient than *Fahrenheit 451* about the role of high art in totalitarian states, it may be because Bradbury is to Burgess what American slave narratives are to religious tracts. Reading looks more heroic when proscribed than when prescribed, just as book burning poses a greater threat when it implies indifference than when it bespeaks hostility.

Even in liberal democracies, print is now less often sold than given away for the purpose of selling something else. By one modern scholar's calculations, more unsolicited advertising passes through the UK mail than individually addressed letters; and thanks to Metro International (founded in 1995), in many cities fewer newspapers are now bought than given away.[9] Ubiquity ensures invisibility: we continue to think of sale as the norm, and gift as the exception. But "gifts" is too kind a term for paper foisted upon readers who have never consented to receive it, let alone demanded. It's telling that the English language lacks even a noun that would encompass spam sent electronically, leaflets handed out on the street, and circulars delivered through the post. Certainly important differences divide these genres. The cost of some materials is borne by the distributor, of others by third parties: in the United States in 2010, newspaper revenue is 87 percent advertising, 13 percent sales. And paper catalogs retain resale value that electronic spam lacks—unlike the fiber-optic cable through which it's transmitted. Nonprinted nonmatter like Viagra ads stand opposite the pure materiality of Victorian wastepaper. Yet the absence of any umbrella term points less to the heterogeneity—far less to the insignificance—of these genres than, on the contrary, to the threat that their combined volume poses to unspoken assumptions about how print works.

In decoupling who reads from who pays, free print challenges three tenets of what could be called "bookish liberalism": that acquiring a book implies choosing it; that owning a book implies an intention to read it; and that virtual encounters with an author distant in space or time can release readers from the constraints of their own social position.[10] Giving free print its due would result in a different economic history in which disposal and storage would upstage production and distribution, and a different cultural history in which—far from enabling mobility or independence—the book would become a prop for commemorating one's forebears, deferring to the judgment of one's elders, and accepting favors from one's betters. Lest that sound too gloomily Foucaultian, let me rephrase in positive terms: taking free print seriously would allow us to recognize forms of individual and collective creativity that otherwise remain invisible, because their inventiveness operates at the level of the book's circulation, not its content.

Measuring by what has survived, we call the nineteenth century the age of the novel; but if we counted instead what was produced, the Victorians might look more like a people of the tract. One historian calculates that in the first seven years of its existence, the Religious Tract Society (founded by an interdenominational group of Evangelicals in 1799) had distributed two million titles; between 1840 and 1849 that number went up to twenty-three million (Fyfe, "Commerce and Philanthropy" 166; St Clair 569). Another estimates that the Society for Promoting Christian Knowledge (its Anglican counterpart) unloaded eight million in 1867 alone (R. Altick 100; see also Fyfe, "Commerce and Philanthropy," and St Clair 350–54). And another adds that individual RTS titles could sell over a million copies (Ledger-Lomas 327). These modern scholars at least take the individual volume as their unit of measure; in 1837, Frances Trollope despaired of any such precision, simply referring to "tracts, so numerous that it would be impossible to give their measure or their value by any other calculation than that of their weight" (191).

For Trollope, the proliferation of tracts provided the most visible instance of the more general proliferation of books. In keeping with Simon Eliot's periodization of book production overall, which dates a "distribution revolution" to 1830–55, followed by a "mass production revolution" (1875–1914), it was clear even at the time that the scale of tract societies' operations owed less to new production technologies than to new distribution channels, including middle-class female volunteers and working-class male hawkers (S. Eliot, *Some Patterns and Trends in British Publishing* 107). If you can't beat 'em, join 'em: tracts aped the very chapbooks they were meant to drive out of business, not only in verbal style and material format but also in their retail model. In 1796, a group of Evangelical gentlemen went slumming in order to investigate how best to feed their Cheap Repository Tracts into existing systems of peddling: as Hannah More wrote to Zachary Macaulay, "Mr Henry Thornton and two or three others have condescended to spend hours with the hawkers to learn the mysteries of their trade" (Z. Macaulay 129). A radical like Mayhew could hardly have done more.

As Leslie Howsam has shown in rich detail, the British and Foreign Bible Society devoted considerable ingenuity to wresting bibles out of older free distribution channels and into something that mimicked market mechanisms, and the same held true for tracts (Howsam; see also Stott 207).[11] Yet that transaction was rarely as simple as a sale outright. The RTS lent tracts for a fee, while the Bible Society collected on an installment plan (R. Altick 100–103). And more fundamentally, even those books not given away were subsidized through volume discounts. Some Cheap Repository Tracts list "Price One Penny, or 4s. 6d. 100" (Kelly 154); from 1796 onward, the text of each tract was printed in two formats, distinguished by paper quality as well as by price structure. As Kevin Gilmartin describes, "profits from the more expensive version were used to subsidize the distribution of cheaper editions, reinforcing the different roles played by different sorts of readers" ("'Study to Be Quiet'" 513).

As a result, neither "sale" nor "gift" adequately describes the distribution mechanisms that tract and bible societies elaborated. It's true that the operations by which books changed hands (what Howsam calls the "Bible transaction") weren't required to produce a net profit: both subsidized the retail cost through complex and graduated volume discounts. Yet both also strove to couple the outflow of books with the inflow of money, not only to avoid "pauperizing" recipients or undervaluing their own wares, but also because they understood giving and taking payment—the very transactions whose anonymity secular economists at the same time were theorizing by contradistinction to the personalized nature of the gift—as an opportunity for face-to-face encounters and personal accountability.[12]

Widely distributed but rarely read, innovative in their dissemination but clichéd in their composition, tracts fit awkwardly into Darnton's communications circuit—if only because they so well illustrate Adams and Barker's point that the book's life neither begins at the moment of writing nor ends at the moment of reading. Tracts link author and reader less strongly than giver and taker: a virtual meeting of minds becomes the pretext for a face-to-face transaction between unequals (whether in gender, class, or age). The tract did not create what Roger Chartier calls "communities of readers," much less what Stanley Fish calls "interpretive

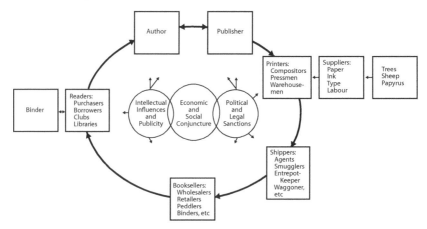

Figure 5.1. Darnton's Communications Circuit.
From *The Kiss of Lamourette: Reflections in Cultural History* by Robert Darnton. Copyright © 1990 by Robert Darnton. Used by permission of W. W. Norton & Company, Inc.

communities"—if only because to be interpreted, tracts would actually have had to be read. What it did create were relationships between givers and takers (as well as among givers themselves), bonds less cosy and more conflictual than the term "community" implies. As the next chapter will argue in more detail, tracts pose an equally sharp challenge to Benedict Anderson's model of national community, not only because their circulation encourages their handlers to differentiate themselves from (instead of, or as well as, identify with) one another, but also because they draw their significance from face-to-face interactions within a single household or parish rather than from a virtual "community in anonymity."

When the hero of More's "Tom White, the Postilion" "sent home for his Bible and Prayer book, which he had not opened for two years, and which had been given him when he left the Sunday school," the narrator takes the opportunity to digress into an address to a very different implied reader: "And here let me remark what encouragement this is for rich people to give away Bibles and good books, and not to lose all hope, though, for a time, they see little or no good effect from it. According to all appearance, Tom's books were never likely to do him any good, and yet his generous benefactor, who had cast his bread upon the waters, found it after many days." What's true of the tract's characters also holds for its audience: the "remark" interpellates a giver of books, not a reader of them (More, *Cheap Repository Tracts* 225). In fact, from

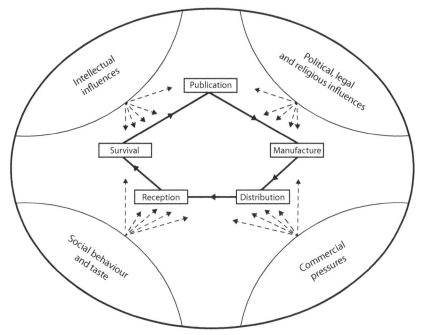

Figure 5.2. Adams and Barker's Life Cycle of a Book.
Thomas Adams and Nicolas Barker, "A New Model for the Study of the Book," *A Potencie of Life: Books in Society* (London: British Library, 1993), 14. © British Library Board (2708.e.2071).

the beginning tracts seem less interested in representing reading than in representing giving: thus Hannah More's *Sunday School* takes as iconic the scene of a book changing hands. No less than it-narratives, religious tracts make changes of ownership the moment of highest drama. Maria Charlesworth's *Book for the Cottage* inserts the title of another tract within its plot, where a poor woman's "visitor did not stay long there, but left a little book with her, called, 'Cecil's Visit to the House of Mourning'; the rector's lady had sent it for her, thinking that if she could read it, it might bring a word of comfort and instruction to her heart" (268–69). The economics and iconography of book distribution alike confirm Susan Pedersen's claim that "the real success of More's tracts is to be found less in their conversion of the poor than in their effective recruitment of the upper class to the role of moral arbiters of popular culture" (109).[13] The formal corollary is that where middle-class secular novels marginalized the self-referentiality that had been their eighteenth-century

THE

SUNDAY SCHOOL.

I Promifed, in the Cottage Cook, to give fome
account of the manner in which Mrs. Jones fet
up her fchool. She did not much fear being
able to raife the money, but money is of little
ufe, unlefs fome perfons of fenfe and piety
can be found to direct thefe inftitutions. Not
that I would difcourage thofe who fet them up,
even in the moft ordinary manner, and from
mere views of worldly policy. It is fomething
gained to refcue children from idling away their

P 6   Sabbath

Figure 5.3. Hannah More, "The Sunday School," 1798.
Hannah More, *Cheap Repository Tracts; Entertaining, Moral, and Religious [v.1]*
(London: Sold by F. and C. Rivington, no. 62, St. Paul's church-yard, 1798).

predecessors' stock-in-trade, religious tracts (even those whose fictional stories borrowed novelistic conventions) inscribed self-referential images of their own circulation in order to illustrate the providential spread of the Word.

Eighteenth-century philanthropists assumed that existing associations—like that linking clergyman's wife with parishioners, or squire's daughter with tenants—could form a conduit for bibles or blankets. In that setting, goods could be given away because the implicit repayment—improved behavior—could be checked up on. In contrast, as Howsam points out, "not many commercial and professional people like the members of the early Committee [of the British and Foreign Bible Society] enjoyed a direct relation with those whom they wished to covert" (39). In the modern city, a credit economy no longer made moral sense; donating bibles to strangers would have been risky, but selling bibles on the installment plan provided an excuse for return visits. Paying for a book week after week (whether by giving money or by accepting visits) turned out to be just as entangled in ongoing relations of trust, guilt, and obligation as any gift giving ever was.[14] Rather than being slotted into preexisting social relationships, the transfer of books became the occasion for inventing them.

Did social networks serve to transmit books, or books serve to connect human beings? The former model underpins a manual for district visitors which explains that "The first idea they learn to associate with you is, that the Bible has to do with you, and you with it: your visit is a signal for the mother to read to you, or you to her, and they must be silent the while. By degrees, as reason dawns, they gather somewhat of the meaning of what passes, the sound of words and names becomes familiar, and a feeling of interest, perhaps of love, is first linked with the precious volume, because of its inseparable association with you" (Charlesworth, *The Female Visitor to the Poor* 80). Yet in the opposite direction, another manual predicts that "you will find this exchange of tracts a good introduction to your visits, especially at first. It will give time for the little awkwardness that is sometimes felt on both sides to pass away" (Nixon 49). Perhaps the distinction is too stark to begin with: as we saw in the previous chapter, the meaning of books resided not only in their printed content but in the history of gifts and exchanges to which their manuscript markings bear witness.

A freethinker like Henry Mayhew saw hypocrisy in the enlistment of books as a pretext for interpersonal transactions, whether philanthropic or commercial: "tract-sellers . . . are regarded by the other street-traders as the idlers, beggars and pretenders of the trade" (Mayhew, *London Labour* 289)—which is not to say that tracts were not also the mirror image of radical leafleting on the other side of the political spectrum.

The confusion troubled even more orthodox middle-class observers: for surreptitious subsidies to misrepresent gift as sale was as bad as for the charity bazaar and the missionary basket to blur the boundaries between donation and purchase.[15] (Charlotte Brontë refers to such a basket as "a monster collection of pincushions, needle-books, cardracks, work-bags, articles of infant wear, &c.&c.&c., made by the Christian ladies of a parish, and sold perforce to the heathenish gentlemen thereof, at prices unblushingly exorbitant" [*Shirley* 134].) The next chapter will examine in more detail why a genre whose own distribution depends as heavily on the market as does the novel's might express particular distaste for pseudocommercial transactions, whether in the form of Dickens's and Thackeray's satire on tract societies or Trollope's extended send-up in *Miss Mackenzie* of a charity bazaar.

More surprisingly, some within the Evangelical press itself echoed free-thinkers' accusations of unfair competition. Missionaries, in fact, were the first to identify books as the locus of two problems that some twenty-first-century commentators imagine unique to digital media: whether content can be distributed more effectively through gift, sale, or rental, and what role social networks should play in that transmission. Even the native colporteurs paid by mission societies to hawk pamphlets complained that charitable English ladies who distributed tracts for free were under-cutting their customer base: information overload was compounded by what can only be called distributor overload (Manning). Hawkers present themselves, too, as torn between the desire to give tracts away and the duty to charge a price: the same colporteurs who describe subsidizing tracts out of their own pockets are also capable of self-consciously resist-ing the temptation, like one who confesses the urge "to return the 50 centimes, and thus enable this good woman to free herself from her debts, paying for the New Testament out [of my?] own funds; but knowing by experience that the Lord regards [* illegible] sacrifice that is made for His name, I felt that I ought not to [deprive?] this female of a great blessing which is in store for her" (*Fifty-Sixth Report of the British and Foreign Bible Society* 18). This model invested paying for a book with as much significance as reading it: where the secular press represents the book's imbrication in the market compromising its textual value (think back to Ranthorpe, Mr. Brownlow, or David forgetting how much his books cost), the Evangelical press takes the commercial transaction as a guaran-tor for a more than commercial value.

Distributors of tracts and bibles faced a double bind, as worried that recipients would value the book for the wrong reasons as that they would refuse to value it at all. Or more precisely—since in times and places where paper is scarce, no book lacks market value—they worried that its pages would be perceived either as something more than a carrier of text, or something less. As something more, because Protestants' eagerness to distribute bibles could be interpreted as idolatry. Instead of "Behold the Book, fall ye down and worship it," Alexander Duff insisted, converts must learn to "behold your God revealing himself through the medium of his written word; fall ye down and worship him" (quoted in Viswanathan 51). But also as something less, because the fear that natives would exalt what should be a "medium" to a "fetish" or subject it to "pharasaical idolatry" coexisted with Duff's equally sharp anxiety that free books might decompose into wastepaper: that pages of the Bible could be spotted wrapping goods in the bazaar, or worse (Viswanathan 51). Even Protestant missionaries acknowledged that "many a tract and many a Bible portion is torn or used for window-panes" ("Foreign Missions at Home" 231). One describes residents of a New York boardinghouse blaming him for wanting to "'leave some waste paper,' as they denominate tracts and Bibles" ("The Sailors and Their Hardships on Shore" 99).[16] More surprisingly, distributors themselves could play on the equation of tracts with rubbish, as when Canon Christopher was witnessed "carrying to the General Post-office in St Aldate's a large waste-paper basket full of tracts addressed to Members of the University" (Reynolds and Thomas 233).

In 1879, the assistant secretary of the British and Foreign Bible Society enjoined his listeners to

> ask any Missionary from India or China what he thinks of flooding those empires with Scriptures, regardless of the power of the people to read and understand, and he will reply that the Bible distributed after this fashion will be sure to be dishonoured, will be collected for waste paper, or at best be valued as a charm, and a true thirst for God's Word be postponed. So has it been in lands nearer home. (C. Reed 235)

According to one report of the Christian Vernacular Education Society in Allahbad, too, "It was objected at the Conference that to give 50 per cent discount in North India, where tracts and Scriptures are large and cheap, might lead to their sale as waste paper. This, it must be admitted, is no imaginary danger. One-third was recommended on Scriptures and other religious books, with 25 per cent on school books and picture books. In Madras the sale of Scriptures as waste paper is largely guarded against by printing them on pages of so small a size that they are almost useless for bazaar purposes" (Brodhead 33). Paradoxically, books that are too valuable face the greatest risk of destruction. A contributor to the *Missionary Register* complained that the average native accepted a bible for no better reason than "that he may store it up as a curiosity; sell it for a few pice; or use it for waste paper."[17] "Storing up," "wasting," "selling": abuses of the book are easier to name than the proper use—reading—which, in missionary publications as much as in Trollope's novels, dwindles to a vanishing point.

The right price proved frustratingly elusive: low enough to under-cut competing secular publications, but high enough to prevent bibles' being bought for resale as waste. One clergyman in India remarks that "although the price charged is extremely small (less than a halfpenny a copy), yet it is sufficient to prevent the book being sold for waste paper, and it appears, therefore, to be a test of the sincerity of the desire felt by the purchasers to read it" ("Circulation of the Scriptures" 309). In Hyderabad in 1860 the Reverend A. Burn reports, "In the beginning of the year (March last), believing that our objects would be best served by getting our books read at any rate, we lowered all the prices, setting only a value on them sufficient to prevent their being bought for waste paper." Another missionary gives the following cheerful testimony from Lahore:

> I have been delighted with the success the attempt to sell books has met with. I had two men engaged in the work last month, who suc-ceeded in realising 7 rupees, 4 annas, 2 pice for the books sold. We have affixed very low prices, only six pice for a single Gospel; this is no more than the value of the waste paper, which is all we can hope to get. The books, in my opinion, are more valued and read than they were before, and the demand appears to be steadily on the increase. I do not think that they are often bought for the paper and binding now, as the Urdu books are more in demand than the Hindi and Gurmukhi, which would not be the case, if they were purchased for that purpose. (*Fifty-Sixth Report of the British and Foreign Bible Society* 120)

In "lands nearer home" (as Reed might have put it), Henry Mayhew notes in 1850 that "'the gentlemen who manage the Ragged School . . . make [the children] presents of Testaments and Bibles' (I find by the Reports that they are sold)" (Mayhew, *Voices of the Poor* 15).

One missionary in China, a member of the British and Foreign Bible Society named Dyer, attacks what he calls distribution "by free gift," which too many of his colleagues advocate "because they consider the Gospel should be without charge." In passing, he alludes to what seem to be standard tricks of the (non)trade, such as blackmailing shopkeepers into buying bibles:

> Advantage is taken of a man's fear of loss, as for instance, the staying in a shop unduly long. The crowd that presses in to see the foreigner may cause the proprietor to purchase, in order to get rid of him, lest any in the crowd should steal. To take advantage of this would, of course, be exceedingly wrong, and the result that might be expected, a bias against, or even hatred of the book.

Just as wrong is "any statement concerning the books which would lead to their being looked upon as a kind of charm." Yet in a veiled reference to wastepaper, Dyer adds that when books are sold "there is much less fear of any misuse" of them; but distribution "by loan" has advantages of its own, since a missionary visiting to exchange a Bible "portion" for the subsequent installment can ask, "Understandest thou what thou readest?" (118–20). Replacing sale by loan also means replacing a punctual transaction with an ongoing social relationship—a point as important to Victorian missionaries as it is to marketers of digitized music or e-books today.[18]

The word "waste" haunts every account of Evangelical book distribution. The same *Missionary Register* article adds that "an indiscriminate distribution of the Scriptures, to everyone who may say he wants a Bible, can be little less than a waste of time, a waste of money and a waste of expectations." Macaulay denigrated Arabic and Sanskrit texts by estimating that "twenty three thousand volumes, most of them folios or quartos, fill the libraries or rather the *lumber rooms* of this body. The Committee contrive to get rid of some portion of their vast Oriental stock by giving books away. But they cannot give so fast as they print. About twenty thousand rupees a year are spent in adding fresh *waste paper* to [this] hoard." (T. B. Macaulay "Minute on Indian Education" 170; my emphasis). At best good for "waste paper," these "books are of less value than the paper on which they were printed was when it was blank." Just as he contrasts education in native languages (which he claims students must be bribed to undergo) against English-medium education (for which students pay), so Macaulay contrasts the Arabic and Sanskrit books—which circulate either as gifts or as waste—with the English-language books that change hands for money. Circulation through the market thus comes to be aligned with secular content: the absence of both taints Eastern literature as much as Christian tracts. "They cannot give so fast as they print": in the hands of the son of Zachary Macaulay, one of the original backers of the Religious Tract Society, the "free gift" becomes indistinguishable from trash.

When Dyer declared that "the people are much more likely to value what they have paid for," he wasn't simply noticing that owning doesn't necessarily imply reading; more strongly, he speculated that strategies conducive to the former can discourage the latter. In this hypothesis, giving books away free isn't just ineffective, but positively harmful. If we took seriously the trope of tract as medicine, we could see Evangelical publishers as the forebears of those twenty-first-century public health researchers who debate whether drugs distributed for a nominal fee are likelier to be effective than those given free (Shiv, Carmon, and Ariely; Cohen and Dupas). In both cases, the debate plays out most sharply with goods that flow from center to periphery; and in both cases, the belief that a particular category of good is too important to put a price on collides with the observation that end users are more likely to vest trust where they've already invested money.

Yet as those researchers have also argued, in the opposite direction, a high resale value can also risk diverting the subsidized goods from the uses for which they were distributed. Christian missionaries in India discovered that the paper of their books and tracts "could be put to a range of practical uses—from wrapping medicines to lining baskets carrying scented oil bottles, and even flying kites" (Ghosh 43).[19] Even within Britain, the Cheap Repository Tract Society found hawkers eager to distribute subsidized publications because nontextual uses guaranteed their resale value: as William St Clair points out, "the chapmen's insistence from the start that the tracts should be printed on soft paper, rather than the smoother book paper favoured by the promoters, suggests that they were well aware of the ignoble fate to which they would soon be consigned" (354). And a tract-writer herself describes a child heaping "a few sticks and cinders on a spark of fire communicated to the torn lead of a Bible; given probably by some compassionate visitor to the wretched woman [her stepmother], who received it for the sake of the accompanying shilling, and then used it as so much waste-paper" (Tonna 161). Subsidized or donated books run a particular risk of nontextual use: compare the outrage that erupted in winter 2009 when the British newspaper *Metro*—itself a free giveaway devoid of market value—reported that old-age pensioners were buying up secondhand books from charity shops because they were cheaper than coal to burn (Erwin).[20]

The logical extension of Dyer's argument for distribution by installments was to misdescribe gifts as loans: Howsam shows that although the BFBS was founded on the principle of never giving books away, the cholera epidemic of 1831 prompted the realization that dying people might not be able to pay for books. The compromise solution consisted of cheap copies marked "loan stock," which were never in fact reclaimed out of fear that they could become a vector for spreading cholera. The BFBS judged it unwise to collect books loaned "among that class of person who from their circumstances & habit of life will be generally found most exposed to the ravages of this and other diseases many of which are highly infectious & likely to prove communicable by books that have been for some time in their possession" (65–66).

Even Dyer acknowledges that the book's base material properties can paradoxically ensure the transmission of its content: "Be it that the book remains unread, that it is used by the women for putting their silks in. What then? Is it impossible or unlikely that in the course of time some visitor should come in, take up the book, read?" (121) Yet as the Reverend J. L. Nevius points out in a response to Dyer's paper, "the promiscuous sale of the Bible, and pressing it upon those who do not want it," can defeat its own ends: "Mr. Dyer's paper speaks of the special trials and insults to which Bible agents are exposed from those who hate them and their work. It is well for us to enquire whether much of this abuse may not be a direct consequence of disregarding the specific command of our Saviour: 'Give not that which is holy unto the dogs, neither cast ye your pearls before swine, lest they trample them under their feet and turn again and rend you'" (Dyer 130). One tract by Mrs. Sherwood responds specifically to that threat, acknowledging that when its Christian heroine gives a book to an Indian swindler trying to propitiate her, "perhaps my reader may be inclined to ask in this place, 'Was not this well-meaning lady overhasty in not stopping to consider somewhat more the person on whom she was bestowing this precious gift, and was she not putting herself in the case of one who *casts his pearls before swine?*'" The reader is instructed, however, that a different quotation is more relevant: "*cast thy bread upon the waters, and thou shalt find it after many days*" (M. Sherwood 135).[21]

Resale value, as we'll see in chapter 7, can preserve as well as destroy. In *London Labour and the London Poor*, the pious remark that "Mr. Mayhew is . . . afraid that the distribution of [religious] tracts among the profligate is a pure waste of good wholesome paper and print" is later undercut by the speculation that "could the well-intentioned distributors of such things . . . see what is done with the papers they leave, they would begin to perceive, perhaps, that the enormous sum of money thus expended . . . might be more profitably applied."[22] It's only because used paper is always good for a scatological laugh that Mayhew can leave "what is done" with tracts to the imagination.

One bible distributor whose constituency included the dustmen of the Paddington Dust Wharf District worried about the resale value of her wares: "A Jewess who I saw was very intelligent" tells that author "that if I *gave* my Bibles to the Jews they would, the next moment, sell them." "I was to tell my friends that she, a Dutch Jewess, told me so in kindness. The Jews knew that they had the truth, and were not like ignorant Christians, bowing down to images of wood and stone and kissing them, &c."[23] The missionary is reassured, however, when "in several of their rooms I saw a picture of Moses holding the two tables of the Law." The vertical paper bearing an image of a vertical stone becomes continuous with the horizontal pages that the Christian distributes. Yet in a world where paper is put to relentlessly material uses, as in the house of a weaver whom L.N.R. visits where "the holes in the ceiling were pasted up with newspapers," "bowing down" to the bible is no less dangerous than treating it as one more item to be pawned.

If the bible-publishing industry reduced a priceless text to a commodity, prize books posed the opposite problem: how to deal with objects that should be subject to the laws of the market but on which no end user would spend her own money. Often described as the first category of objects to be marketed specifically for gift giving—Victorian annuals sport titles like *The New Year's Gift*—books were given by middle-class parents not only to their own children but to their social inferiors, whether in the form of austere tracts or showy volumes (Nissenbaum 143). Although the practice itself was hardly new, the spread of schooling (by 1870, 3.5 million pupils were enrolled in English Sunday schools) ensured that "reward books" were increasingly produced and marketed as such (Ledger-Lomas 338; Laqueur 113–18). From 1810 onward, the RTS catalog categorized certain titles as "reward books to the children of Sunday-schools," where good-conduct tickets could be exchanged for tracts that not only represented characters' good behavior but also attested to their owners' (Reynolds 191; Bratton 17).

As the downward spread of schooling expanded the reach of prize giving, reward books became especially prominent in poor households, where they constituted a higher proportion of total books owned. Charlotte Yonge lamented that "the usual habit is to choose gay outsides and pretty pictures, with little heed to the contents, but it should be remembered that the lent book is ephemeral, read in a week and passed on, while the prize remains, is exhibited to relatives and friends, is read over and over, becomes a resource in illness, and forms part of the possessions to be handed on to the next generation." She adds that "weakness and poverty of thought should be avoided, especially as these books may fall into the hands of clever, ungodly men, and serve to excite their mockery" (*What Books to Lend and What to Give* 10–11). Yonge's advice emphasizes that books owned by working-class people will pass through more phases of existence than do books designed for their betters. Like the narrator of *The Story of a Pocket Bible*, the hypothetical prize book is owned first by a child and then by an adult, first by a believer and next by a scoffer. To target a readership too precisely (as we'll see in a moment) was to ignore how many times books changed hands.

Yonge's stricture on "gay outsides" reminds us, too, that the very tracts that encouraged their wearers to don plain but serviceable clothes were themselves made of cheap paper, expensively bound: unlike for most other books, the outside accounted for a higher share of production costs than the inside (P. Scott 220). In an age where books were expected to critique the consumerism that they had once exemplified, prize books (shoddily written and gaudily bound) posed an even greater embarrassment than tracts (whose production values at least matched the poor quality of their literary style). In fact, the protectiveness toward shabby books that we saw in the previous chapter finds its mirror image in recipients' eagerness to rid themselves of fancy presentation copies: Arthur Ransome points out that "sumptuous volumes are always easiest to part with; a ragged, worn old thing, especially if it is small, tugs at our feelings, so that we cannot let it go, whereas a school prize or an elegant present—away with it" (142). Remember Mozley's declaration that "the book thus influential came to [the child] by a sort of chance, through no act of authority or intention": to receive a book from a teacher or parent strips reading of its transgressive force.

The problem wasn't just the production value of prize books, but also their mode of distribution. Books designed as gifts for middle-class children punctured the myth of the self-made reader that we saw in chapter 3. Presented in front of an audience rather than devoured in secret, reward volumes disjoined reading from self-determination. Gender and class alike could determine whether books were chosen by, or forced upon, the reader: one middle-class woman, for example, remembers that her older brothers' "[book] prizes fell into my keeping, handed to me in disgust" (Hughes 61). And age combines with those two categories to ensure that a grown-up man like the brickmaker in *Bleak House* should feel particularly infantilized by having his reading matter selected by a "School lady, Visiting lady, Reading lady, Distributing lady":

> Have I read the little book wot you left? No, I an't read the little book wot you left. There an't nobody here as knows how to read it; and if there wos, it wouldn't be suitable to me. It's a book fit for a babby, and I'm not a babby. If you was to leave me a doll, I shouldn't nuss it. (132)[24]

Part of what makes Mrs. Pardiggle disturbing is that she's anything but (in the words of the previous chapter) a "silent messenger": far from slipping into places where a human body can't penetrate, the book she proffers provides an occasion for rather noisily embodied humans to poke into other humans' houses. Equally significant, however, is that the brickmaker's parody of the catechistic form of tracts rejects the double degradation of being associated with children and females. No "cottage woman," Yonge notes, "likes a book manifestly for children lent to themselves"; yet Yonge's own annotated catalog of books suitable for giving away remarks of one title, "May be useful where children or servants fear a haunted house" (*What Books to Lend and What to Give* 8, 24). "Children" are interchangeable with "servants" not simply because both are ignorant and superstitious—a commonality that results in part from servants' telling *oral* ghost stories to their masters' children—but because both depend on others to select and supply *printed* books for them. The involuntary listening to stories echoes the involuntary receiving of books: children "should never be terrified with nonsensical stories of ghosts,"

Ann Taylor writes in *The Present of a Mother to a Servant* (1816), "unless you wish to render them as unhappy about such things as you may have been, and perhaps still are" (quoted in Steedman 235). The middle-class adult's past is the servant's present—when it comes to books as much as to oral traditions. The working-class man who has books thrust upon him stands opposite the middle-class child who seizes his own books independently of any curriculum. On the one hand, Mrs. Pardiggle's tracts; on the other, the romances in the Copperfields' lumber room.

"CLASS LITERATURE"

Free print reminds us that where booksellers fulfilled the orders of middle-class adults, most readers—whether young, female, or working-class—had to submit to the literary direction of others. The same belongings that allowed financially independent adults to stamp their identity onto bookcases and sofa tables were tainted for middle-class children by association with the teacher or parent, and for working-class adults by association with the district visitor. Tracts, reward volumes, and advertising circulars all contradicted print's claim to individualize its users.

The Protestant overtones of that claim may help explain why tract societies show such ambivalence about bunching readers into market segments. On the one hand, Evangelical groups pioneered niche marketing long before commercial advertisers began to copy them. One of the earliest sets of instructions for tract distribution explains that they should be "adapted to various situations and conditions . . . When an address is particular, and directed to a specified situation, it comes home to the man's bosom, who sees himself described . . . Hence the propriety and necessity of Tracts for the young and for the aged, for the children of propriety and of affliction" (Bogue 13). According to a later chronicler, tracts were variously crafted "for seamen and soldiers; others for person in particular situations, as prisoners, attendants on pleasure fairs and races, patients in hospitals, the sick, etc. . . . Thirty-two tracts most suitable for the aged are in large type" (Jones 118). Jones's history is itself divided into sections such as "Books for the Young," "The Road Labourer," "The Hospital Patient," "The Showman." "Jewish tracts" meant tracts for Jews, not by them. Even Jones's tables adding up numbers of books sold are broken down by audience:

Emigrants 641,639
Soldiers, sailors etc. 1,622,661
Sabbath-breakers 1,059,590

Prisoners 106,303
Hospitals 60,924
Workhouses 68,836
Railroad men 447,407
Foreigners in England 50,742

The RTS annual reports break grants down into tracts for "Soldiers, sailors, rivermen," "Patients in hospitals," "railway labourers," "foreigners in England," "fairs," "races," and so on. In this sense, tract societies blazed the trail that newspapers would follow in dividing their material into a fashion section for women, a comics page for children, and a sport section for men. The directiveness of tracts, which explicitly addressed particular readers, stood opposite the studious neutrality of the bibles distributed by the BFBS, whose lack of notes or commentary was designed to avoid slotting their implied reader into any particular sectarian identity.

On the other hand, the Religious Tract Society worried that publications targeted too precisely could draw invidious distinctions. Remember the narrator of *The Story of a Pocket Bible* "recounting the duties of the various classes of persons to whom I had messages to deliver, and among other things I had something to say to servants. 'Servants,' I said, 'obey in all things your masters.'" (Sargent 38). Yet that narrator's reluctance to be pulled off the shelf by the duster-wielding servant provides a reminder that what's presented as a principle of inclusion (servants get to have their very own books, all to themselves) can also signal exclusion (servants are not to share their masters'). In 1851, the prospectus for a new RTS magazine declares explicitly that

> Avoiding the pernicious principle of creating a distinct literature for each of the different sections of society, there will be no ostentatious parade of condescension in the choices of topics or the mode of treating them; but animated by feelings of pure catholicity, "THE LEISURE HOUR" will seek to utter sentiments which shall meet an equally quick response in the parlour and the workshop, the hall and the cottage . . . the whole forming a miscellany aiming to be highly attractive in itself, and one which the Christian parent and employer may safely place in the hands of those who are under his influence. ("Prospectus for the Leisure Hour")

It remains unclear whether the resident of hall and parlor is expected to read the magazine herself, or merely to place it in others' hands. What is clear is that the influence of texts on their readers is doubled here by the influence of "parents and employers" on children and workers—no matter whether their authority derives from class or age.

Two decades later, a magazine article by Charlotte Yonge identified market segmentation as the most pernicious innovation of modern publishing. Yonge lumps books "for children or the poor" under the category of "what may be called class-literature," whose development she dates to the beginning of her century. Her neologism registers the historical specificity of a model in which "every one writes books *for* some one: books for children, books for servants, books for poor men, poor women, poor boys, and poor girls. It is not enough to say 'Thou shalt not steal,' but the merchant must be edified by the tale of a fraudulent banker, the schoolboy by hearing how seven cherries were stolen, the servants must be told how the wicked cook hid her mistress's ring in the innocent scullerymaid's box" ("Children's Literature: Part III" 450).[25]

In her book of advice for parish visitors, *What Books to Lend and What to Give*, Yonge insists more explicitly that "there is no reason against giving details about persons in different stations of the life from that of those who received them, and in fact they are often preferred . . . A book labeled 'A tale for—' is apt to carry a note of warning to the perverse spirits of those to whom it is addressed" (12). Her own list of books recommended to read aloud, lend, and distribute in parish work includes a chapter on "drawing-room stories," that is, representations of life in a higher rank—although a mixed message is conveyed by the very fact that this chapter is separate.

None of this is to say that religious tracts had any monopoly on the attempt to align the identity of characters with the identity of readers. Secular publishers were quicker to recognize, however, that aspiration could trump identification. Yonge's critique echoes Margaret Oliphant's observation about cheap literature:

> If any one supposes that here, in this special branch of literature provided for the multitude, anything about the said multitude is to be found, a more entire mistake could not be imagined . . . An Alton Locke may find a countess to fall in love with him, but is no hero for the sempstress, who makes her romance out of quite different materials; and whereas we can please ourselves with Mary Barton, our poor neighbors share no such humble taste . . . It is not because their own trials are shadowed—their own sentiments expressed—their own life illustrated by the fictitious representation before them, that our humble friends love their weekly story-telling. When the future historians of this century seek information about the life and manners of our poorer classes, he will find no kind of popular print so entirely destitute of the details he seeks as are those penny miscellanies which are solely read by the poor. (207)

Oliphant herself slots class into the place occupied by age in Dr. Johnson's famous remark

> "Babies do not want . . . to hear about babies; they like to be told of giants and castles, and of somewhat which can stretch and stimulate their little minds." When in answer [Hester Thrale] would urge the numerous editions and quick sale of Tommy Prudent or Goody Two Shoes: "Remember always (said he) that the parents *buy* the books, and that the children never read them." (Piozzi 14)

Johnson anticipates Oliphant and Yonge in linking two apparently unrelated phenomena: the coincidence of readers' status with characters', and the gap between buyer and end user.

Historical hindsight makes it easier to see a higher-order association: that the structure of arguments about children's identification with literary characters can be seamlessly slotted into arguments about working-class readers' identification with literary characters. The missing link, as Dickens's bricklayer perceives, is that neither audience chooses its own reading. When Charles Knight accused "the learned and the aristocratic" who "prattle about bestowing the blessings of education" of "talking to thinking beings . . . in the language of the nursery" (*Passages of a Working Life During Half a Century* 243), he was referring not just to the style of tracts but also to their distribution method. Like children's books, tracts addressed a double audience: the buyer didn't coincide with the reader. And in both cases, the distributor disciplined rather than fulfilling the end user's desires.[26]

Like the adult who forces books about goody-goody children upon young readers, the tract-distributor who tries to wean servants away from aspirational romances of high life takes the moral value of reading to inhere not just in the content of the text itself (in which case children could just as easily become virtuous by reading about sanctimonious adults) but in the degree of overlap between readers' and characters' identity. For twelve-year-olds to read about twenty-year-olds, or maids about marquises, was to engage in aspirational escapism; for bad children to read about good children or godless barrow women to read about pious barrow women counted instead as a spiritual exercise.

Yet a logic that put the "self" back into self-improvement inevitably came into conflict with the secular defense of fiction reading that valued identification precisely as an escape from egotism through the enlargement of the sympathies. "It is the endeavour to hold up a mirror to each variety of reader of his or her way of life," Yonge continues, "as if there were no interest beyond it, and nothing else could be understood or cared for, that we think narrowing and weakening. If it be true that imagination is really needful to give the power of doing as we would be done by, surely it is better to have models set before us not immediately within our own range" ("Children's Literature: Part III" 450).

Yonge shrewdly diagnoses two contradictory assumptions that structure the genre. First, a logical contradiction, because the same readers who are expected to identify across species (in texts featuring talking animals) or even with inanimate objects (in it-narratives) are presumed incapable of crossing the finer lines separating children from adults, men from women, or rich from poor. Second, a moral contradiction, because the metaphor of "hold[ing] up a mirror" points to the narcissism inherent in an act supposed to foster selflessness. "When sympathizing with the heroine let it be with *her*, not with yourself under her name," warned the children's writer Mrs. Molesworth. "Nothing is more dwarfing and enervating than to make all you read into a sort of looking-glass" (454).[27]

ASSOCIATION COPIES

The moral ambiguities of both forms of other-directedness catalyzed by books—defining one's identity by imitation of characters', or by association with other readers'—find their fullest expression in a novel whose scale, audience, and doctrinal stance (or lack thereof) might appear to place it worlds apart from these cheap tracts. The new copies of the three volumes of *The Mill on the Floss* that went on sale for a guinea and a half in 1860 presented their readers with representations of books being read, dropped, given, bequeathed, resold, inscribed, and sworn on; they also contribute one installment to Eliot's lifelong exploration of what will happen to written matter after its owner's death. *Romola* centers on the question of whether to obey the promise given to a dying father to keep together the library that he spent his life collecting. (The question projected onto Renaissance Florence was a live one at a moment when great aristocratic libraries were being dispersed into public hands and across the Atlantic.) Casaubon hopes that an analogous promise, once again exacted by an older man from a younger woman, will allow notebooks to outlive their owner.[28] In contrast, *The Mill on the Floss* asks whether to keep or break an oath of vengeance—also sworn *to* a dying father, but in this case sworn, just as crucially, *on* the family bible. The book forms a "chain" both in the sense of connection, and in the sense of constraint, drag, or dead weight.

Three questions, then. When text pulls apart from book—when the words that preach forgiveness take the form of a talisman on which vengeance can be sworn—which will win out? What relation does a text establish among its successive users—either between the dead and those who inherit their books, or between the current reader and the traces of their predecessors that linger in the form of inscriptions, dog-earing or even dirt? And why should the materiality of the book swim into focus at the moment of its owner's death—or, conversely, why should looking at books conjure up the thought of their dead owners?

Ranthorpe would appear to find his double in Maggie, who seizes on an *Imitation of Christ* with its "corners turned down in many places," in which "some hand, now for ever quiet, had made at certain passages strong pen and ink marks, long since browned by time" (G. Eliot, *The Mill on the Floss* 303).[29] In fact, her asceticism is associated just as strongly with the book's material form as with its textual content: even before she reads a word of Thomas à Kempis, Maggie prefers the "little, old, clumsy" copy of *Imitation of Christ,* "for which you need only pay sixpence at a book-stall," to the tastelessly illustrated annuals that only a peddler like Bob Jakin (the auctioneer's down-market double) would call the "bettermost books" (303, 295).[30] When Maggie "read[s] where the quiet hand pointed," however, she merely replaces one version of bibliophilia by another: the materialism that registers clean pages and lavish bindings by the fetishism that prizes a book for the hands through which it has passed. We flinch when Tom answers Maggie's lament for the auctioned-off books by asking, "Why should they buy many books when they bought so little furniture?" (252). The juxtaposition of the two nouns sounds as oxymoronic as Jewsbury's coupling of noun with verb: "a few splendidly bound books *furnished* the heavily carved rosewood table." Yet the novel's own homology between inscribed books and initialed linens levels those terms, reducing books to keepsakes (namesakes of Rosamond Vincy's annual) while elevating housewares to expressions of identity.

The Mill on the Floss thus makes association copies—books whose value, like that of a religious relic, derives from the history of their transmission rather than from the text that they contain—a model for the circulation of nontextual possessions.[31] In the process, Eliot flattens any distinction between books and humbler objects: chairs, teapots, sheets. Where almost every other Victorian novel—including some by Eliot herself—use sofa-table books to exemplify the evils of the market, *The Mill on the Floss* turns a specific sector of the book trade into a template for a larger reflection on the embeddedness of objects within human relationships.

No less than her flower-pressing aunts and bible-signing father, Maggie understands the book as a means of preserving memories. And she, too, locates this preservative power not in its printed contents but in the traces that past owners have left of their own presence: "Our dear old Pilgrim's Progress that you coloured with your little paints . . . I thought we should never part with that while we lived . . . the end of our lives will have nothing in it like the beginning!" (252).[32] Painted woodcuts mirror scribbled flyleaves: the authorial hand that describes the heroine as fair or blonde matters less than the hand-colored illustration that brings to mind a beloved brother.

Where Maggie calls down Tom's wrath by mourning the auctioning off of her inscribed books, what Mrs. Tulliver regrets is the sale of her trousseau, the "things wi' my name on 'em" (215). In *The Mill on the Floss,* books form only the last addition to a long list of inside-out objects whose value derives less from their contents than from their owners' signature. That category includes the hand-colored *Pilgrim's Progress* and the family bible whose flyleaves are inscribed but whose printed content goes unread, but it also encompasses the monogrammed teapot too good to be dirtied with tea and the initial-embroidered sheets too fine for a living body to lie between. Like the pristine pot or folded sheets, Mr. Tulliver's confusion of a "good book" with a "good binding" repeats the Dodsons' determination to empty containers of their contents—whether Mrs. Glegg's "brocaded gown that would stand up empty, like a suit of armour," or the display of pristine jelly-glasses that Mrs. Tulliver invokes in response to Mrs. Deane's offer of jelly for the invalid: "There's a dozen o' cut jelly-glasses upstairs . . . I shall niver put jelly into 'em no more." Even when Aunt Pullet finally progresses from the "very bright wardrobe, where you may have hastily supposed she would find the new bonnet" but which holds only a key, to the second wardrobe in which the hat is actually hidden, the object inside turns out to be less interesting than its paper covers: "The delicious scent of roseleaves that issued from the wardrobe made the process of taking out sheet after sheet of silver-paper quite pleasant to assist at, though the sight of the bonnet at last was something of an anticlimax" (127, 95, 96).

To judge books by their covers, as Mr. Tulliver does, is to make them an exemplar of, rather than an exception to, that perverse logic. The petals inside the Dodsons' unread bible form a mirror image of other characters' obsession with the outside of books: their bindings (in Defoe), flyleaves (in the bible), illustrations (in *Pilgrim's Progress*), marginalia (in the *Imitation of Christ*), and (via an interest in metadata such as the author's name) their title pages. In the end, the symmetry that levels books (meaningful but impractical) with household goods (insignificant but utilitarian) sharpens into a chiasmus: as the Gospel dwindles to a material surface on which oaths are written or flowers pressed, "well-cured ham at one's funeral" takes over its sacramental function (G. Eliot, *The Mill on the Floss* 285). By seeking in the book trade a model for its own understanding of the human associations embodied in things, *The Mill on the Floss* subordinates the one attribute that distinguishes books from baser commodities—their linguistic content—to the attribute that they share: that is, their mode of circulation. The only difference is that where Tom reduces books to a subset of "furniture," Eliot lifts your ugly furniture (as *Middlemarch* will put it) into the serene light of science (G. Eliot, *Middlemarch* 264).

If you open a one-volume reprint without regard to Eliot's characteristic alternation of "historical" (narrative) with "doctrinal" (sententiousness), the page that will lie flat is the one containing a description of the "quarto Bible open at the fly-leaf" in which Tom Tulliver will inscribe his oath of vengeance. Maggie herself draws attention to the blasphemy of making a book that preaches forgiveness into a repository for revenge: "'O father, what?' said Maggie, sinking down by his knees, pale and trembling. 'It's wicked to curse and bear malice.'" (G. Eliot, *The Mill on the Floss* 280). The disagreement between father and daughter is bibliographical as much as theological. Maggie's conspicuously oral contraction ("it's wicked to bear malice") calls attention to the looseness with which she paraphrases the words that we can recognize as originating in the Gospels—even if Mr. Tulliver doesn't. Maggie understands the bible as a container for truths that remain stable from one copy to another, one edition to another, even print to voice. Mr. Tulliver, on the contrary, treats it as a thing that can be owned, inscribed, held, and sworn on.[33] In relocating meaning from textual content to paratextual margins, *The Mill on the Floss* also replaces metaphorical intimacy with a virtual author by metonymic intimacy with other handlers.[34]

To measure the force of the tension between a paratext that prescribes vengeance and a text that proscribes it, compare *The History of a Family Bible. A Tale of the American War* (1851). Narrated by a young minister who transcribes the family records on the flyleaves of cottagers' bibles, this moral tale quotes from their manuscript paratexts as much as from their printed texts. The story ends with a family's bible being seized by the bailiff, as its owner cries "spare me but this book—my father's book—our family Bible—the book I brought from America on purpose for him—his and my mother's comfort in their latter days. Spare me this precious book." When the bailiff insists, the owner replies,

> "I'll have a part of it; I'll have some memento of those I so fondly love," and before he was aware of what I was about, I had cut the thread that fastened on the linen cover, and grasped that in my hand with the violence of despair. No sooner were the lids disclosed, than two bits of paper folded perfectly smooth fell on the ground."

These turn out, of course, to be Bank of England notes inscribed by the narrator's parents. One is marked "your father's ears are never deaf"; the other, "when sorrow overtakes you, seek your Bible." Like the bible, then, the banknote signifies doubly: through its official content (the printed text, the formulaic language of a financial instrument) and through a personal inscription alluding to the hands through which the object has passed. The latter literalizes the former: when her granddaughter exclaims, "look at the treasure you have found," the narrator replies, " 'tis true that literal banknotes may not be given as in this case but what is far better than gold and silver is the certain portion of all believers" (Best 138–40). In Mr. Tulliver's case, however, no such dovetailing: text and paratext contradict, rather than reinforcing, one another.

The tension between print (the Gospels) and manuscript (the Family Record) reflects a struggle between literally writing oneself into the book and imaginatively putting oneself in its characters' place. And as writing on the flyleaf replaces reading in the book, self-inscription replaces selflessness.[35] We all know that Eliot expected texts to give their readers practice in empathizing with fictional Others—in what "The Natural History of German Life" calls the "extension of our sympathies." Yet her characters invest most heavily in texts where reader coincides with writer, and writer becomes protagonist: "It's got everything in [it]—when I was born and married" (G. Eliot, *The Mill on the Floss* 273).[36] Even the title page of *Holy Living and Holy Dying*, oddly, becomes indistinguishable from the flyleaf of the bible: "Mr Tulliver felt somehow a familiarity with that great writer [Jeremy Taylor] because his name was Jeremy" (G. Eliot, *The Mill on the Floss* 21).

Jeremy—but why always Jeremy? Perverse where we expect it to be pious, the novel insists that reading doesn't involve losing oneself in a text so much as finding oneself in it.[37] That swerve away from empathy culminates in the scene in which Mr. Tulliver orders Tom to inscribe in the bible what sounds like a summary of the book that we ourselves are reading:

> "Write as your father, Edward Tulliver, took service under John Wakem, the man as had helped to ruin him, because I'd promised my wife to make her what amends I could for her trouble, and because I wanted to die in th'old place. Where I was born and my father was born. Put that i' the right words—you know how . . .
>
> "Now let me hear what you've wrote," said Mr Tulliver. Tom read aloud, slowly. (281)

Tulliver's shift from the third person ("took service under the man as had helped to ruin him") to the first ("because I'd promised") mimics the model of reading that governs *The Mill on the Floss* as a whole. To subsume "him" into "I" is to displace the hope that the novel will enlarge sympathy by the acknowledgment that whatever text one reads turns out to be about the self.

Maggie's replacement of plot by Scott reminds us that characters "dreaming over a book" often turn out to be less interested in its content than in a story about its author—which becomes in turn a story about its reader (18). You'll remember that Scott appears as a secondary character in Maggie's own fantasized *autobiography*, not as the author of a text representing lives distant from her own. For Maggie as for her father, the printed book functions less to contain a story than to occasion one. Whether handwritten (Mr. Tulliver's family bible) or purely mental (in the case of Maggie's "vision"), that story always concerns the self. In fact, Maggie can recognize another's handwriting only in the process of remembering a flyleaf inscribed with her own name: "At last, Bob brought her a letter without a postmark—directed in a hand which she knew familiarly in the letters of her own name: a hand in which her name had been written long ago in a pocket Shakespeare which she possessed" (522). While a name handwritten in a printed book brings the family bible to mind, the term "familiar" recalls more closely the language that had described Jeremy Tulliver's identification with Jeremy Taylor. Just where we expect Maggie's empathy to contrast with her father's egotism, we're reminded that the daughter and the father share the inability to read anything but their own names.

More sophisticated than her father, Maggie resists "applying" texts to her own life—whether reading the *Imitation of Christ* as a how-to manual or identifying with the dark Scott heroine in preference to the blonde. Yet even as the *Imitation of Christ* inspires Maggie to deny herself the luxury of a mirror, the book itself becomes a medium for the "sort of looking-glass," as Mrs. Molesworth would put it, that makes the heroine "yourself under her name." With a crucial difference: the figure with whom Maggie "sympathizes" isn't the heroine *in* a text, but rather an earlier reader *of* a text. By the same token, where her father wants to imagine that his namesake Jeremy speaks uniquely to him, Maggie understands herself as only one of a long line of readers. The "low voice" and "quiet hand" that she discerns in her secondhand copy of the book belong to a previous owner, not to the author, let alone to a character represented in its pages. Communion with earlier readers can undo the covert narcissism of identifying with fictional characters or recognizing oneself in authors' names.

A copy of *The Mill on the Floss* that Emily Dickinson's family bequeathed to Harvard's Houghton Library bears two traces of use: like Mr. Tulliver, Susan Dickinson has written her name on the flyleaf; and like the Dodsons, someone has pressed a fern leaf between the leaves of the book. Within the novel, the value of readers' marks remains more ambiguous. Where the family bible has been held out of the auction (where its markings would in any case have lowered its price), the *Imitation* acquires an extraeconomic value precisely by changing owners. Mr. Tulliver's bible proves as difficult to alienate as his wife's monogrammed linens—one "spoiled" by writing in the margins, the other by stitching in the corners. Both offer a countermodel to Maggie's hunger for a virtual community of readers: for her parents, objects are valuable only to the extent that they can be imagined as never having been in, and never destined to pass into, others' hands.

CHAPTER 6

The Book as Go-Between: Domestic Servants and Forced Reading

THE PREVIOUS CHAPTER located the meaning of tracts in the interactions that they represent, but also in the relationships they establish: relationships of difference between giver and reader, relationships of similarity between character and reader. That second meaning may be more readily available to book historians than to literary historians: a quarter century ago, a magisterial analysis of the literary representation of servants never asked what servants read (Robbins). If tract distributing punctures the myth that makes reading an expression of individual choice, it also threatens the hope—or staves off the fear—that the shared act of reading will break down social barriers. The relationships that reading vehicled were asymmetrical, conflictual, and frequently asynchronous. Even, or especially, where a single text such as the Bible was shared among different audiences, the different physical formats of the book and different temporalities of reading restored social differences.[1]

Throughout the nineteenth century, three spaces became staging grounds for the tension between commonality and hierarchy: the public library (where successive borrowings of a single volume linked patrons across classes); the home (where the fear that masters and servants could access each other's corrupting books mirrored the hope that masters could inflict godly books on their servants); and the margin (where the trace of previous readers' hands bridged the living with the dead). These spaces all differ from one another, of course: the first two stand on opposite ends of a public/private spectrum, and both of them contrast in scale with the third. The home and the library were (and are) often contrasted as spaces for reading, whether one is imagined as a retreat from the other's civic openness or a refuge from its capacity to spread disease. Yet in both settings, books undermined social distinctions not through their content but through each user's knowledge that the object in his or her hands had been handled by someone else before, and would eventually fall into someone else's.

If eighteenth-century subscription lists inscribed the reader in an economically and sexually homogeneous public, borrowers' names at the back of a library book did just the opposite. As literacy became more evenly distributed, the fear that texts might corrupt their readers gave way to the fear that books might blur social distinctions among their handlers. The tracts that represent servants' reading (and dusting) give way, after 1850, to discussions of books' circulation among public-library patrons of different social classes; yet even after the rise of the public library and the fall of domestic service, social commentators continue to use servants' reading as a front for critiques of the middle classes whom they themselves address.

This chapter asks in conclusion why secular fiction devoted so much space to jokes about tract distributing. Where tracts imitate the formal conventions of the same novels with which they competed, mid-Victorian novels almost obsessively represent characters distributing (though rarely reading) tracts. Yet tract distribution was only one among several practices that the secular press used to figure questions about the relation between supply and demand. The experiences of being handed a tract, read aloud to, and tricked into mistaking printed advertisements for personal letters, all provided the novel with mirror images for its own claim to be freely chosen.

By satirizing intrusively personal forms of charitable and familial transmission, the novel made a virtue of a traditional accusation against it: that its commercial distribution and solitary consumption made the novel an antisocial genre. Mid-Victorian novels made both the vertical relationships that bound givers of books to their more or less willing recipients, and the horizontal relationships that linked each user of a book with others who had read or at least handled it before, foils for the kinds of solipsism that they represented fiction itself enabling—whether in the heroic mode of a Dickensian child being beaten for reading in the lumber room, or the bathetic mode of a Trollopian wife hiding from her husband in the pages of an unread book.

DUSTERS AND READERS

If reading is synonymous with individualism (as Victorian secular genres as diverse as the bildungsroman and the political treatise assume), what to make of its logistics? The historical record only occasionally yields reminders that, for example, Frances Hamilton relied on a manservant to return and pick up her books from the local circulating library (Steedman 75). Even apart from books' dependence on multiple agents to recommend them, sell them, clean them, maintain them, sort them, retrieve them, and dispose of them, the affordability of printed matter has traditionally depended on amortizing costs by sharing, whether in the form of resale (the secondhand trade), of loan (be it public library or private book club), or more complex arrangements in which newspapers were handed down along a chain of users. In an age when books (rather like cars or college textbooks today) were bought with an eye to resale value, every owner who calculated how much demand there would be for his possession should misfortune lead him to the auctioneer or the pawnbro-ker was also a reader picturing those other readers who would succeed him (Goolsbee). In the opposite direction, the margins of the page could form a repository for the traces of earlier readers, whether owners of books or borrowers of library stock. Yet those traces were not necessar-ily articulate or even intentional: in post-1850 public libraries especially, concerns about jam smears and mucus upstaged meetings of mind by contamination of bodies.

Closer to home, masters and servants faced the challenge of reconcil-ing shared access to bookshelves with differential uses of their contents. (This is relevant, of course, only to that subset of servant-employing households whose masters were literate and owned books.) Most simply, how might you get your wife or maid to dust your books, without allow-ing either to read them? The narrator of Henry James's 1892 tale "Brook-smith" observes that "a certain feeling for letters must have rubbed off on [the servant] from the mere handling of his master's books, which he was always carrying to and fro and putting back in their places" (77); and as late as 1935, the protagonist of Canetti's *Auto-da-Fé* hesitates to lend his housekeeper a book, on the grounds that "it was true that she dusted the books every day and had not yet injured one of them. But dusting and reading are different. Her fingers were coarse and rough . . . A hard binding can naturally stand rougher handling than sensitive pages . . . Charity is all very well indeed, but not at other people's expense. Why should the books have to foot the bill? . . . They are defenceless against the uneducated" (38–39). As books are anthropomorphized, servants be-come dehumanized; as books become companions, humans inhabiting the same space are banished to arm's length.

Not literally, of course; on the contrary, the problem is precisely that books lay within the reach of servants' arms. Just as men's newspapers and women's novels repelled others physically present (as we saw in chapter 2), men's newspapers and women's novels also created bonds with physically absent or socially distant others who had touched the same object. When Alfred Austin called the sensation novel "that one touch of anything but nature that makes the kitchen and the drawing-room kin" ([Austin] 424), he echoed another reviewer's claim that Mary Elizabeth Braddon "may boast, without fear of contradiction, of having temporarily succeeded in making the literature of the Kitchen the favorite reading of the Drawing-room" (Rae 204). That claim itself responded to the "likeness" that the novel itself establishes between Lady Audley and her lookalike maid, who "knew enough of the French language to be able to dip into the yellow-papercovered novels which my lady ordered from the Burlington Arcade" (M. E. Braddon 104).

Sauce for the gander: Thackeray asks the reader of his *Roundabout Papers* in the *Cornhill* to "suppose you ask for your newspaper, and Jeames says, 'I'm reading it, and jest beg not to be disturbed.'" The narrator of *The Newcomes*, too, introduces a digression with the observation that "'our John finishes reading the newspaper before he answers our bell, and brings it to us'" (*Roundabout Papers* 85; *Newcomes* 428).[2] The problem here is not just the fact of different classes handling the same object, but, more specifically, the order in which they do so: the traditional logic in which the newspaper descended along a social chain as it aged (or in which a maid inherited her mistress's castoffs once they went out of fashion) is upended in the scenario of the valet getting first crack at his master's newspaper.

I say "handling the same object" instead of "reading the same text" to register that shared reading of a single text—in particular the Bible, as we saw in the previous chapter—was sanctioned as long as each audience owned it in a different format or a different binding. The particular text chosen was often less significant, in fact, than the distinction between books given by masters to servants (or forced by masters upon servants) and books that servants themselves took the initiative to buy, beg, borrow, or steal. Tract societies reassert the social order threatened within the home by book borrowing, and outside the home by the rise of the public library. They do so not by denying that books have the power to link masters with servants, but by substituting unilateral giving for secret sharing. Where Thackeray's secular fiction represents valets thrusting their way into the *Times*'s putatively middle-class audience, tracts require their middle-class writers and distributors to think themselves temporarily into the consciousness of working-class readers. Different power dynamics, different temporalities: far from the valet cutting ahead of his master or the maid peeking into a yellowback en route to her mistress, a tract reaches its end users only after being screened by their betters.

Those competing models of the relationships brokered by books collide in Charlotte Adams's *Little Servant Maids*, a tract published by the SPCK in 1848 for 3s. 6d. in a binding suitable for middle-class ladies to give to young servants, with a style to match. Its episodic narrative, structured by a long-suffering lady's hiring and firing of successive servants with the assistance of a tattling upper servant named Martha, culminates in the elderly Martha's being taught to read. Yet no less than *The Story of a Pocket Bible*, this tract also criticizes the two premises on which its own circulation depends: that servants are literate and have access to books. (Both would have applied more fully to the indoor servants who figure in such tracts than to the larger class of farm workers.)

Figure 6.1. Religious Tracts, Cambridge University Library.
By permission of Cambridge University Library.

Mrs. Sewell had, among others, a few handsomely bound volumes, lying on a table in her sitting-room, they were chiefly presents to Frank,
and contained some pretty engravings: these proved a great attraction
to her young servant; and heedless of soiling the binding, or marking the fair pages with dusty fingers, [Caroline] would amuse herself

by looking over them whenever opportunity offered. She knew that she had no business to open these books, and whenever she heard a step approaching she closed them hastily, so that many of the leaves got creased and dog-eared. One small volume of poems she was so delighted with, that she did very wrong, and took into her bed-room, where she kept it concealed under her pillow, to draw forth and study whenever she could do so unobserved . . .

Detection came at last, and with it disgrace, as is the case with most deceitful persons. Martha discovered the volume of poems in its hiding place, and carried it directly to her mistress. This led to an examination of the other books, and Mrs. Sewell was vexed to find how much they had been thumbed and pulled about . . .

"It is very strange," observed Mrs. Sewell to Martha, after she was gone, "that these sort of persons always fix on one's best things to entertain themselves with; there were other books quite as amusing, with plain bindings, lying beside those that are handsomely bound, but they do not appear to have been touched; and there was a book of poems in a paper cover, exactly the same as the morocco one she took away to her bed–room." . . .

Poor people should begin early to teach their children to respect what is costly and ornamental; they should make them learn to look without touching. If this was generally attended to, we should not see in every direction clean white walls scribbled all over with pencil or charcoal, panes of glass windows scratched, benches hacked and cut about with letters. (Adams, *Little Servant Maids* 138–40)

Two contradictions vertebrate this passage. First, the claim that only "these sort of persons" judge a book by its cover collides with the granularity at which the mistress (and the narrator) describe the different books' physical form. The suggestion that the servant should have chosen the "book of poems in a paper cover, exactly the same as the morocco one," anticipates the moment in *The Story of a Pocket Bible* where the servant puts down the high-end narrator and takes up "the very counterpart of myself, indeed, except in mere externals." Different editions of the same text, or even the same edition on a different paper or in a different binding, can reconcile the cultural imperative for different classes to share a common canon with the equally strong imperative to mark status. The British and Foreign Bible Society's annual report for 1840 insists that in cheap bibles, "it is but the casket that is homely, the gem still retains its purity and richness: the peasant, or the peasant's child, when taking the *cheap* Bible in his hand, looks upon the same great truths . . . as meet the eye of the prince when he bends over the vellumed page" (quoted in Howsam 71). Yet by 1860, its catalog meticulously distinguished fifty-nine different formats of its English Bible, at prices ranging from sevenpence to twenty-three shillings depending not only on whether a subscriber's discount was given but also on a dizzying series of permutations and combinations of size, paper weight, edges (plain, gilt, red, or marbled), binding style (limp or circuit), and binding material (morocco, roan, sheep, calf, cloth, enameled cloth, or artificially grained sheepskin) (*Fifty-Sixth Report of the British and Foreign Bible Society*).[3]

Even the most doggedly anti-Evangelical of writers and publishers had something to learn from this trickle-down model. Dickens's strategy of "working the copyrights" by issuing each novel in successively cheaper formats owes more than the creator of Mrs. Pardiggle would have admitted to the tract societies' schemes for repackaging the same content at different price points.[4] At one end of the century, from 1796 onward, each Cheap Repository Tract was issued in two different qualities of paper: one, coarser, sold to hawkers at 24 for 6d., and another, finer, to the gentry at 24 for 1s. 6d. The result was what Kevin Gilmartin calls a "tension between a desire to incorporate every reader and every text within a single print economy, and an insistence that differences of privilege and function within that economy be strictly enforced" ("'Study to Be Quiet'" 512). By 1893, J. M. Barrie could reverse this trope for comic effect, by imagining a master becoming embarrassed to read his selected Landor once he discovers that the maid owns the *Collected* (Barrie)—itself an updating of the eighteenth-century servant Elizabeth Hands's joke:

> Quoth Madam, I have it;
> A Scripture tale?—ay—I remember it—true;
> Pray is it i'th' old Testament or the new?
> If I thought I could readily find it, I'd borrow
> My house-keeper's Bible, and read it to-morrow.
>
> ("A Poem, on the Supposition of the Book
Having Been Published and Read," 261)

The Bible, in particular, needs to remind readers of their place while including them in a larger religious community; not only does its numbering system unmoor citations from the pagination specific to any edition or even any language, but it's precisely because its contents are never outdated that its flyleaves can be enlisted to record the passage from one generation of owners to the next. In Mrs. Sherwood's popular tract "Little Henry and his Bearer," the observation that the text of the Bible remains stable across different formats is undercut by the space devoted to leather bindings and silk coverings:

> One day Henry came into the lady's room, and found her opening a box of books. "Come," said she, "Henry, help me to unpack these

books, and to carry them to my bookcase." Now, while they were thus busy, and little Henry was much pleased to think that he could make himself useful, the lady said, "These books have different kinds of covers, and some are larger than others, but they all contain the same words, and are the book of God . . . You shall have any one of these books you like best."

Henry thanked the lady with all his heart, and called Boosy in to give his advice whether he should choose a book with a purple morocco cover, or one with a red one. (M. Sherwood 14)

As another moralist warns, "Children, particularly, should never suffer themselves to be tempted by the rich outside of the book: often a worthless production shines in gold, whilst many a moral and useful work appears in a plain and simple cover ("Little Jack of All Trades").

The second contradiction is that we might expect Caroline's "dusty fingers" to connote industry rather than laziness: the reason they are dusty, after all, is that she has been dusting the sitting-room table.[5] Mrs. Sewell's injunction against handling reverses the more predictable logic of *The Story of a Pocket Bible*, where the servant is expected to dust without reading—to use her hands instead of her eyes. Unlike smelling, looking is not in any simple sense opposed to reading; in fact, the explicit imperative to "look without touching" oddly inverts the implicit order to touch (that is, dust) without looking (that is, reading).

As an activity that requires books to be handled by those who (for reasons of gender or class) are not supposed to read them, dusting makes anxieties about shared access particularly visible. Mrs. Beeton enjoins housemaids that even when the books are in a locked case, "every corner of every pane and ledge requires to be carefully wiped, so that not a speck of dust can be found" (990). In *The Enemies of Books*, however, the bibliophile William Blades asks in the pseudojocular tone beloved of fin-de-siècle male collectors:

> Why need the women-folk (God forgive me!) bother themselves about the inside of a man's library, and whether it wants dusting or not? . . . When your books are being 'dusted,' don't impute too much commonsense to your assistants; take their ignorance for granted . . . Your female 'help,' too, dearly loves a good tall pile to work at, and, as a rule, her notions of the center of gravity are not accurate, leading often to a general downfall. (118–20)[6]

Book collecting found its mirror image in tract distributing: one an assertion of wealth and taste, the other an act of social condescension and aesthetic slumming; one associated with bachelor dandies, the other with do-gooding spinsters; one aimed at accumulation, the other at dispersal. Yet even though written by, for, and about women, *Little Servant Maids* seems to side more with Blades's misogyny: in a tract whose practical hints are devoted to the ways servants should touch—and dust and scrub and polish—every surface around them, writing and reading form the only context in which servants' hands are imagined to defile, not to clean. That contradiction can be resolved only through a bifurcation of the hand: neat fingers polish and iron, dirty thumbs crease and dog-ear. The thumbprint that illiterates use to replace a signature gives way to a thumb smear that semiliterates use to mark their territory, or rather the domestic territory that they service but don't possess.

When a character in *Little Servant Maids* covers a glass of beer "with a dirty-looking tract," the generic designation provides a bridge between the pages that we're holding and the world represented within them (Adams, *Little Servant Maids* 15). Yet where the SPCK's book is designed to link mistresses with servants, Adams's text represents a world in which the potential for papers to connect their successive handlers must be held at bay. The proper relations that the book enables find their mirror image in the improper relations that it represents. Inhabiting spaces traversed by both masters and servants, depending on the former to select and pay for it but on the latter to carry, tidy, and dispose of it, written matter inevitably posed choices about how to balance privacy against convenience. (Do you tell your IT administrator your password?)

The scene of Caroline thumbing what's on the sofa table dramatizes traditional debates about whether to lock the bookshelf—themselves back-formed from questions about how best to secure the places where food and drink are kept. Remember that the noun "safe" designated a chest for holding meat long before it came to refer to the storage of money; remember, too, that the late nineteenth century saw the invention of the tantalus, which allowed decanters—like books in a glass case—to remain visible while preventing servants from accessing their contents.[7] The difference, however, is that books were not included among "vails" or perquisites: while leftover food or hand-me-down clothes provided a socially accepted (if often disputed) bond between masters and servants, printed matter spoiled less quickly than food and went out of fashion less quickly than clothes. Even old newspapers tended to be passed along at gradual removes, rather than changing hands all at once from servant to master.[8]

Some exceptions involve legacies upon a master's death; others, rejects that masters never wanted to read even when new, as when one servant remembers that she was given only "'goody books,' deportations from the parlor."[9] In Mrs. Sherwood's story of the same name, Susan Gray recounts that her mistress "left me her Bible and Prayer-book, and a black stuff gown and petticoat to wear as mourning for her"; *The Story of a*

Pocket Bible makes one wonder about the binding. Gladstone (hardly a typical reader or master) assembled a library of SPCK volumes for his servants, though he allowed them to borrow freely from his own collection as well (Windscheffel 114). In general, however, newspapers are to hand-me-down clothes as books are to servants' livery: even when new, they would not be used by masters. Where American slave narratives make literacy both symbol of, and means to, freedom, contemporaneous British tracts make receiving books a sign of servants' dependence. And where American slave narratives blame masters for forcibly withholding books, secular British novels blame mistresses for forcibly distributing them. It speaks at once to Harriet Martineau's eccentricity, conscientiousness, and obsession with unwanted paper that she worried about the effect on servants of her own discards, complaining of the "curious assortment of religious books and tracts sent to me by post . . . too bad in matter and spirit to be safe reading for my servants; so, instead of the waste-basket, they go into the fire" (*Autobiography* 111).

Books, in other words, fit uneasily within the accepted models for managing the tension between the need for servants and masters to handle the same objects, and the need to avoid direct contact between masters' and servants' bodies. Adams's disgust at the maid who touches her mistress's volumes transposes from manuscript to print the traditional insistence that letters be presented on a tray—more specifically, that the moment when the servant's hand puts the paper onto the tray be separated in time and space from the moment when the master's hand takes it off. Caroline's successor, too lazy to walk upstairs to take a letter, asks the messenger to throw it down to her in the area:

> Now Becky had been told not only not to look out of the area door when there was a ring or knock, but she had been especially charged never to let messengers throw down their notes or letters, a thing which has a very bad appearance, besides often dirtying them, so that they are unpleasant for a mistress to handle. She remembered all this, but being in a hurry to finish her toast before Martha's return, she determined to disobey orders.
>
> The note was from a lady recently married, and had a silver border, and a delicate white seal. The area was wet and dirty, for it was the month of November; still Becky thought she should avoid all risk of detection by holding out her apron to catch the note; so she bid the boy throw it down; but he made an awkward cast, and it fell in the dirt. Becky picked it up and began wiping it, but she only made matters worse, for her fingers and apron were all black from the fire she had been poking about to make her toast. (Adams, *Little Servant Maids* 214)

The fear of servants' dirtying books is common enough: a later Wesleyan tract for servants, for example, advises its readers, "if, through inadvertence, you have allowed a drop of ink to fall on linen, or a drop of tallow from the candle on the leaf of a book, the more speedily you do all you can to remedy the evil, the less the mischief which will ensue" (Smith 202). And another tract figures a mistress telling the narrator to "never handle a book carelessly . . . never take a dirty duster to books. Have always a clean one for them . . . Never come to books with hands dirty with work"; later, the maid illustrates her remarks about slatternly households by describing "books the worse for their handling" (Charlesworth, *The Old Looking-Glass* 89, 118).

Servants' unworthiness to own books has less to do with their inability to understand their contents, then, than with their refusal to handle them properly. The hero of *John Hartley* warns another servant boy that, "as mother used to say, by often wetting the corners, the paper gets rotten, and the edges tear off, and the [prayer]book looks shabby" (Adams, *John Hartley* 163). Yet Adams extends this problematic to middle-class children as well: in *The Useful Little Girl*, when an adult "sent the young ladies a book of prints to look at, desiring, at the same time, they would be careful of it, as it was a borrowed book," a girl grabs the book from her cousin, spilling broth on the book in the process: "in vain was the print carefully wiped and then dried, a large stain remained" (Adams, *The Useful Little Girl* 75). And in another of Adams's tales for and about middle-class children:

> A servant passing at the moment, Peter seized upon a single cup of coffee that remained on the tray, and was going to drink it, when Hugh snatched it from his hand, declaring that his brother had already had two cups, while he had had none. Peter attempted to regain his coffee, but Hugh jerking it away, the cup was overturned, and a great part of the contents spilled on Claude's beautiful prints. In a moment Hugh closed the book, and hastened to another part of the room, wholly indifferent to the fate of the engravings, and only anxious to escape detection. (Adams, *Boys at Home* 182)

In the remainder of the chapter, defacing the book leads to telling a lie: to lose respect for books is to lose the self-respect that underpins virtue.

In *Little Servant Maids*, however, bodily traces are linked more specifically with secrecy. The echoes of Bluebeard (with ink replacing blood as a mistress replaces a husband) reflect not only the SPCK's ambivalent stance toward chapbook fairy tales, but also the oddly sexual register in which servants' contact with written matter is represented. (Again, what makes the book sexual is not any pornographic content, but simply the fact that Becky hides it under her pillow.) Becky's successor is ordered to "look at the chairs—the backs all covered with dust, so that you might write 'slut' with your finger on every one of them" (Adams, *Little Servant Maids* 375). A few years later, *Punch* would picture a mistress who complains that she can write her name in the dust being answered, "Lor, mum, so you can! Now I never had no edgercation myself!"[10] The joke appears less funny if you remember how many American slave narratives represent their narrators learning to write by scratching their letters in the dirt.

Cleaning is the only form of inscription permitted to servants—a negative of the marks left on the white note by Becky's sooty fingers. And where ashes take the place of artist's charcoal, reciprocally ink becomes dirt as soon as a servant touches it:

> When Becky had done so, having nothing else to do, or rather, not choosing to do anything which she was not immediately bid, she pulled the inkstand towards her, and, leaning over the table, began to mark the first letters of her name on the window-seat. She could write but very little, and only on a slate, so that being unused to a pen, she filled it too full of ink, and made large blots on the clean white paint; these she from time to time removed by wiping her hand over them, and then cleaned her hand by rubbing it down her pincloth. (Adams, *Little Servant Maids* 219)

"Dirty books" are no metaphor. When Paget's antiquixotic novel *Lucretia* attacks "novels which some straitlaced folk would call the dirtiest in every sense, as being the loosest in their morals, and the most greasily thumbed by a discerning public," "dirt" joins "inflammatory" and "volume" as words whose literal sense Paget substitutes for a more common figurative meaning (Paget 104). The traditional fear that the book's content will inflame a servant's desires is crowded out by the fear that the book's binding will tempt her to disobey. Just as Greenwood imagines *Jack Sheppard* inspiring apprentices to steal goods that they can use to pay for a copy of *Jack Sheppard*, so what Caroline steals is the book itself, not (as in so many accounts where reading makes servants covetous) some ribbon or bonnet described within its covers. *Little Servant Maids* stops short, too, of imagining Becky peeking into her mistress's correspondence: no worse fault than the dirtying of paint and cloth. The servant's grimy hand becomes a leitmotif linking the turned-down corners of a book with the graffiti scratched on the wall: the white page has more in common with white cloth and white walls than with the whitey-brown paper that wraps food. In an inversion of the it-narratives that make handling books gently a predictor of kindness to women, children, animals, and slaves, here respect for books predicts respect for masters; the maid

A SOFT ANSWER.

"Susan, just look here! I can write my Name in the Dust on the top of this Table!"   "Lor, Mum, so you can! Now I never had no Edgercation myself!"

Figure 6.2. "A Soft Answer," *Punch*, 30 November 1895, 258.

who inscribes her response to the text is also the one who answers back to her mistress.

Taken cumulatively, these scenes conflate an emerging anxiety about the proper relation of reading to marking (do marginalia betray caring too much about the text or being too careless of the book?) with an older debate about whether servants' literacy should be passive or active. The internal discussion in the early years of the SPCK about whether poor children should be taught to write or only to read gives way here to a plot that positions writing where we would have expected reading to appear—marking the outside of the letter in place of prying into its contents. At some moments, that is, a servant's meddling with her mistress's books looks similar to eavesdropping on conversations; at others, it bears more resemblance to breaking a china plate.[11] Or maybe a better analogy would be taking a swig from your master's glass or trying on your mistress's bonnet behind her back. The image of a servant leaning on her broom figures alike in "One Thing at a Time," where the other hand holds a book, and in and "My mistress's bonnet," where the other hand is trying out a muff.[12] Sometimes, in contrast, what seems to be being stolen is not an object at all, but the user's time or attention. In engaging in what Walmart now dubs "time theft," maids like Hannah or Caroline go back on the tradition in which (as we saw in chapter 3) an industrious apprentice like Richardson could boast of reading on his own time, by the light of his own candles. And when the mistress in a different tract tells the servant, "Never spend your time for work in looking into books," she reminds us that the same activity at which middle-class children are praised as "working hard" takes servants the same age away from their legitimate labor (Charlesworth, *The Old Looking-Glass* 90).

The phenomena known as "Sunday reading" and "airplane reading" remind us that books take on different functions depending on when and where they're read. Any scholar knows that a bible means something different in a library (even a Gothic one like Sterling Memorial, whose vaulted reading rooms helped me begin this project) than in a church. And religious tracts taught their readers that a bible does something different when read by a maidservant perched on a ladder with a duster in her other hand, than it does when read by the same maidservant in her own room in a plainer binding.

In suggesting that *Little Servant Maids* represents books being handled in ways different from those that it invites itself, I don't mean to gloss over the inconsistencies in its own cues about who is expected to buy and who to read. The age of its implied readers remains in doubt, and, with it, the question of whether the book is made to be given in the school or in the home which is also a workplace. (Remember Yonge's reference to books from which "children and servants" will benefit.) As often as the text addresses servants, some narratorial asides suggest that this role lies in its readers' future:

> If servants generally would follow this good rule, how much of what is disagreeable and inconvenient would be spared to their masters and mistresses, and from how much temptation, and often misery, they would save themselves. Many children who read this tale will be too young or too inexperienced to feel the full importance of this observation, but let them bear it in mind and act upon it, and some day or other they will be aware of its value. (Adams, *Little Servant Maids* 49)

On the other hand, any such speculations about the implied reader (and by extension, the implied giver) are complicated by the confusion of age with social class diagnosed by Charlotte Yonge, echoing the bricklayer in *Bleak House*.

The *Leisure Hour*'s assumption that the influence of "parents" is interchangeable with that of "employers" nostalgically invokes the early modern model in which (as Lawrence Stone and Bruce Robbins have described) families treated their own children as sources of labor and hired servants as objects of paternalistic care: how-to manuals for servants, in fact, were known as "babies' books" (Robbins 150). In exhortations of "children should remember" and references to "the children who read her history," the referent sometimes seems to be a young girl already in service, as when the text acknowledges that, by definition, anyone whose hands it reaches is likely to differ from its untaught protagonist:

> Jessy had had few advantages of education; probably much fewer than any of the children who read her history. They most of them have been taught "to do their duty in the state to which it has pleased God to call them"; and such conduct as this untaught girl's would in them be inexcusable. They all know and feel that it would be very wrong to leave a place without giving notice; but let such children ask themselves if they have not been guilty of conduct which was as wrong in them as Jessy's was in her. (Adams, *Little Servant Maids* 86–87)

The reference to "the advantages of education" reveals the ambivalence at the heart of *Little Servant Maids*: the saintly old servant who is taught to read by her mistress's kind son is first impressed when a charity-schoolgirl entering her service arrives bearing a bible and prayer book that have been given to her as a school-leaving present, but then dismayed when the girl glosses her possession of these fine objects as a "right" rather than a gift (Adams, *Little Servant Maids* 185).

Here as in *The Story of a Pocket Bible*, the content of the text tugs in the opposite direction from its material form. Reading a bible or tract under the wrong circumstances can cancel out its message, as when *The*

Pawnbrokers' Gazette reports that "one of the very poorest and most wretched class of prostitutes" tries to pawn "a very large and handsome illustrated Book of Common Prayer"; suspecting it was "improperly come by," the pawnbroker refuses to advance money on the volume and reports the prostitute to the police, but she testifies that when "a perfect gentleman" in a "house of ill repute" discovered that he had no money on him, "he said that sooner than be dishonorable, he would give her the book in question, which was worth two guineas, and no doubt some pawnbroker would lend her 10s. on it."[13]

When tracts represent the reading of characters from the same social class as their own implied reader, it's often hedged with qualifications about the exact circumstances under which that reading is acceptable. Just as the decision of the servant in *The Story of a Pocket Bible* to buy her own cheaper bible cancels out her earlier sin of reading the book she's supposed to dust, so Hannah More makes clear that her working-class characters don't steal their reading time from working hours: "As her mother hated the sight of a book, Hester was forced to learn out of sight. It was no disobedience to do this, as long as she wasted no part of that time which it was her duty to spend in useful labour . . . Hester would not neglect the washing-tub or the spinning-wheel, even to get on with her catechism, but she thought it fair to think over her questions while she was washing and spinning" (More, *Tales* 104).[14]

More makes what we now call "multitasking" exemplify the house-wifely virtue of efficiency. Yet the dangers of that activity are illustrated in a later tract where "Jane has left off her business of sweeping and dusting the room, and is looking at the books which were placed on the table. Very likely it may be some good book which will do her no harm. Perhaps it has pictures in it of pleasant places far off; but whatever it may be, it would seem better if Jane finished her work first, for if her mistress finds the room not ready, she will see that the maid has been idling. Perhaps Jane has never heard the rhyme: 'One thing at a time, and that done well / Is a very good rule as many may tell' " ("One Thing at a Time" 194). The illustration shows a broom stuck through the back of a chair and a duster thrown carelessly down upon the seat, while Jane stands absorbed in the book she is holding.

For male servants as much as for middle-class girls, reading is stigmatized when it occasions oblivion to others, especially to superiors: we know that the eponymous pageboy in Adams's tract *John Hartley, and How He Got on in Life* has gone to the bad when "he often suffered [the cook] to call to him before he would stir, and then he would look up from a newspaper he might perhaps be reading, cast his eyes over it again before laying it leisurely down" (128). Conversely, reading is praised when it involves interpersonal exchanges, whether in the form of giving books

Figure 6.3. "One Thing at a Time," *Making the Best of It*, n.d.
Little Wide-Awake: an Anthology of Victorian Children's Books and Periodicals
in the Collection of Anne and Fernand G. Renier (Cleveland: World Pub. Co.,
1967), 194.

to others or reading aloud to them, as when the same tract describes John
burning his fellow servant Alice's licentious books and offering instead to
read his sister's school-leaving prize aloud to the other servants:

> The tale was a very interesting one, well and powerfully written. It was
> the history of the work, trials, and temptations of a young woman in

service . . . The time allotted for reading only allowed of half the tale being gone through, and Alice, contrary to her expectation, became so deeply interested in it that she begged John to go on with it after family prayers were over. He, however, said that . . . it was against rules to sit up after their master and mistress had dismissed them for the night. She then begged him to let her have the book to finish to herself upstairs in her room.

"No, Alice," said the cook; "even you, I think, would not be guilty of such misconduct. Call to mind the order mistress gave you on first coming, 'never to burn a light in your room longer than was necessary for getting to bed.'"

"Very well," said Alice, "I give in; I must try and be like good Jane in John's book, and obey orders."

. . . The story was finished by John, but it was not once hearing it that satisfied Alice. Again and again she begged him to read to her parts or the whole. (118–20)

Good books crowd out bad, just as the fire into which the latter are cast to burn doubles the candle not used to read the former by. Here as in conduct books and it-narratives, the lesson learned from hearing the good book read aloud is precisely not to read (or, at least, not to read to oneself)—to be able to bear interruption and suspense. But unlike in those middle-class books, what's set up for imitation is not only a certain rhythm of reading, but also a certain modality of acquiring: what makes the book good is not just its contents, but the fact of its having been given as a school prize. More specifically, given to John's sister: in a boy's acceptance of a feminized book, the logic of Molly Hughes's brother scornfully handing his reward books down to her is reversed.

The immorality of reading at the wrong place and time is limited neither to servants nor even to working-class characters. Hannah's crime of reading books when she should be dusting them is anticipated by an 1850 didactic fiction in which a middle-class girl's absorption in a book causes a servant to catch her forgetting her own light housework:

Once or twice she became so interested in the books which she found in the book-case, that the servant came in to set the table for breakfast before she had finished. She found Sally with one knee resting on the book-case, the dusting-cloth on the floor, and she so deeply absorbed in a tale as to have forgotten every thing else. Her father came into the room, walked quietly towards her, and laying his hand on her shoulder, asked what she was reading. She replied, "I have commenced a story which so much interests me, that I am sorry I shall have to stop." Her father asked her if she would not enjoy it more after her work was finished, advised her to defer reading it till evening, and let him partake with her of the enjoyment. The anticipation of reading this to her beloved parent was enough for Sally. The book was soon replaced, and the more commonplace business of life reverted to. (*The Useful and the Beautiful* 60–61)

A knee on the bookcase instead of a duster on the books, a dust cloth on the floor instead of a body kneeling: the topsy-turvy logic of the scene is corrected only once the book is put back into the proper place (a shelf), the proper time (the evening), and the proper operation (reading aloud).

When the fear that a book will distract women from their domestic tasks extends beyond servants, it comes to reflect a tension between the self and the social, as much as between work and leisure. Think back to the middle-class women whose novel handling distracts them from cleaning their houses or acknowledging their husbands. A different Jane's daydreams of "pleasant places far off" are prompted by her idling with books that John Reed reminds her belong to someone else: "You have no business to take our books; you are a dependent, mama says . . . I'll teach you to rummage in my book-shelves" (C. Brontë, *Jane Eyre* 17). In fact, the same pastiche of *Jane Eyre* that asked just how a persecuted dependent would lay her hands on the books needed to furnish her imagination also noticed that language enough to repeat it for a third time, making a daughter of her employer's family say of the governess: "she has not the run of the house, to go about it as she likes; she has no business in the library" (Wood 100).

No less than the middle-class secular fiction that we saw in chapters 1 and 2, religious tracts equate picking up a book with asserting a self. The difference is that *Little Servant Maids* makes the narrator, rather than an unsympathetic character like John Reed, the source for the comment that Caroline "had no business to open these books." Once "having business to take" or "no business to open" books (note that neither text speaks of whether the servant has "business to read" them) becomes a synecdoche for membership in the middle class, the opening scene of *Jane Eyre* comes to look like less like a psychological meditation on readerly interiority than like a prefiguration of the narrower social questions that critics like Elizabeth Rigby responded to: the application of John's claim that dependents have no right to pick up books to subcategories like "orphan," "servant," "governess," and "village schoolmistress."[15] Yet whether that individualism is attacked (in the conversion narrative) or endorsed (in the bildungsroman), both genres pit reader against family. The only difference is that as tracts continue to define "family" as an economic unit joining masters with servants, novels place that in tension with the modern sense of a nuclear household where gender and age replace class as sources of difference.

SOCIOLOGICAL REVOLUTIONS

If we took a long historical view, we might expect that as one model of family replaced the other, domestic service would cease to be the locus of anxieties about the transitive powers of shared books. It's true that after 1850, the library arguably replaced the household as the site where middle-class ladies could funnel subsidized books into the hands of their inferiors—with the difference that library books involved turn taking across classes. And it's also true that fears that books might become a vector for dirt migrated from housekeeping manuals to library science guides. The dirty thumb of *Little Servant Maids* reappears half a century later in Marie Corelli's remark that

> the true lover of books will never want to peruse volumes that are *thumbed and soiled* by hundreds of other *hands*—he or she will manage to buy them and keep them *as friends* in the private household . . . A little saving on drugged beer and betting would enable the most ordinary mechanic to stock himself with a very decent library of his own. To borrow one's mental fare from Free Libraries is a dirty habit to begin with. It is rather like picking up eatables dropped by someone else in the road, and making one's dinner off *another's leavings*. One book, clean and fresh from the bookseller's counter, is worth half a dozen of the soiled and messy *knockabout volumes*, which many of our medical men assure us carry disease-germs in their too-frequently *fingered* pages. (9; my emphasis)

So far, so conventional: reading is represented as both competitor to, and enabler of, betting; both competitor to, and adjunct to, eating. What's more original is Corelli's assumption that relations among readers compete with relations to books instead of enabling them. Where tracts function as both cause and effect of interpersonal connections, Corelli uses the metaphorical "friend" that is a book to displace actual friendships among human beings: the library must link reader to book, not one reader to another. In the same way, kindness to the metaphorical children that are books justifies unkindness to very real children: in *The Enemies of Books*, the Emersonian dictum that "the surest way to preserve your books in health is to treat them as you would your own children" is followed by the remark that "a neighbour of mine some few years ago suffered severely from the propensity, apparently irresistible, in one of his daughters to tear his library books . . . A single 'whipping' effected a cure" (Blades 32, 131).[16] Beat your daughter and she'll stop beating up your books.

The library described by Corelli looks as socially promiscuous as the home described by Adams (servants literalized Corelli's metaphor every

Figure 6.4. Book Disinfecting Apparatus, 1890.
Thomas Greenwood, *Public Libraries: A History of the Movement and a Manual for the Organization and Management of Rate-Supported Libraries* (London: Simpkin, Marshall & Co., 1890), 495.

time they "made their dinner off of another's leavings"), the page as overpopulated as the reading room. Corelli is hardly alone in identifying books as vectors of contagion. The eighteenth-century policy of dipping in seawater the bibles on which shipmasters swore that they had disinfected the contents of their ship—the text guaranteeing purity, the book spreading contagion—gave way to the act of 1910 that forbade the kissing of bibles during the taking of oaths (Rickards and Twyman 122; Watson 485). Between those dates, the library reformer Frederick Greenwood campaigned on behalf of what he called the "book disinfecting apparatus":

> At Dundee [the] apparatus . . . consist[s] of a sort of closed cupboard made of ordinary tinplate, with a lid at the top, a wire shelf half-way up, and a little door at the foot . . . At Sheffield they tried a system of heating the books in an oven to the temperature of boiling water, and that at the same time they should be exposed to the vapour of carbolic acid. It is claimed that this plan does not injure the binding or cause the books to smell of carbolic acid for very long afterwards . . . The simplest and best arrangement which has yet been introduced is . . . a metal fumigator made from 16th wire gauge sheet iron, with angle iron door-supports and side-shelf rests. Its weight is 3 cwt 1qr. and the cost of it was £5 10s. Compound sulphorous acid is burned in a small lamp, and a very little suffices to disinfect the books . . . The

shelves should be perforated in order to allow of a free circulation of the fumes of the acid. (T. Greenwood 494–95; see also Roberts; Black, "The Library as Clinic")

This antivirus hardware did little to stem the fear that particular genres of library book were being read on the sickbed and then put back into circulation. "I always sympathize," says the narrator of a Rhoda Broughton novel, "with the woman, who, when she went to a circulating library, asked them to give her a dull novel, because she thought it was less likely to have been thumbed and read by convalescent scarlet fevers and mumps" (*A Beginner* 7). And Corelli's metaphor of book as food is compounded by a metonymic association of books soiled by food in Arnold Bennett's exhortation to

> go into the average good home of the crust, in the quietude of 'after-tea,' and you will see a youthful miss sitting over something by Charlotte M. Yonge or Charles Kingsley. And that something is repulsively foul, greasy, sticky, black. Remember that it reaches from thirty to a hundred such good homes every year. Can you wonder that it should carry deposits of jam, egg, butter, coffee, and personal dirt? You cannot. But you are entitled to wonder why the Municipal Sanitary Inspector does not inspect it and order it to be destroyed. (105)

As the traditional fear that "unhealthy" texts could "poison" their readers was literalized by the worry that book-objects could spread disease, older concerns about the communion of a reader with a text gave way to newer ones about readers' contact with one another. This fear could involve manuscript as easily as books: in Maria Edgeworth's *Patronage*, when the Duke receives Lord Oldborough's letter sealed with a wafer (which is against etiquette in a letter to a superior), he "flung the note immediately to his secretary, exclaiming: 'Open that, if you please, Sir—*I wonder how any man can have the impertinence to send me his spittle!*'" (83). But it becomes especially sharp in the case of mechanically reproduced multiples, addressed to no one and exposed to "from thirty to one hundred" thumbs.

This is not to say that books ceased to be *compared* to poison: the Obscene Publications Act of 1857 (20 & 21 Vict. c.83) was prompted by a trial for the sale of pornography that coincided with a debate in the House of Lords over a bill aiming to restrict the sale of poisons, and that prompted the chief justice to term pornography "a sale of poison more deadly than prussic acid, strychnine or arsenic." But once the medical threat posed by books replaces the moral threat posed by texts, circulation comes to look more dangerous: what connotes health and prosperity when it applies to blood or money seems, in the case of books, to cause disease or (as we'll see in the next chapter) reflect bankruptcy. Where books circulating within the household were accused of making gender trump class (the valet peeking into his master's newspaper, the maid swiping her mistress's novel), here cross-gender contamination posed a greater threat. One early twentieth-century librarian joked about a "male borrower (holding out to Lady Assistant the latest novelty in bookmarks)": "Please, Miss, I wish you would tell some of your lady readers not to leave their fringe-nets in the books. I found a hairpin in my last book, and a fringe-net in this one, and my wife is getting a bit suspicious" (Coutts 142). Transitively, the book that I touch after you've touched it blurs the boundary between my body and yours; the fin-de-siècle cult of the uncut page provides a physical barrier against this danger.

The bookmark that stands in for the reader's place-holding finger makes the book a conduit between successive bodies.[17] And the erotic charge of book borrowing points to the peculiar intimacy that shared handling could broker. In *The Children of the Abbey* (1796), for example, marginalia first establish and then commemorate a romantic relationship:

> When alone within it, she found fresh objects to remind her of Lord Mortimer, and consequently to augment her grief. Here lay the bookcase he had sent her. She opened it with trembling impatience; but scarcely a volume did she examine in which select passages were not marked, by his hand, for her particular perusal. Oh! what mementoes were those volumes of the happy hours she had passed at the cottage . . . The night waned away, and still she continued weeping over them. (Roche 34)

Half a century later, in *Shirley*, a student writing in a tutor's book or even a tutor marking the student's notebook makes the contact between hand and page stand in for the contact between one body and another:

> "I never could correct that composition," observed Shirley, as Moore concluded. "Your censor-pencil scored it with condemnatory lines, whose signification I strove vainly to fathom."
>
> She had taken a crayon from the tutor's desk, and was drawing little leaves, fragments of pillars, broken crosses, on the margin of the book.
>
> "French may be half-forgotten, but the habits of the French lesson are retained, I see," said Louis: "my books would now, as erst, be unsafe with you. My newly-bound St. Pierre would soon be like my Racine: Miss Keeldar, her mark—traced on every page."
>
> Shirley dropped her crayon as if it burned her fingers. (C. Brontë, *Shirley* 461)

Across the Atlantic, Margaret Fuller uses an annotated book to compensate for the absence of the beloved, writing that some guests were announced. She went into another room to receive them, and I took up her book . . . I opened where her mark lay, and read merely with the feeling of continuing our mutual existence by passing my eyes over the same page where hers had been. (35)

In figuring the book as a bridge, these scenes reverse Trollope's deployment of the book as a wedge. (Writing of "the pleasure of reading out of the same book with" the beloved, Collins asks, "When is your face so constantly close to hers as it is then?—when can your hair mingle with hers, your cheek touch hers, your eyes meet hers, so often as they can then?" (*Basil* 103). Public libraries extended such triangular intimacies to strangers. Marginalia, valued earlier in the century as proof that reading involved strenuous production rather than idle consumption, was embargoed by the new public libraries, which saw readers' hands as wandering, dirty, or even capable of spreading disease.

And yet, it would be too simple to claim a one-to-one swap between pre-1850 admonitions to reading servants and post-1850 caricatures of public library patrons. For even over the course of two centuries that saw equally drastic changes in the nature of domestic service and in modes of book distribution, novels representing master-servant relations continued to prompt the middle-class reader to worry about running into his own servants in their audience. A century before *Little Servant Maids*, the prefatory letters to *Shamela* include Parson Tickletext's recommendation, "Pray let your Servant-Maids read it over, or read it to them," and his correspondent's reply that he "stand[s] excused from delivering it, either into the hands of my Daughter, or my Servant-Maid." A century in the other direction, every juror is charged to imagine a copy of the second most famous English novel about master-servant sex passing from his hands to those of a dependent differentiated by class, age, and gender:

> You may think that one of the ways in which you can test this book, and test it from the most liberal outlook, is to ask yourselves the question, *when you have read it through*, would you approve of your young sons, young daughters—*because girls can read as well as boys*—reading this book. *Is it a book that you would have lying around in your own house? Is it a book that you would even wish your wife or your servants to read?* (*Lady Chatterley's Trial* 4; emphasis mine)

While the reference to "your wife" conveniently ignores the presence of three women on the jury, the reference to "your servants" (plural) makes the remark more aspirational than descriptive: perhaps jurors were flattered by the assumption that they were wealthy male householders.

Michael Warner has argued that in the United States, "black illiteracy was more than a negation of literacy for blacks; it was the condition of a positive character of written discourse for whites. By extension, print-ing constituted and distinguished a specifically white community; in this sense it was more than a neutral medium that whites simply managed to monopolize" (*The Letters of the Republic* 12). Here, too, readers' iden-tity was formed through differentiation from other readers; this chap-ter should perhaps be called "Masters' Reading," Not "Servants'." Dif-ference, but also self-recognition: for the normative reader (adult, male, middle-class) could think about reading only via a detour through other audiences. Readers discussed were distinguished from readers addressed: thus men were instructed on how to control their daughters' reading, or middle-class philanthropists informed about the reading habits of me-chanics, or jurors asked to speculate about the effects of a book on their maid. An analysis of working-class reading habits can even be couched in a language designed to remind its own readers of their class and gender, as when a Latin tag concludes an 1844 pamphlet polemicizing against postal reform: "let any man only consider the known correspondence of his own servants: *sufficit una domus*" (*Administration of the Post Office* 195).

One magazine responded to Martineau's *Manchester Strike* (1832) not by assessing its appeal to the review's own readers, but rather by imagin-ing those middle-class magazine-readers distributing it to their employ-ees: "If the masters knew their own interest, this little work would be circulated by tens of thousands among their labourers; and the philan-thropist who feels the deplorable state of society in Manchester, could not spend a year better than in devoting himself to the circulation of its ideas and pictures" ("Review of Illustrations of Political Economy"). Neither the reader of the book review nor the buyer of the book is imagined to be the end user of the *Illustrations*: their role is to distribute the text, not to read it. Even more strikingly, the legitimate reader is asked, not to introspect about his or her own response to the stories, but rather to speculate about someone else's—defined either in contradistinction to his (as when the impressionable servant stands directly opposite the dispas-sionate judge) or by analogy with it (as when the reviewer both predicts and ventriloquizes the taste of his own readers). Once choosing replaced reading as the expression of working-class selfhood, "placing [books] in the hands of those who are under his influence" replaced reading as the mark of upper-class virtue. And when both vie for the power to decide what the weaker party will read, it's hardly surprising that conflict should result.

When Yonge repeats the warning to mistresses that their books can fall into the hands of servants, she draws the opposite conclusion from this fact than Charlotte Adams or the Chatterley prosecution do: "As for servants, it really is needless to try to select books for them, considering the cheapness of novels, and their easy access to all we have in the house"

("Children's Literature: Part III" 451). More specifically, although I have found no evidence that she read *Little Servant Maids*, Yonge is careful to decouple the acknowledgment that servants can touch any book from the prediction that they will dirty every book:

> I believe it is a great mistake to have a special library of "books adapted for servants." There is nothing they so dislike, or that is so unlike themselves, as the model Thomases and Marias in books, except, perhaps, that literature in which little nursery-maids convert all the children, while the nurse drinks wine in the pantry, and hides her lady's jewels in their boxes. Remember that the servants *can*, if they choose, read any book of yours they like, and that many of them have been well educated. Tell them, therefore, freely what you think is pleasant reading and give them a turn of a book from your box, if it is suitable. They are no more likely to soil it than you are. (C. Yonge, *P's and Q's, or, the Question of Putting Upon* 199)

The warning against model Thomases and Marias fits neatly enough with Yonge's critique of "class-literature's" pigeonholing of readers. What's more surprisingly is that the author of didactic literature often seen herself as a conservative goody-goody rubs her middle-class readers' noses in the power of books to cut across social barriers—whether by dragging masters down to the level of servants (middle-class readers are now the ones accused of soiling their own books), by invoking principles of fairness ("turn" taking), or by reminding masters that a shared domestic space implies shared access to texts ("remember that the servants *can*, if they choose, read any book of yours they like"). This last could, of course, sound like a reactionary threat (servants can also, if they choose, spit into whatever food of yours they like), which the tacked-on "if it is suitable" can hardly defuse by invoking mistresses' superior judgment at the eleventh hour.

When Yonge remarks in another set of hints for district visitors that "whatever *wholesomely* interests our own households may well be sent into the club-room," working-class readers become a proxy through which aspersions are cast on the purity of her own middle-class readers' tastes (C. M. Yonge, *What Books to Lend and What to Give* 9). The implication that her readers' own interests may not bear scrutiny humbles their pride as thoroughly as (and more deliberately than) the unlucky tract-distributor who, by mistake, "handed the tract" 'To An Unfortunate female' . . . to a respectable lady" (Jones 174).[18] Brontë plays on the same problem of address when Jane Eyre, handed a "thin pamphlet sewn in a cover" containing "'An account of the awfully sudden death of Martha G——, a naughty child addicted to falsehood and deceit,'" offers to make the book over to Mrs. Reed's own daughter. *Vanity Fair* makes the vertical logic of tract distribution even more explicit when Lady Southdown backs down from her plan to thrust a tract on the wealthy Miss Crawley, deciding that Miss Crawley's downtrodden companion and servants can take her place as the unlucky recipients:

> "Emily, my love, get ready a packet of books for Miss Crawley. Put up 'A Voice from the Flames,' 'A Trumpet-warning to Jericho,' and the 'Fleshpots Broken; or, the Converted Cannibal.'"
> . . . By way of compromise, Lady Emily sent in a packet in the evening for [Miss Briggs], containing copies of the "Washerwoman," and other mild and favourite tracts for Miss B.'s own perusal; and a few for the servants' hall, viz.: "Crumbs from the Pantry," "The Frying Pan and the Fire," and "The Livery of Sin," of a much stronger kind. (Thackeray, *Vanity Fair* 335)

Lady Southdown, we've learned already, "launched packets of tracts among the cottagers and tenants, and would order Gaffer Jones to be converted, as she would order Goody Hicks to take a James's powder, without appeal, resistance, or benefit of clergy" (333). Like medicine, tracts hover between categories: understood by one party as a gift and the other (whether children or the poor) as a burden, they cure but also disgust. This isn't to say that the economically powerful figure couldn't also be emotionally vulnerable: anyone who's ever leafleted on a street corner—or taught an English class—knows how powerless he or she is to make anyone read anything.

To read about bad readers is comfortable; to be addressed as one of them, less so. In a surprising number of late nineteenth- and early twentieth-century essays, a moral failing (the soiling of books, for example) is first attributed to working-class others but then brought home to the conscience of the essay's own readers. In an essay titled "Cheap Literature" in the middle-class *Contemporary Review*, for example, Helen Bosanquet wonders whether it is quite fair to judge penny dreadfuls by their effect on readers: "We ourselves should resent having our recreative reading judged by a moral standard." Even Bosanquet's attack on cheap novels' "narrowing and morbid" obsession with marriage is qualified by an afterthought: "Of course, I am aware that the same criticism applied to the majority of novels placed in the hands of girls of the more educated classes." And after mocking the puffs that penny magazines supply for themselves, she adds, "But might we not say much the same of the readers who make up their Mudie lists with such touching confidence in the verdict of 'Athenaeum' and 'Spectator'?" (680). Another article, "Reading and Readers," in 1893 similarly declares that "human nature is pretty much the same wherever we go, and the record Mr Hileken of Bethnal Green and I can give would shame that of many a West-End knight and dame" (191). (G. F. Hileken was the librarian of the Bethnal Green Free Library.) The rueful tone drags the author and his readers down to the level of those they observe.

Even the addressee of a proto-sociological tract rarely maintained complete distance from the styles of reading under discussion. When Florence Bell set out to survey the reading habits of a manufacturing town in 1911, she ended up turning her gaze back on her own readers: "On finding what were the results of the inquiry made respecting reading among the workmen, a similar investigation was attempted among people who were better off, and the result of this inquiry among those whom we may call 'drawing-room readers' is curiously instructive" (250). In describing workers' reading to a middle-class audience, Bell appeals over and over to her readers' own experience for confirmation. Thus "There are a certain number of born readers among the workpeople in the town described, as there are, happily, in every layer of society . . . But . . . reading, perhaps, is not as prevalent in any class of society as we think" (203–4). Or, on working-class women's reluctance to take time away from daily tasks for reading: "Is there not some truth in this view? Even among the well-to-do this idea persists a great deal more than one would at the first blush admit" (237).

As novels like *Northanger Abbey* forced readers to recognize themselves in the romance-reading misses being satirized, so periodicals made the working-class people whose reading habits they investigated into a mirror for their own middle-class public. The common act of reading cut cross moral distinctions in one case, social distinctions in the other. And if reading books "made kitchen and drawing-room kin," so did distributing them. A Religious Tract Society pamphlet that praises the best tract-distributors for being "as clever as Fagin was at sliding their silent messengers into people's pockets without attracting notice" erases any distinction between missionaries and common thieves (N. Watts 13). The scene of Evangelical gentlemen "spending hours with the hawkers in order to learn the mysteries of their trade," too, emblematizes the danger that the books designed to shore up class distinctions ultimately blur them.

In fact, the fear of maids' peeking into their masters' bookshelves proves surprisingly reversible: if servants can be led astray by their masters' books, young ladies can also be corrupted by accepting forbidden books from a governess, as in Maria Edgeworth's "Mademoiselle Panache" (1801) or Sewell's *Laneton Parsonage* (1846–48). Just as Jeames's or Caroline's act of illicitly borrowing books cancels out the innocuousness of their content, so the badness of the romances described here is doubled by the sinfulness of the process by which they're acquired—whether lying to parents, hiding the volume, borrowing money to pay for it, or making oneself vulnerable to blackmail. Accepting a book from a servant thus forms the modern, print-culture equivalent of the older nightmare that your children might listen to ghost stories told by an old nurse (Steedman 50; Trumpener, *Bardic Nationalism* 211). From a servant, or even from other inferiors: in a plot that Mrs. Sherwood constructs to illustrate the dangers of preferring low and vulgar friends to one's own sisters, the offer of a book is what tempts the culprit to jump the fence separating her family's property from the neighbor's ("Intimate Friends" 234).

Conversely, writers signify proper reading not only by naming the book's title, but also by mentioning how it reached the servant's hands. A good maid in *The youth's magazine or evangelical miscellany* "was very soon lost in her book, Miss Thornton's present to her when she left home. She had read the Pilgrim's Progress through. Her father had an old copy, full of quaint pictures; but this was a large and handsome edition, beautifully illustrated" ("The Bunch of Keys" 331). In one model, good books trickle down; in the other, bad books percolate up. In both, their meaning lies as much in how they're transmitted as in what they represent. Where secular writers cast books as a refuge for the powerless, didactic literature represents access to books as dependent on social relationships, and domestic power struggles playing out in tugs-of-war over books.

I spoke earlier of "cementing or severing relationships . . . by giving and receiving books or by withholding and rejecting them." It may not have been clear there how little those pairs map onto each other: to withhold a book is not necessarily to sever a relationship any more than to give a book is to cement one. On the contrary, anonymous market transactions in a bookshop often grant freer access to books than does the more intimate censorship carried out by personal acquaintances like parents or masters. Yet the intimacies that sometimes blocked access to books also enabled it. Associated in theory with individual liberation, in practice reading not only reflected but created social dependence. This is not to say that those models were mutually exclusive. One 1824 critic of female bible distributors asked, "is it correct, particularly in females, to go from house to house, and sow seeds of discord between husbands and wives, parents and children, masters and servants?" (Martin 50). The echo of Diderot (himself echoing Matthew 10:35) makes clear that a tract can corrupt as much as any Richardson novel—as long as its giver stands outside the household.

Autodidacts' memoirs celebrated literacy as a means for individuals to rise above their social class of origin; middle-class bildungsromans focalized the delicate child narrator withdrawing from an insensitive family; cartoons satirized men hiding from wives or fellow commuters behind a newspaper. Protestant tracts like *Claude the Colporteur* made Catholic countries the site of bookish heroism, turning readers from sedentary goody-goodies into manly risk-takers: like the adults who serve as blocking figures in the bildungsroman, the Catholic authorities make reading narratable.[19] Yet in Britain—to state the obvious—reading was learned in schools run by the church, then by the state. And even outside of formal education, the provision of something to read depended on complex networks of gift, sale, loan, exchange, even theft. What was true for middle-class adults held even more strongly for those who depended on others' voices to spell out text, others' judgment to select titles, others' money to provide copies, and others' permission to read them.[20]

THE TRACT AND *THE MOONSTONE*

If social commentary invited its middle-class readers to displace their self-criticism onto working-class counterparts, light literature established its own entertainment value by contradictions to tracts. More specifically, by mocking tract-distributors for harassing a captive audience, secular novels and magazines congratulated themselves on being freely chosen and paid for. Although the RTS entitled its internal history *The Romance of Tract Distribution*—as if the wanderings of books, like foundlings, led providentially toward a happy ending—the secular press more often chose a satiric mode. In 1843, for example, Thackeray reprinted in *Punch* a note from the *Times*:

> 'The Agents of the Tract Societies have lately had recourse to a new method of introducing their tracts into Cadiz. The tracts were put into glass bottles, *securely corked*; and, taking advantage of the tide flowing into the harbour, they were committed to the waves, on whose surface they floated towards the town, where the inhabitants eagerly took them up on their arriving on the shore. The bottles were then uncorked, and the tracts they contain are *supposed to have been* read with much interest.'

Thus far verbatim; but Thackeray goes on to reply in the persona of the "Regent of Spain . . . and of the Regent's Park," calling the "manoeuvres of the Dissenting-Tract Smuggler (*Tractistero dissentero contrabandistero*)" worse than any Jesuit arts.

> Let *Punch*, let Lord Aberdeen, let Great Britain at large, put itself in the position of the poor mariner of Cadiz, and then answer. Tired with the day's labour, thirsty as the seaman naturally is, he lies perchance, and watches at eve the tide of ocean swelling into the bay. What does he see cresting the wave that rolls towards him? A bottle. Regard-

" SHERRY, PERHAPS,"

Figure 6.5. "Singular Letter from the Regent of Spain," *Punch*, 1 December 1843, 268.

less of the wet, he rushes eagerly towards the advancing flask. 'Sherry, perhaps,' is his first thought (for 'tis the wine of his country). 'Rum, I hope,' he adds, while with beating heart and wringing pantaloons, he puts his bottle-screw into the cork. But, ah! Englishmen! fancy his agonising feelings on withdrawing from the flask a Spanish translation of 'The Cowboy of Kennington Common,' or 'The Little Blind Dustman of Pentonville.' ([W. Thackeray], "Singular Letter from the Regent of Spain")

The joke doesn't require much exaggeration. "The Cowboy of Kennington Common" is hardly more alliterative than real RTS titles like *Lucy the Light-Bearer* (1871), *Claude the Colporteur* (1880), or *The Book-stall boy of Batherton* (1872). And RTS reports did in fact associate temperance with thirst for the Word:

Eight men went in a punt to convey goods to a ship in the river; while they were on board, the captain asked them whether they would have a glass of rum each, or a book; they all chose a book, and said they did not drink rum. Tracts were given, which were gratefully received. (Jones 599)

The tract society that devised the message in a bottle was punning on the same confusion of word with drink that allowed the magazine to publish miscellanies like *A Bowl of Punch* (1848)—or Lady Southdown, in *Vanity Fair*, to "pitilessly dose [Miss Crawley's household] with her tracts and her medicine" (Thackeray, *Vanity Fair* 351). Even secular reformers could borrow the analogy: although Rowland Hill acknowledged grudgingly that "the wish to correspond with their friends may not be so strong, or so general, as the desire for fermented liquors," he persisted in predicting the effects on the revenue of the proposed penny post system by extrapolating from taxes on beer and wine (93).

Books competed with drink both for poor people's time (did one go to the pub or stay at home with a book?) and their money (as when a temperance tract exults that "the variety of useful books they can afford to purchase, at once proclaims how much has been gained since the public house has been abandoned" (Best 13). The same Evangelicals who countered requests for bread with offers of a tract worried that free bibles would be sold or pawned to pay for gin. (Lower resale value gave tracts an advantage over bibles in this respect.) In 1860, one bible-distributor responded to those who "shut the door in my face, and said, 'We want bread instead of Bibles'" by "invit[ing] her poor neighbors to devote the pence too often squandered in gin to the purchase of the precious Word of God, which is the 'water of life' to all who drink its refreshing streams" (Ranyard 69). As a result, philanthropists remained unsure whether to link books with food metaphorically or metonymically: should tracts be piggybacked onto gifts of food and blankets, or substituted for such tangible handouts? In Trollope's *Castle Richmond*, the recurring pun that renders as "mail" the Irish characters' pronunciation of the "meal" provided at soup kitchens during the Famine places debates about the public distribution of food in parallel with debates about the public distribution of letters. Yet if free food and free bibles alike disturbed the workings of the market, how to make sense of the need for advertising circulars and other forms of free print to stimulate consumer appetites?

Whether middle-class triple-deckers or penny serials, novels characteristically represent tracts as filler—a poor substitute either for a different genre of book (usually a novel), or for nonbookish objects like food, drink, or paper money. Thus when a canting character in Rymer's radical *The White Slave* leaves a piece of paper on the heroine's father's snowy grave: "Is it a bank note?" "No, this is a religious tract." Later, an Evangelical lady who has just denied the starving heroine food exclaims: "You would have been supplied with tracts, provided you kept them clean, and returned them" (34).

In invoking *Crusoe*, the Religious Tract Society's message in a bottle takes us back to a paper scarcity and a readerly solitude as far as possible from the modern overabundance of both literature and fellow readers. If the RTS's hawkers and chapmen mimicked the way that romances *were* distributed, its publicity stunt in Cadiz mimicked the way that romances *represent* the distribution of texts. Reciprocally, and obsessively, the Victorian novel represented the distribution of tracts. The two genres provided one another with mirror images: when characters in tracts read novels or characters in novels reject tracts, the stakes are intellectual-historical (Evangelical fiction opposite the epic of a world without God) as much as formal (both torn between narrative and didactic modes). But also affective: if the novel was the competitor that the tract was trying to beat at its own game, the novel returned the compliment by making the dullness of tracts a foil to its own pleasures.[21] And finally, economic: both widely distributed, neither much respected, tract and novel achieved their common ubiquity via opposite routes—one anonymously or even surreptitiously bought and rented, the other forced on readers through face-to-face relationships.

Like Trollope's or Thackeray's embedded newspapers and novels, embedded tracts perform an antiquixotic function: everywhere present in the hands of characters, but nowhere read.[22] Mrs. Jellyby and Mrs. Pardiggle emblematize the imbalance between supply and demand. In fact, when the latter "pull[s] out a good book as if it were a constable's staff," it's not enough for every character in *Bleak House* to refuse to read the tract. Adding insult to injury, characters are dragged in from another novel to second that refusal: "Mr. Jarndyce said he doubted if Robinson Crusoe could have read [the book], though he had had no other on his desolate island" (133). The mounting absurdity of those counterfactuals—if a nineteenth-century genre had already existed, if a copy had made its way to Crusoe's island—drives home the contrast between two models of transmission: one, viral, that organically diffuses an old romance across economic lines; another, top-down, in which middle-class adults bribe or bully others into owning (if not reading) demographically tailored fictions.[23]

Bleak House is one of three mid-Victorian novels that couple Robinson Crusoe's name with an imaginary tract. *Hard Times* contrasts Coketown workers' eagerness to "[take] De Foe to their bosoms" with their reluctance—echoing the bricklayer's in *Bleak House*—to read "leaden little books [written] for them, showing how the good grown-up baby invariably got to the Savings-bank, and the bad grown-up baby invariably got transported" (Dickens, *Hard Times* 65). Not until *Crusoe* reappears for a third time in *The Moonstone* (1868), however, does a nineteenth-century novel stage an all-out battle between an earlier romance and the modern tract. As every reader will remember, the First Period of Collins's novel takes *Robinson Crusoe* as its intertext, the Second a tract-distributor as its narrator. Armed with a "little library of works . . . (say a dozen only)," its narrator, Miss Clack, throws tracts in windows, slips them under sofa cushions, stuffs them into flower boxes, sneaks them into the pockets of dressing gowns, and thrusts them into the hands of a cabdriver. "If I had presented a pistol at his head, this abandoned wretch could hardly have exhibited greater consternation" (W. Collins, *The Moonstone* 214): Miss Clack at once echoes Thackeray's mock-epic description of Lady Southdown "before she bore down personally upon any individual whom she proposed to subjugate, fir[ing] in a quantity of tracts upon the menaced party (as a charge of the French was always preceded by a furious cannonade)," and anticipates the logic of the journalist who remarks in 1899 that "the gratulations of a successful tract-writer may be only on a par with the boasts of a soldier who knows he has killed 150 of the enemy because he has fired 150 rounds," and adds: "it may be doubted if as high a proportion of tracts as of Mauser bullets reach their billets" (Thackeray, *Vanity Fair* 335; Ogden). Even Rowland Hill refers to advertisements for books as "random shots"—a metaphor that survives today in our term "mailshot" but was literalized more directly in the 1944 German experiment firing postcards in rifle grenades over France (89; Rickards and Twyman 11). The tone in which Miss Clack's struggles are chronicled is hardly more comic than a biographer's description of Canon Christopher: "On one occasion the younger members of his family thought they had prevented such untimely and embarrassing activity—as they conceived it—at a Governor's garden party in the Isle of Man. All his pockets had been quietly emptied, and those in the secret set out with a light heart. They were presently disillusioned by the sight of Christopher offering tracts. More alert than they had supposed, he had taken the precaution of stuffing a few packets into his elastic-sided boots" (Reynolds and Thomas 233).

The tract, then, is to the Second Period what *Robinson Crusoe* was to the First. As if to recapitulate the historical shift from chapbook to tract, a romance treated as if it were a bible gives way to a tract that feebly imitates the conventions of fiction. The bibliomancy that allows Betteredge to "[wear] out six stout copies" through the "wholesome application of a bit of ROBINSON CRUSOE" paves the way for Miss Clack's equally instrumental and equally discontinuous reading habits. Like Betteredge, she understands reading as both aleatory and combinatory, prefacing her quotations with the claim to have "chanced on the following passage" and wresting out of context snippets that "proved to be quite *providentially* applicable to" the recipient.

Miss Clack herself sees Bruff's reading not as identical to hers, however, but its competitor: when she describes Bruff as "equally capable of reading a novel and of tearing up a tract," she imagines the two genres fighting over a limited pool of readers' attention (W. Collins, *The Moonstone* 9, 193, 215). William Wilberforce's famous declaration that "I would rather go render up my account at the last day, carrying up with me The Shepherd of Salisbury Plain than bearing the load of [the Waverley novels], full as they are of genius," reminds us that one virtue of tracts is negative: they fill the time that might otherwise be spent reading novels (Rosman 188).

Novel and tract fight to occupy the same space twice over. In the editorial frame, Franklin Blake rejects the "copious Extracts from precious publications in her possession" that Miss Clack tries to insert into her "Narrative," forbidding her to slip quotations from tracts into the novel that we are reading: "I am not permitted to improve—I am condemned to narrate." The converse is more literal: within the "Narrative" Miss Clack tries to slip novels inside tracts. "On the library table I noticed two of the 'amusing books' which the infidel doctor had recommended. I instantly covered them from sight with two of my own precious publications" (W. Collins, *The Moonstone* 237, 24).

The image of a tract covering up a novel figures the uncomfortable proximity of the two genres. On the one hand, tracts represent novels competing with devotional texts. The narrator of the *Adventures of a Bible* reports, "I am sorry to say that Jane, was by no means pleased with the good woman's gift; but pouted, and said she would much rather, had it been left to her own choice, have a story book to amuse her" (12). And *Gilbert Guestling, or, the Story of a Hymn-Book* (published by the Wesleyan Conference Office in 1881) features a character moved to buy the hymnbook after reading Brontë's parody of a Dissenting Sunday school in *Shirley* (Yeames). Here as so often, More's Cheap Repository Tracts set the tone for their successors. In "Mr Bragwell and his Two Daughters," the title characters

> spent the morning in bed, the noon in dressing, and the evening at the Spinnet, and the night in reading Novels . . . Jack, the plow-boy, on whom they had now put a livery jacket, was employed half his time in trotting backwards and forwards with the most wretched trash the little neighborhood book-shop could furnish. The choice was often left to Jack, who could not read, but who had general orders to bring *all the new things, and a great many of them.*

The quantitative language in which More's tract describes novels anticipates the quantitative language in which Frances Trollope's novel evokes tracts. Just as reductively, "Mr Bragwell" lumps novels together with other, humbler, consumer goods: the title character complains that "our Jack the Plowboy, spends half his time in going to a shop in our Market town, where they let out books to read with marble covers. And they sell paper with all manner of colours on the edges, and gim-cracks, and powder-puffs, and wash-balls." One daughter rejects a good match because "he scorned to talk that palavering stuff which she had been used to in the marble covered books I told you of" (More, *Works* 133, 31, 41). Not for nothing did Bishop Porteus speak of More's "spiritual quixotism" (quoted in Pedersen 87): the tract's representation of novel renting and novel buying at once repeats the eighteenth-century novel's embedding of romance reading and prefigures the nineteenth-century novel's satire on tract distributing.

On the other hand, middle-class novels are themselves haunted by the fear of turning into tracts—as when Herbert Pocket refuses even to call Pip by a name that "sounds like a moral boy out of the spelling-book, who was so lazy that he fell into a pond, or so fat that he couldn't see out of his eyes, or so determined to go a-bird's-nesting that he got himself eaten by bears who lived handy in the neighborhood" (Dickens, *Great Expectations* 165). The tracts represented in *The Moonstone* share the pseudo-orality of the frame narrative ("There was a precious publication to meet her eye, or to meet her hand, and to say with silent eloquence, in either case, 'Come, try me! try me!'") as well as its strategy of mixing the domestic with the sensational (W. Collins, *The Moonstone* 223). Titles like "Satan in the Hair Brush" and "THE SERPENT AT HOME" parody the sensation novel's own obsession with "those most mysterious of mysteries, the mysteries which are at our own doors" (H. James, "Miss Braddon" 593).

Like tracts, too, sensation novels not only represent the home, but pervade it. Compare Trollope's assertion (also in 1868) that novels are in "our library, our drawing-rooms, our bed-rooms, our kitchens,—and in our nurseries," with Miss Clack's campaign to "judiciously distribute [tracts] in the various rooms [Rachel] would be likely to occupy" (Anthony Trollope, "On English Prose Fiction" 108; W. Collins, *The Moonstone* 224). Earlier, Miss Clack attempted to ensure that a tract lies in wait "in every room that [Lady Verinder] enters," an ambition that requires her to imagine—like the author of some domestic novel—what activities her victim will engage in over the course of an ordinary day:

> I slipped [one] under the sofa cushions, half in, and half out, close by her handkerchief, and her smelling-bottle. Every time her hand searched for either of these, it would touch the book; and, sooner or later (who knows?) the book might touch HER . . . In the drawing-room I found more cheering opportunities of emptying my bag. I disposed of another in the back drawing-room, under some unfinished embroidery, which I knew to be of Lady Verinder's working . . . I put a book near the matches on one side, and a book under the box of chocolate drops on the other . . . But one apartment was still unexplored—the bath-room, which opened out of the bed-room. I peeped in; and the holy

inner voice that never deceives, whispered to me, "You have met her, Drusilla, everywhere else; meet her at the bath, and the work is done." (223–25)

Miss Clack's strategy recalls the *Quarterly*'s observation that novels "come to us when we are off our guard, and gain their place and position before we have fairly begun to discuss them" ("Recent Novels"). Both reach their readers by stealth, in the places of daily life rather than in the dedicated space of classroom, church, or library. Yet no classroom teacher who spends her life "assigning" works of literature—including, in my case, *The Moonstone*—can escape recognizing herself in Miss Clack.

The tract doesn't just resemble *The Moonstone*; in the opposite direction, it resembles the moonstone. The Evangelical spinster who sneaks into bedrooms to deposit books mirrors the Evangelical suitor who sneaks into bedrooms to remove jewels. Miss Clack's reverse shoplifting anticipates the Religious Tract Society's invocation of Fagin—or another RTS anecdote in which an atheist flees a coach to escape the exhortations of an Evangelical passenger, but "as he got down, the pocket of his coat gaped open, and, unperceived, his fellow-traveller quickly and quietly dropped into it Mr. Blackwood's little book, *Eternal Life*" (N. Watts 7).

What would it mean to model book distribution after pickpocketing? The parallel between Miss Clack and Mr. Ablewhite makes tracts look as anomalous as the eponymous diamond: it's in people's interest to acquire books and jewels, but to get rid of tracts and moonstones.[24] Just as Sir John's bequest of the diamond turns out to be motivated by malice masquerading as kindness, so Miss Clack's and Lady Verinder's attempts to force tracts onto one another reflect aggression misrepresented as generosity. The niece disguises printed matter as personal letters, copying out extracts by hand to elude Lady Verinder's suspicions; the aunt, in turn, sends back the tracts disguised as a "legacy." Once tract and moonstone exemplify a circulation driven by push rather than pull, then Christian charity becomes hard to distinguish from Hindu curse. And when printed matter is imagined as burden rather than boon, it becomes more appropriate for mistresses to give their young maids than for a poor relation to thrust upon her aunt.

The two objects that *The Moonstone* places in parallel—printed matter so worthless, and exotic talismans so potent, that both need to be unloaded onto someone else—would eventually converge in the magic book that forms the subject of M. R. James's 1911 short story "Casting the Runes." To the extent that such a convoluted story can be summarized, its plot involves a scholar who gets into another researcher's "bad books" by returning his conference proposal and finds himself on the receiving end of papers containing an ancient curse. Those papers pursue both indoors

and out: a billboard on the tram, "a handful of leaflets such as are distributed to passers-by by agents of enterprising firms . . . thrust" into his
hand on the street, "a calendar, such as tradesmen often send," received
by post, a paper that he never dropped returned to him in the British
Library reading room. As the rarest of manuscripts becomes interchangeable with mass-produced junk mail, the claustrophobic intimacy of the
library (where a scholar can penetrate the secrets of the anonymous reviewer at the next desk) becomes indistinguishable from the home whose
door is now breached by a letter slot that any tradesman can penetrate,
and the street where every passerby is exposed to billboards and leaflets
(M. R. James 131, 42).[25] As the plot that begins with an unwanted scholarly paper ends with unwanted paper of a more literal and dangerous
kind, free print becomes reenchanted—as if a tract could bear a Hindu
curse, or Martineau's junk mail could not only inconvenience but poison
its recipient.

The more tracts look like the moonstone, however, the less they look
like *The Moonstone*. Even if, as I suggested a moment ago, the two genres
share their osmotic distribution mode and their formal strategy of mixing the domestic with the wild and the oral with the printed, they stand
opposite each other in one crucial way. Those who love a novel want to
hold onto it (even, like Betteredge, wearing out multiple copies); those
who love a tract want to give it away. Miss Clack unloads books as surreptitiously as *The Moonstone*'s own readers acquired them: according
to one observer, "even the porters and boys were interested in the story
and read the new numbers in sly corners, with packs on their backs" (W.
Collins, *The Moonstone* xxxviii). The force-feeding of tracts throws into
relief the hunger for sensation fiction.

Such hostility to religious tracts is hardly peculiar to the novel: Thackeray could vent his hatred just as easily through the medium of the sketch.
But the representation of tract reading, or more precisely of the refusal to
read tracts, took on a particular charge in this context, because the novel
slotted the tract into the position traditionally occupied by a different
embedded genre, the romance. Once fictional characters stop reading romances and start handling tracts, two things change. First, the novel can
no longer define its own seriousness in contradistinction to the frivolity
of the genre embedded within it. Second, embedded romance and embedded tract imply diametrically opposed models of how books reach their
users. Framed by mass-distributed novels, embedded romances harked
back to an earlier age of serendipitous individual finds: new novels rented
out by circulating libraries represent dusty manuscripts whose price can
be measured only in eyestrain. What distinguished frame narrative from
framed genres in *Don Quixote* was that the latter never represented buying and selling; but what distinguishes embedded genres in the gothic

is that they're never bought or sold themselves. As we'll see in the next chapter, found manuscripts obey an economics of scarcity: the candle gutters before the roll is finished, or the page is torn midsentence. In contrast, the tracts represented in mid-Victorian novels embody an economics of surplus: everywhere distributed and nowhere read, they substitute painful duties for guilty pleasures. That shift helps explain the novel's particular animus toward nonmarket forms of distribution, whether tract societies, missionary baskets, or bazaars—all competitors to the secular, commercial networks on which its own dissemination depends.

Force-Reading

The standard biography traces the genesis of Miss Clack to two Evangelical tract-distributors who targeted Collins during a vacation at Broadstairs with his mistress and illegitimate daughter (N. P. Davis 226). When juxtaposed with the other forms of information overload cataloged in the previous chapter, however, *The Moonstone*'s representation of tracts becomes harder to reduce to a localized expression of anticlericalism. In that context, its logic looks not just biographical but also generic (aimed at the Evangelical press's ambivalent use of fiction) and economic (aimed at a particular mode of distribution).

Particular but also pervasive: for what Collins take tracts to exemplify isn't just religiosity, but *any* social arrangement that licenses one person to force a printed text onto the consciousness of others. His later novel *Poor Miss Finch* instances "that particular form of human persecution which is called reading aloud" in a character (unsurprisingly, a clergyman) who "inflicted his accomplishment on his family circle at every available opportunity":

> Of what we suffered on these occasions, I shall say nothing. Let it be enough to mention that the rector thoroughly enjoyed the pleasure of hearing his own magnificent voice. There was no escaping Mr. Finch when the rage for "reading" seized on him. Now on one pretense, and now on another, he descended on us unfortunate women, book in hand; seated us at one end of the room; placed himself at the other; opened his dreadful mouth; and fired words at us, like shots at a target, by the hour together. Read what he might, he made such a noise and such a fuss over it. (W. Collins, *Poor Miss Finch* 132)

The military metaphor echoes *The Moonstone*'s comparison of Miss Clack's tract to a highwayman's pistol.

The listener doesn't always lack power, of course, any more than the recipient of a free book does. The servants whom eighteenth-century gentry reduced to human audiobooks (standing behind their master's chair with a book during meals or hairdressing) found themselves forced, a generation later, to listen to their masters read aloud godly books. Daughters, too, were enlisted alternately to read at their fathers' sickbeds and to listen to their fathers reading family prayers. In fact, by broadening the audience being read to from "family" in the sense of biological kin to "family" in the sense of "household," family prayers placed wives and children closer to servants than to the master of the house. Or, in some cases, to its mistress: when Trollope lumps Lady Amelia and Mr. Gazebee together with the footman—all three put to sleep by Lady Rosina's reading—the speaker is clearly the winner; yet his novels draw little distinction between what it feels like to be read aloud to and what it feels like to have someone else read silently in your presence (Anthony Trollope, *The Small House at Allington* 386).

In classroom or church, the audience is captive, the reader its captor. Florence Nightingale extended that logic to the drawing room in a series of rhetorical questions:

> Don't you feel, when you are being read to, as if a pailful of water were being poured down your throat, which, but that it comes up again just as it goes down, would suffocate you? Very few swallow it at all; fewer still digest it. Many people like to read aloud; but how many can bear being read to without going to sleep? Yet *everybody* can't be reading aloud . . . What is it to be "read aloud to"? The most miserable exercise of the human intellect. Or rather, is it any exercise at all? It is like lying on one's back, with one's hands tied and having liquid poured down one's throat. (Nightingale, *Cassandra* 714, 213)[26]

Compare a young girl's diary entry from 1875: "Such a detestable evening, Grobee actually made a fuss because I was writing my journal while the reading was going on. He not only makes us read a dry, stupid old book (which we do willingly to please him) but forbids those who are not reading to do anything which prevents them from listening. It is really too bad. Why should we be forced to listen when we don't want to? . . . No one may move or speak a word the whole evening—it is most dull" (Troubridge 125).

Such accounts of domesticity draw on a tradition of comic writing that stretches back to Jane Collier's *Art of Ingeniously Tormenting* (1753):

> The same indifference, also, you may put on, if [your husband] should be a man who loves reading . . . If, for instance, he desires you to hear one of Shakespeare's plays, you may give him perpetual interruptions, by sometimes going out of the room, sometimes ringing the bell to give orders for what cannot be wanted till the next day; at other times tak-

ing notice (if your children are in the room), that Molly's cap is awry, or that Jackey looks pale . . . If you have needle-work in your hands, you may be so busy in cutting out, and measuring one part with another, that it will plainly appear to your husband, that you mind not one word he reads. If all this teazes him enough to make him call on you for your attention, you may say, that indeed you have other things to mind besides poetry. (89)

To listen is to submit to another's power; like Trollope's characters going to sleep, the wife resorts to passive resistance. Yet even Nightingale— whose leading questions can themselves be experienced as coercive— acknowledges that listening can also be experienced as parasitism: "women like something to tickle their ears and save them the trouble of thinking, while they have needlework in their hands. They like to be spared the *ennui* of doing nothing, without the labour of doing something" (Nightingale, *Cassandra* 74).

On the one hand, Elizabeth Hamilton's 1800 *Memoirs of Modern Philosophers* contrasts the selfless Harriet's reading aloud at a sickbed with Bridgetina Botherim's selfish insistence on reading silently to herself even in a roomful of other people (Hamilton and Grogan 176, 72, 84).[27] On the other, *Poor Miss Finch* proportions female listeners' "suffering" to the male reader's "pleasure." In its indifference to content ("read what he might"), *Poor Miss Finch* echoes Nightingale's sense that the text itself is almost incidental to the power relations created between reader and listeners.

On the other hand, reading aloud can constitute torture just as easily as being read aloud to. In Broughton's *Second Thoughts*, a father whose daughter arrives at his sickbed prepared to read him "something . . . a little serious" amuses himself by reminding her of her duty to read whatever she is asked (in this case, a French gossip column) seated "under his direction, exactly opposite him, where he can nicely observe every shade of expression, every nervous blush and mortified contraction that passes over her face" (Broughton, *Second Thoughts* 41–43, 54). The violation of a girl's innocence is even more explicitly sadistic in the late-Victorian pornographic novel *"Frank" and I*, where the narrator's first reaction to unmasking "Frank" as a cross-dressed girl is to take out a Mudie's subscription. More specifically, after Frank is discovered to be Frances, the narrator discovers as well that the most pleasant sequel to flogging her is to lie in an armchair listening to her read aloud . . . *The Moonstone* (*"Frank" and I* 43). As a middle-aged male landowner orders his teenaged female dependent to read a trashy novel, the power dynamics of the tract distributing represented *within* Collins's novel take on a secular and sexual charge.[28]

The analogy with music may give a subtler sense of the contradictory connotations of reading aloud: girls' piano playing, too, could be coded either as self-importance (Mary Bennet being told that "You have delighted us long enough. Let the other young ladies have time to exhibit") or self-abnegation (Sophia Western playing "Bobbing Joan," rather than her beloved Handel, to her drunken father) (Austen, *Pride and Prejudice* 69). By 1934, when *A Handful of Dust* ends with Tony Last being kept alive on condition that he read Dickens aloud to his captor (like David to Steerforth?), the inventive Scheherazade is reduced to a mechanical drudge. "He had always rather enjoyed reading aloud and in the first years of his marriage had shared several books in this way with Brenda, until one day, in a moment of frankness, she remarked that it was torture to her" (Waugh 292).[29] The "torture" of being read aloud to prefigures the torture of reading aloud: ontogeny recapitulates phylogeny as the paterfamilias's voice gives way to something more like the prison workforce that today recites scripts on customer service lines. When the contrast is between who speaks and who listens, the reader-aloud looks like the powerful figure, inflicting sounds on his public as autocratically as Miss Clack inflicts objects on hers; but when the question is rather who decides what book he wants to hear and who executes those orders, then the reader-aloud becomes as powerless as the child or woman or bricklayer whose reading material is chosen by their betters.

Such harangues don't need a book as prop; when Sydney Smith "dreamt I was chained to a rock and being talked to death by Harriet Martineau and Macaulay," neither nightmare figure held a book (Kemble 65). The two contrasting models of reading aloud do, however, exemplify the larger question of whether reading (in whatever medium) expresses selfhood or dependence. If tract distribution enforces the latter model (while apotropaically representing the former) and novel reading enacts the former (while systematically satirizing the latter), the postal debates with which this chapter began can be understood as a face-off between these two models, with reformers championing correspondence as the medium of individual enlightenment and conservatives exposing it as a generator of mass markets.

In this context, a third possible explanation for Collins's obsession with tract-distributors emerges: Miss Clack transposes the effects of Rowland Hill's new postal regime into a religious register. I've suggested that hand-distributed tracts and junk mail delivered by post together emblematized the shift from paper scarcity to paper surfeit; more specifically, both prompted concerns about the relation of print to manuscript, broadcast to personal communication, and written matter to nontextual objects. The difference between tracts and bulk mail lay less in their content (after all, both drew on a rhetoric of persuasion) than in their distribution methods: in one case, a reformed postal system driven by modern liberal principles that glorified virtuality, impersonality, and mediation; in the other, a nostalgic (though equally reformist) network of face-to-face relationships.

This is not to say that the two enterprises were antagonistic. In one direction, tract distribution provided a model for postal distribution: when Cobden offered to subsidize a cheap edition of Hill's pamphlet *Post Office Reform*, the model of the Cheap Repository Tracts must have come to mind (Robinson 273). Rowland Hill the postal reformer was named after the Rowland Hill who helped to found the Religious Tract Society; reciprocally, the post provided a channel for tracts. One chronicler of the RTS rejoices that "in these days of penny-postage blessedness, in almost every letter we write we can proclaim the glad tidings of mercy, by inserting an eight-paged tract" (Jones 258). A prospectus for a system of "Evangelization by book-post" explains that

> the post, we thought, is a neutral agent, often spreading evil, but capable of spreading good, why then not make use of it to scatter the seed of truth in all directions? Doubtless many papers so distributed find their way to the wastepaper basket, without obtaining the favor of a reading; but, after all, there must be a real benefit in so expensive a distribution, since so many tradespeople persevere in it. (Dardier 318)

Commercial circulars provide a generic model for religious tracts as well as a template for their distribution. Henry Mayhew had more on his side than shock value when he lumped the religious tract together with the commercial advertisement, describing "sham indecent" packets stuffed with "a religious tract, or a slop-tailor's puff" (*London Labour and the London Poor* 1:241). Both genres developed an ambivalent relation to the market: where the religious tract tried to beat commercial chapbooks at their own game, the advertising circular sold goods by giving printed paper away for free.

While tract societies disguised gift as sale and mass-duplicated books as personal "messengers," the same advertisers who took over the Christian language of the "free gift" presented broadcast communications as point-to-point.[30] As one early twentieth-century observer pointed out, a society overloaded with paper creates the temptation to misrepresent the mass-distributed as personal communications and vice versa: "One of the most curious recent developments in the graphic arts is the effort of advertisers to make printed matter look like typescript, while the authors of books that are not in sufficient demand to warrant publication are seeking a typescript that will look like print" (Binkley 526).

The distinction between personal letters and bulk mail becomes especially fraught at a moment where print—once charged with brokering the meeting of an individual reader's mind with an author's—comes instead to insert each reader into a mass public. In fact, the statistics about the number of new readers mirrored equally oft-cited calculations about the number of new books. The mechanical production of newspapers spread metonymically to their readers: in 1883 Northcliffe attributes his market to the fact that "the Board Schools are turning out hundreds of thousands of boys and girls annually who are anxious to read" (quoted in R. Williams, *The Long Revolution* 196). The excess of print implied an excess of fellow readers.

I am a reader, you are a public, they are a market. The commercial nature of the book is best displaced onto others—preferably of a different rank and gender. Even when middle-class men were forced into awareness of who had handled a book before them, the question of whose hands it would fall into later was easier to avoid. Part of what distinguished working-class and female readers, in contrast, was that they never had the luxury of ignoring what would happen to a book after it left their hands. Those after-uses form the subject of my final chapter.

CHAPTER 7

The Book as Waste: Henry Mayhew and the
Fall of Paper Recycling

OVER THE COURSE of this study, the physicality of print has swum into focus at extremes: in the case of books that are especially expensive (bibliophilic collectibles) or especially worthless (free advertising circulars, subsidized religious tracts); among subcultures especially bookish (antiquarians, collectors) or especially bookless (the illiterate, the heathen); with books considered especially sacred and timeless (the Bible) or especially profane and ephemeral (newspapers, almanacs, novels); at the beginning of their life (manufacture) or the end (pulping).

Taking this last case seriously would mean replacing the traditional question "what is a text" by "when is a text?" In an age of taxed paper, reading constituted only one point in a cycle: beginning its life as rags no longer worth wearing, the page dwindled back into paper once its content was no longer worth reading. In the wood-pulp era, only bibliographers continued to notice the prehistory and afterlife of legible objects. But even bibliographers need limits. Can the study of printed books stop short of forestry and the secondhand clothing market? Does the interpretation of graffiti require expertise in brickmaking? In the opposite direction, how far downstream should reception theorists venture: to the archive, the depository, the Dumpster?

By the turn of the twentieth century, one modern scholar reminds us, "most of the paper used in Britain was not used for printing. Of what was printed, most was thrown away" (McKitterick, *The Cambridge History of the Book in Britain* 63). Modernity can be defined not just by what's produced, but by what's discarded, and when. Until the second half of the nineteenth century, most reading matter was made from old rags, and much of it went on to be recycled in turn. Newspapers were handed down a chain of households as their contents staled; letters were torn to light a pipe; broadsheets pieced out dress patterns or lined pie plates or wiped shit. In their passage from hand to hand and use to use, loose sheets accreted scars and bruises as telling as any it-narrator.

To think about the transmission of paper is to think about the contingent, the unmentionable, and the mundane. Much of the vernacular Chinese fiction now extant has reached our hands by accident, unearthed from tombs or stumbled across in the backing material for other books (Zeitlin 254).[1] In Europe, the same "secondary causes" that destroy books have preserved pages. Some of those unintended consequences are bibliographic (binder's waste), others more vulgarly domestic (trunk linings) (Adams and Barker 31). In Han China, "paper was probably used for wrapping before it was used for writing" (Needham et al. 122); in Britain as late as 1911, the *Encyclopedia Britannica* continued to define paper as "the substance commonly used for writing upon, or for wrapping things in." Where pages can make readers forget hunger, as in so many accounts of prison reading, paper serves as a reminder of the need to ingest and excrete. Or at least, *did* serve as such a reminder, because this chapter will suggest that two phenomena that usually get explained in terms of the rise of electronic media in the late twentieth century—the dematerialization of the text and the disembodiment of the reader—in fact have more to do with two much earlier developments. One is legal: the 1861 repeal of the taxes previously imposed on all paper except that used for printing bibles. The other is technological: the rise first of wood-pulp paper (in the late nineteenth century) and then (in the twentieth) of plastics.

More specifically, I'll suggest that working-class users' relation to books in general, as well as middle-class users' relation to books that they wish to abject, proceeds not just by omission but also by commission: in the first case, a negative quality (illiteracy) goes together with a positive one (expertise in the use value and exchange value of different weights, textures, and colors of paper); in the second, not only refusing to read a book, but also determining to smear it by association with some culinary use, to besmirch it more literally with one's own excrement, or to call its author's gentlemanliness into question by passing his books along to one's servants. Conversely, to exempt a page from base uses is to exalt its textual contents: a sixth-century Chinese scholar-official noted that "paper on which there are quotations or commentaries from the *Five Classics* or the names of sages, I dare not use for toilet purposes" (Needham et al. 123).[2]

Nothing more embarrassing than a book past its read-by date. What's no longer worth reading, however, still needs to be handled, if only to make room for more. And at the opposite level of abstraction, what's no longer fit to read may remain good to think with—if only because the moment where the book's shelf life diverges from the text's calls into question the relation of words to things. Such thinking, I want to suggest, stands at the heart of Henry Mayhew's *London Labour and the London Poor*, the loose, baggy ethnography of the urban underclass that swelled out of a messy series of media: eighty-two articles serialized in the *Morning Chronicle* (October 1849–50) provided the raw material for freestanding weekly numbers published between December 1850 and February 1852, which in turn were expanded, revised, and collected in volume form in 1861–62.

It may seem perverse to conclude a project that opened with realist novels like Trollope's and Dickens's by turning to a text that only its harshest critics have described as fiction. But precisely because *London Labour* decouples the realist mode from fictionality, Mayhew provides the starkest possible test case for the question with which this book opened: why have book historians drawn a disproportionate number of their case studies from the realist novel? Like the novel, *London Labour* foregrounds practice over theory; Mayhew's study of the informal economy, too, aspires to dignify the everyday. But he borrows novelistic techniques—including the pileup of metonymic detail that has led so many readers to praise Mayhew as "Dickensian"—only to turn them against the novel. *London Labour* repudiates fictionality, that is, by means of characteristically realist tics—including, most crucially for my purposes, an obsessive staging of moments where verbal structures pull apart from material objects.

After-Uses

This chapter asks why Mayhew's "cyclopaedia of the industry, the want, and the vice of the great Metropolis" so encyclopedically catalogs the uses to which used paper can be turned.[3] In the city that *London Labour* describes, books and newspapers never stand still: they're sold to fishmongers, to middlemen who distribute them to fishmongers, to a flypaper manufacturer, and even to a member of what the narrator terms the "sham indecent trade," whose sealed packets, advertised as "not [to] be admitted into families," turn out to be stuffed with "a lot of missionary tracts and old newspapers that [the vendor] got dirt cheap at a 'waste' shop."[4] Wrapping, wadding, padding, lining: why so much attention to paper, so little to the page? The answer that a middle-class contemporary might have given, for reasons that we'll see in the following section, is that Mayhew's informants aren't literate. *London Labour* contradicts that hypothesis not only in its content—which represents them reading, among other things, previous installments of *London Labour*—but in its form, which attributes the reading of texts and the recycling of papers to the same agents.[5] As a result, those actions are distinguished not by who performs them (gentlemen read, street urchins recycle) or even by which genres or media invite them (bibles are for rereading, newspapers for recycling), but rather by successive moments in the life cycle of the *same* piece of printed matter. And even that minimal distinction gets broken down as Mayhew replaces the conventional time line in which wrapping follows reading by a counternarrative in which food packaging gets resurrected as legible text.

What, then, if we were to replace "illiteracy" with a more positive term? "Orality" might be an obvious candidate, given how central speech is not just to the informal economy described in *London Labour*, but to its own use of the interview. Books signify bankruptcy, if only because the wastepaper described is as likely to consist of financial and legal manuscripts as of printed books. One waste-seller offers Mayhew

> railway prospectuses, with plans to some of them, nice engravings; and the same with other joint-stock companies . . . Old account-books of every kind. A good many years ago, I had some that must have belonged to a West End perfumer, there was such French items for Lady This, or the Honorable Captain that. I remember there was an Hon. Capt. G., and almost at every second page was "100 toothpicks, 3*s.* 6*d.*" I think it was 3*s.* 6*d.*; in arranging this sort of waste one now and then gives a glance to it. (2:114)

Contrast that memento mori with the busy street vendor who tells Mayhew that "it's all headwork with us" (2:24)—by which he means that he operates, as we now say, "off the books." The more lifeless the papers that Mayhew describes, the more vivid the voices that he quotes: padded packets provide a foil to street cries, wastepaper to oral interviews. In this analysis, the uncanny immediacy that *London Labour* produces would be thrown into relief against the backdrop of dead media.[6]

A third possible explanation is more reductive: you could say that paper gets resold in Mayhew's London simply because everything does. Readers today—at least in the developed world—will be even more struck than middle-class Victorians were by the ubiquity of reuse in *London Labour*. Unable to afford either to buy new things or to discard old ones, his informants lack the luxury of ignoring the past and future of their possessions. When we're told of one man that "his dress could not so well be called mean as hard worn, with the unmistakable look of much of the attire of his class, that it was not made for the wearer" (2:65), the narrator's self-correction encapsulates a characteristic stutter step. Mayhew begins by gesturing toward the possibility of pricing an object on the basis of the amount of labor or quality of materials that went into its original manufacture; but he goes on to upstage that logic by a competing explanation that bases price on the position that the object occupies in a chain of successive owners and uses. As Suzanne Raitt points out, the term "waste" invokes the history of an object—or at least the process of which it's a by-product—as near synonyms like "rubbish" and "litter" do not (73). More specifically, narrative structures Mayhew's providential model of the market in which, far from exhausting or depreciating objects, circulation animates them and invests them with fresh value.

In this analysis, "after-uses" would provide Mayhew with a lever to topple books from their taxonomic pedestal—to simultaneously defamiliarize and deflate printed matter by lumping it among a long list of humbler commodities that lose value as they pass from hand to hand. (Such a list would encompass every consumer good that turns up in the pages of *London Labour*, with the possible exception of women's stays [2:29]: in Mayhew, no such thing as what Igor Kopytoff calls "terminal commodities," those that make only one journey from production to consumption [75; compare M. Thompson 9].) The question of whether books stand outside the market becomes a test case of whether anything at all stands outside the market. Anything, or even anyone: the value of used paper provides a measure for the value of the human beings who sell it: one "dealer in 'waste' (paper)" "had been brought up as a compositor, but late hours and glaring gas-lights in the printing-office affected his eyes, he told me; and . . . a half-blind compositor was about of as little value, he thought, as a 'horse with a wooden leg' " (1:289).

From wastepaper to blind person to lame horse: as so often in Mayhew, what sounds like a hyperbolic analogy will turn out in retrospect to have been a perfectly serious cross-reference, because the price of the goods and services derived from horse carcasses will form the subject of a tabular breakdown in the next volume (2:9). Books, persons, and horses are all expected to form privileged categories, exempt from base uses. (Or at least, this is the case in England; the French, Mayhew reminds us, are less sentimental about their horses. And even in England, as we've seen, books not only resemble horses but are made from their dead bodies.)[7] Yet each exemption dwindles with age: what can't be eaten turns out to be not horses, but young horses; what can't be pulped turns out to be new books, not books *tout court*. The same aging that most poignantly humanizes books (in it-narrative, at least) also reduces them most ruthlessly to objects.

If books begin their life as an exception, then, they end up exemplifying the rule. Mayhew's uncertainty about whether to place books in parallel with, or contradistinction to, other kinds of object prefigures the tension between internalist accounts of print culture—those that emphasize what sets books apart from other commodities—and those that draw on nonbibliographical analogies, situating debates about copyright in the context of pharmaceutical patents or reducing the history of authorship to a subset of the history of branding. Even structurally, the tension between exceptionality and typicality can be measured by the placement of paper within *London Labour*. The volume devoted to the resale trade opens—before moving briskly along to secondhand backgammon boards and used mattress ticking—with a set piece describing "a body of men in London who occupy themselves entirely in collecting waste paper."

It is no matter what kind; a small prayer-book, a once perfumed and welcome love-note, lawyers' or tailors' bills, acts of Parliament, and double sheets of the Times, form portions of the waste dealers' stock . . . [M]odern poems or pamphlets and old romances (perfect or imperfect), Shakespeare, Molière, Bibles, music, histories, stories, magazines, tracts to convert the heathen or to prove how easily and how immensely our national and individual wealth might be enhanced, the prospectuses of a thousand companies, each certain to prove a mine of wealth, schemes to pay off the national debt, or recommendations to wipe it off [a bad pun?], auctioneers' catalogues and long-kept letters, children's copybooks and last-century ledgers, printed effusions which have progressed no further than the unfolded sheets, uncut works and books mouldy with age—all these things are found in the insatiate bag of the waste collector. (2:9)

The breathlessness of Mayhew's syntax levels generic "kinds" into undifferentiated "matter." But the reduction of absorbing reading to absorbent paper doesn't necessarily imply a social fall, because Mayhew conflates wrappers with readers. "Some of the costermongers who were able to read," the narrator tells us, "or loved to listen to reading[,] purchased their literature in a very commercial spirit, frequently buying the periodical which is the largest in size, because when they've 'got the reading out of it,' as they say, 'it's worth a halfpenny for the barrow' " (1:26).

What might it mean to "get the reading out of" a newspaper? Before wood pulp, esparto grass, and other raw materials began to replace rags in the decades following the publication of *London Labour*, the obvious answer would have been that paper outlasts its contents.[8] That quality isn't unique to paper, of course. Mayhew notes elsewhere that brass doorplates fetch a fraction of their original value when they fall into the hands of scrap metal dealers after their owners' death: there, too, the value of the material medium paradoxically hastens the erasure of the text (2:10). And recent media theorists have emphasized the gap between the life expectancy of hardware (slow to break down in landfills) and software (replaced on an increasingly short cycle) (Sterne 25; Parks).[9] There's something especially poignant, however, about measuring the ephemerality of a text against the adaptability of a book, because that contrast inverts the traditional hope that words will survive the surfaces on which they're inscribed—whether brass, stone, or marble and gilded monuments, much less paper. Within that tradition, pages transcend the temporal limits that paper embodies. If texts broker a transhistorical meeting of minds, the book—"Poor earthly casket of immortal verse" (Wordsworth 160)—can never break free of a particular location in space and time. Mayhew turns that contrast on its head, pitting the durability of paper against the disposability of words.

Where stone connotes immortality, paper is associated with death: in imperial China, paper "spirit-money" was placed in tombs five centuries before it was used among the living (D. Hunter 207). In pairing the afterlife of paper with the death of text, Mayhew inverts a paradox most succinctly stated by Drummond of Hawthornden in 1711, the year when paper taxes began the climb that would end the year before the volume publication of *London Labour*. "Books have that strange Quality," he observed, "that being of the frailest and tenderest of Matter, they outlast Brass, Iron and Marble" (Drummond et al. 222; West 199). But do they? On the one hand, the high resale value of stone makes it likelier to be erased (Green and Stallybrass 21); on the other, ephemerality makes paper cheap, and cheapness allows each text to be produced in multiple copies that will go on to be stored in multiple locations and transmitted through multiple channels—channels that include the pastry-cook as often as the librarian.

The intertwining of preservation with destruction is not unique to paper: Sumerian clay tablets, for example, often survived thanks to accidents like fire, which baked them into greater durability. Movable type sharpened that paradox, however, for redundancy forms at once an effect of, and a counterweight to, the fragility of printed paper. Texts survive in proportion as books decay. Bibliographers all know that preservation varies inversely with use: not only do small books circulate most widely and reach libraries most rarely, but the genres that get the most handling (such as almanacs) are the hardest for modern scholars to lay their hands on. This paradox is hardly unique to books (printed labels that survive in collections of ephemera usually do so because they were surplus to requirements and sent direct from the manufacturer) or even to print: Jane Kamensky has pointed out that a similar logic governs the survival of clothing: ceremonial articles and those made for infants are preserved in disproportionate numbers (personal conversation). As Thomas Adams and Nicolas Barker note, "popularity tends to operate positively on the text and negatively on the book" (Adams and Barker 33): the more readings a work undergoes, the more reprintings are likely to be produced, but the less likely any given copy is to survive. To which a second paradox could be added: the better preserved a book, the less evidence we tend to have about the ways in which it's been used: the survival of the book itself, in other words, stands in direct conflict with survival of traces of its readers.

William Sherman reminds us that "printed images and texts were part of a dynamic ecology of use and reuse, leading to transformation and destruction as well as to preservation": the more heavily a book was used, the more vulnerable it was to decay (6). The proof is Jan Stock's argument that for popular prints produced in Renaissance Antwerp, "the larger the quantity of impressions made and the larger the number of people they reached, the smaller was the chance of the material being preserved." That so few have survived can be explained, Stock argues, "not through indifference, but because they served their purpose": "cherished to destruction," they may have been "attached to the inside of a travelling case . . . cut up and pasted to a wall as decoration . . . recycled as lining for a book cover or as the flyleaf in a register of archives . . . In the Antwerp city archives, a sixteenth-century woodcut advertising the work of a pin-maker was fortuitously preserved, because it spent centuries serving as wrapping paper for the wax seal attached to a document" (quoted in Sherman 6). In the opposite direction, however, books' "serviceability may partly explain why so few of them have survived. We find printed paper being used . . . for stuffing cracks in chimneys and windows, . . . and as spills to light a fire or a pipe" (Cressy 93).

In Mayhew's own time, the battle lines were clearly drawn between those who saw circulation as life-giving and those who understood it to be life-shortening—or, more specifically, who credited or blamed the library either with putting books into circulation, or with withdrawing them from use. (From it-narratives' characterization of pristine books as "banished" or "imprisoned," it's only one step to imagine them being buried alive.) Because bibliographic debates draw on the language of saving and exchange, it comes as no surprise that economists were among the most prominent participants in early debates about free-library funding: Jevons, for example, acknowledged that free-library books wore out faster than those in private collections, but insisted "how infinitely better it is that they should perish in the full accomplishment of their mission, instead of falling a prey to the butter-man, the waste-dealer, the entomological book-worm, or the other enemies of books which Mr. Blades has so well described and anathematized" (Jevons 30). The image of books as heroes "perish[ing] in the full accomplishment of their mission" makes a valiant attempt to translate ephemerality into a martial language.

Such mock-heroics can do little, however, to counter the less highflown imagery of a 1871 article on circulating libraries which observes that if unpopular books have a short life span (because no one reprints them), popular books die for the opposite reason: too "torn, dirtied, and 'read-to-death'" to serve even as food wrapping, "they will not carry butter; nor will they 'to the trunk-makers.' Their purpose is—for manure!" (Friswell 522). Friswell's two possible destinations for a no longer readable book—butter wrapping and manure—remind us that paper ended its life as an aid to ingestion and excretion. And as it accumulated traces of its successive users' hands, or intestines, the book reneged on its traditional mission of transcending the body.

The secondhand bookstall provides an emblem of mortality: asked by Mayhew why he has no new works, one bookstall-keeper answers that "they haven't become cheap enough yet for the streets, but that they would come to it in time . . . Yes, indeed, you all come to such as me at last. Why, last night I heard a song all about the stateliest buildings coming at last to the ivy, and I thought, as I listened, it was the same with authors. The best that the best can do is the book-stall's food at last. And no harm, for he's in the best of company, with Shakespeare, and all the great people" (1:296).

The metaphor of paper's life cycle makes books look less like the timeless works that they contain than like the mortal beings who read them. This may explain in turn why paper seems to change hands only at the latter's death. "I think we should all be tired" (as Mayhew says) if I repeated a full catalog of examples from the scrap heap of *London Labour*, but here are two:

> "I've had Bibles—the backs are taken off in the waste trade, or it wouldn't be fair weight—testaments, Prayer-books, Companions to the Altar, and Sermons and religious works . . . perhaps a godly old man dies, and those that follow him care nothing for hymn-books, and so they come to such as me, for they're so cheap now they're not to be sold second-hand, I fancy." (2:113)

And a second informant:

> "[I've had] Manuscripts, but only if they're rather old; well, 20 or 30 years or so; I call that old. Letters on every possible subject, but not, in my experience, any very modern ones. An old man dies, you see, and his papers are sold off, letters and all; that's the way; get rid of all the old rubbish, as soon as the old boy's pointing his toes to the sky. What's old letters worth, when the writers are dead and buried?"

Yet what sounds like a rhetorical question turns out to have an answer: "Why, perhaps 1 1/2*d.* a pound, and it's a rattling big letter that will weigh half-an-ounce. O, it's a rattling queer trade, but there's many worse" (2:114). Ghoulishness isn't enough to explain the ease with which Mayhew's informants slip from "old manuscripts" to "old men," from dead letters to dead bodies, from valuing letters at one-thirty-second of a penny to terming their authors "an old boy pointing his toes to the sky."

It's unclear why paper should survive at the expense of its owners, especially in a class where the selling and pawning of goods during an illness or to pay for a funeral leaves little scope for inheritance. (Unlike rich bibliophiles, the poor don't bequeath their books to libraries.) Yet

Mayhew is hardly alone in this association of ideas. Even those Victorian genres that most strenuously celebrated the diffusion of knowledge, whether Useful or Christian, associated circulation with death—both of the book and of its owners.

It-narratives hinge on misfortunes: the deathbed scene during which a bible is willed to a spendthrift, the bankruptcy after which it ends up in a secondhand bookstall, the drinking binge during which it's exchanged for a single dram. This generic convention doesn't contradict what we know about the transmission of nineteenth-century books: closer to home, many of the holdings of our own research libraries are available to scholars today only thanks to the death of their donors. (Once one of many objects whose usefulness outlived their first owner, books are now one of only a few consumer goods that even—or especially—rich persons and institutions buy secondhand.) And that association of ideas shouldn't be entirely foreign to book historians, for whom probate inventories have provided a crucial source.

Even the prospectus for a book fumigator concludes (like Jevons) that contagion is a lesser evil than isolation: "When we have foregone or disinfected our books, . . . killed our cats, declined to use a cab, adopted respirators, and sternly refused to shake hands with our friends, and adopted all the other precautions which are recommended against our microscopical bugbears, will it be worth while to go on living? Happily for our peace of mind the majority of us prefer to take our risk" (T. Greenwood 494). This paean to circulation is confirmed by the history of the volume in which it appears: most of the copies now in circulation on secondhand bookselling Web sites come from library collections. (My own was advertised as "a few marginal blind stamps, library stamp on back of title, *but* a good copy of the entirely re-written third edition of this seminal work" [my emphasis]).

The fear of contagion took on special poignancy, then, when expressed by the most energetic public-library advocate of his generation. Like Gladstone, Greenwood compares private collections with graves: "Oh! ye gentlemen of England, who are said to 'live at home at ease,' if this not worth remembering? . . . By placing your treasures upon [the empty shelves of public libraries] a new lease of life would be given to books you have prized, and it is impossible to say where, along the lines of the generations to follow, they would cease to gratify and enlighten" (T. Greenwood 4). "Books, like coins," he adds, "are only performing their right function when they are in circulation . . . Hoarded up," they remain "only so much paper and leather," but "in a Public Library, books begin to really live" (5). Yet what breathes life into books seems to kill their owners.

If Greenwood's "book disinfecting apparatus" assumes that the sharing of books can cause death and disease, reciprocally owners' death or ruin can set books into motion. One measure of the value that readers attach to books (as opposed to, say, newspapers) is that they enter into circulation only against their owners' will. Where object narratives involving coins take circulation as the precondition of financial health, those narrated by books counterbalance their structural need for change (for something to narrate) with their thematic assumption that books can change hands only in consequence of misfortune. Where financial health is marked by the shucking off of old clothes and furniture to replace them with new fashions, new books enter the house without old books ever exiting (a problem that anyone reading these words must have experienced firsthand).

Where the circulation of blood connotes life and the circulation of coins prosperity, the circulation of books instead announces death or bankruptcy. Both converge in *The Mill on the Floss*, where the books sold at auction after Mr. Tulliver's financial ruin are replaced by the copy of the *Imitation of Christ* in which "some hand, now for ever quiet, had made at certain passages strong pen and ink marks, long since browned by time" (G. Eliot, *The Mill on the Floss* 382).[10] Yet it's never explained why, on acquiring a batch of secondhand books, Maggie (not to mention the narrator) immediately assumes that their owners must be dead—even though Maggie herself still lives after having seen her own books sold at auction.

For Eliot's characters, buying a book conjures up the death of its previous readers as surely as learning to read conjures up one's own mortality: "Better spend an extra hundred or two on your son's education than leave it to him in your will" (G. Eliot, *The Mill on the Floss* 23). Marian Evans's first published poem, "Knowing that Shortly I Must Put off this Tabernacle," appeared in the *Christian Observer* for January 1840. In it, the speaker's death is announced via a farewell to the books that he or she owns:

> Books, that have been to me as chests of gold,
> Which, miser-like I secretly have told,
> And for you love, health, friendship, peace have sold—
> Farewell!
>
> (Deakin 36)

As one article in an antiquarian magazine declares, "It is a peculiarity of bookplates that they bring 'the dead hand' always before the imagination" (Wallis 257). Borrowing the logic of the saint's relic, association copies invest the object with value borrowed from the identities of its human users. And like the saint, the previous owner must be dead.

From there, it's only one step to imagining the book itself as mortal—whether through disuse or overuse. One early nineteenth-century inventor imagined paper manufacture itself as a kind of grave robbing.

In response to the scarcity of linen raw materials for papermaking, his pamphlet (itself printed on a specimen of the inventor's recycled paper) proposes that by an "act of parliament, which prohibits, under a penalty, the burial of the dead, in any other dress than wool, may be saved about 250,000 pounds weight of linen annually; which in other countries perish in the grave" (Koops 75).

Augustine Birrell's 1891 essay "Book-Buying" exemplifies both of these contradictory tenets. On the one hand, books are made to be recycled: "It is one of the boasts of letters to have glorified the term 'second-hand,' which other crafts have 'soiled to all ignoble use.' . . . The writers of to-day need not grumble. Let them 'bide a wee.' If their books are worth anything they too one day will be second-hand. If their books are not worth anything there are ancient trades still in full operation amongst us—the pastry-cooks and the trunk-makers—who must have paper." In this analysis, the wear and tear that debases other commodities ennobles books. On the other, "Book-Buying" associates the circulation of books with the death of their owners. From the past represented in texts, Birrell segues directly to the future in which the book's current owner will be replaced by another: "Alas! the printed page goes hazy beneath a filmy eye as you suddenly remember that Lycidas is dead . . . the 'ancient peace' of your old friends will be disturbed, when rude hands will dislodge them from their accustomed nooks and break up their goodly company" (285, 91).

As early as 1820, Washington Irving dramatized that ambivalence in the essay describing Geoffrey Crayon's visit to Westminster Abbey, which begins by comparing the library to a catacomb, and "authors" (metonymically, books) to mummies—or else to "treasured remains of those saints and monarchs that lie enshrined in the adjoining chapels; while the remains of their contemporary mortals, left to the ordinary course of nature, have long since returned to dust." Yet when the narrator is surprised to find a "hoarse and broken voice" emerging from a small quarto, it tells a different story. "What a plague do they mean," the voice complains, "by keeping several thousand volumes of us shut up here, and watched by a set of old vergers, like so many beauties in a harem, merely to be looked at now and then by the Dean?" Here as so often, books are invested with the literal power of speech at precisely the moment when they fail to metaphorically "speak" to their readers. But also as so often, everything rides on what metaphor you choose: the narrator tells the quarto to thank its lucky stars for being immured in an old library "which, suffer me to add, instead of likening to harems, you might more properly and gratefully have compared to those infirmaries attached to religious establishments for the benefit of the old and decrepit, and where, by quiet fostering and no employment, they often endure to an amazingly good-for-nothing old age" (167–69).

One reason that books can so easily replace financial instruments as the protagonists of it-narratives is that both share the same duality: like coins and banknotes, bibles and hymnbooks derive value at once from their material form (coins are made from precious metals; in this period especially, paper costs account for a high proportion of the production costs of books) and from their power to evoke a reality independent of that material form (a "pretty book" is not a handsome object so much as a tasteful text). The resemblance breaks down, though, once you reach the end of the object's life. For while a well-defined system regulates the moment when worn coins are taken out of circulation, no such thing is true for battered books. This is the breach into which Mayhew steps.

Mortal Pages

Whether martial or scatological, life cycle metaphors overlay the relation of text to book onto a time line. Yet that before and after also correspond to richer and poorer. If wastepaper looms large in the slums that Mayhew describes, the simplest explanation is that the Victorians associated mental operations (such as reading) with the upper classes, manual gestures (such as wrapping) with the lower. If the former reflects the price of new paper, the latter has more to do with the resale value of old—and, by extension, with the lack of materials manufactured specifically for packaging. Today, books themselves are one of the few commodities that, even in the developed world, continue to be packaged in used paper rather than new plastic: what reader of this page has not unwrapped a second-hand book from the outdated local newspaper of an online bookseller in some never-visited small town identifiable only by the real-estate ads? But where we use newspaper to wrap other texts, the Victorians associated it with nontextual and even antitextual contents. Before the invention of toilet paper and paper bags—both first produced for sale in the same decade as *London Labour*—and the even more spectacular rise of plastics, old paper was inextricably linked to food. Its habitats were the kitchen and the privy, its associations with the larder and the body (at best, with Dora Copperfield's curlpapers).[11] In this context, even Mayhew's term "after-use" takes on an anatomical meaning.

The setting in which Mayhew's contemporaries would have expected to come upon such references to grocery packaging was neither ethnographic (as in his costermongers) nor economic (as in Jevons's "butterman"). On the contrary, food wrapping had by the nineteenth century come to occupy a central role in aesthetics, because a long satirical tradition ensured that the easiest way to insult a work of literature was to mention it in the same breath with groceries. The *Monthly Review*'s

1792 attack on Charles Harrington's *The Republican Refuted*, for example, concludes with two lines: "Here, boy! throw this to the great heap that lies there, in the corner, for the cheesemonger: it may be of some use to him, though we can make nothing of it in *our way*."[12] By 1830, Macaulay could demolish Montgomery's *Satan: A Poem* in the pages of the *Edinburgh* by remarking, a propos of nothing in particular, that "the fashionable novels of eighteen hundred and twenty-nine hold the pastry of eighteen hundred and thirty" (T. B. Macaulay, "Mr. Robert Montgomery" 279). A year later, a satirical newspaper dismissed a parliamentary speech with the couplet

> For sale, waste paper lying in a loft—
> Perceval's speech, *particularly soft!*
>
> ("Brevities," 7)

(Nathaniel Ames observed that "sailors paying little or no attention to the 'serious calls' of these 'gospel trumpeters' . . . have quietly handed over to the cook all the tracts which a blind sectarian zeal had intruded upon their notice" (Skallerup 39). In 1844, *Punch*'s parodic "Report of the Select Committee on Parliamentary Petitions" suggested that the dustmen appointed to cart away petitions categorize them by party, so that cheesemongers could avoid losing customers by sending their wares wrapped up in offensive papers: instead, "Conservative petitions [could be] used for sending home Parmesan, Gruyere, and Stilton; the Whig petitions were used for the Cheshire class of cheeses; while the Radical petitions were devoted to the cheaper sorts, including black-puddings, and single rashers of bacon." Along the way, the article restored a series of figurative terms to their literal meaning: learning that the majority of petitions are tabled, the committee proceeds to examine the table; a dustman testifies that the papers in question are both "heavy" and "dry" ("Report of the Select Committee on Parliamentary Petitions").

Reviewers invoked food wrapping to dismiss others' writing, not their own. But authors, too, could smear their own writings by association with household uses, as in *Biographia Literaria*:

> Of the unsaleable nature of my writings I had an amusing memento one morning from our own servant girl. For happening to rise at an earlier hour than usual, I observed her putting an extravagant quantity of paper into the grate in order to light the fire, and mildly checked her for her wastefulness; la, Sir! (replied poor Nanny) why, it is only "WATCHMEN." (Coleridge 1:187)

Coleridge's rueful joke locks text and book into a zero-sum relationship: the more valuable the paper, the more worthless the page. In a culture that dismisses texts as "not worth the paper they're printed on," to register the usefulness of the book-object—how fast it catches fire or how much mutton grease it can sop up—is to assert the uselessness of its contents.[13] In the same vein, the preface to John Mills's *The English Fireside: A Tale of the Past* (1844) apologizes for having boasted that an earlier novel "did not find its way to the trunkmakers' or the chandlers' shops in the time specified for the reception of modern productions of literature." Over the course of a printed object's lifetime, texture replaced text as the source of its value.

If these reviews draw their scatological humor from classical satire, their obsession with domestic uses can be traced to more recent, more religiously pointed forms of biblioclasm. The Reformation—often billed as the triumph of the book—caused manuscripts to be cut up for "sewing-guards, fly-leaves, and wrappers of bookbindings"; in 1536, John Leland complains that iconoclasts cut leaves from ancient manuscripts to clean their shoes (Christopher de Hamel quoted in Sherman 102). And James Simpson has observed that during the dissolution of the monasteries, those into whose hands they fell "reserved of those lybrarye bokes, some to serve theyr iakes, some to scoure theyr candel styckes, and some to rubbe their bootes. Some they solde to the grossers and the sope sellers, and some they sent over the see to the bokebynders" (Simpson, "Bonjour paresse" 258). William Sherman adds that cutting and pasting could just as easily connote reverence: same action, different cause (Sherman 103). The pyres of Nuremberg confer a dignity lacking from the kitchen fire.

Reading, then wrapping: reviewers invoke wastepaper to taunt authors with the ephemerality of the works that were supposed to counterbalance their own mortality. As Fielding's translation of Martial put it, "How many fear the Moth's and Bookworm's Rage, / And Pastry-Cooks, sole Buyers in this Age? / What can these Murtherers of Wit controul? / To be immortal, Books must have a soul" (Gigante, *The Great Age of the English Essay* 170). This may explain why authors themselves so often experience wastepaper as a memento mori, even when its contents happen to be by someone else. In an entry for 4 January 1821—later cut and pasted into a printed book—Byron's diary records,

> I was out of spirits—read the papers—thought what fame was, on reading, in a case of murder, that 'Mr. Wych, grocer, at Tunbridge, sold some bacon, flour, cheese, and, it is believed, some plums, to some gipsy woman accused. He had on his counter (I quote faithfully) a book, the Life of Pamela, which he was tearing for waste paper, etc. etc. In the cheese was found, etc., and a leaf of Pamela wrapt round the bacon.' What would Richardson have said could he have traced his pages from their place on the French prince's toilets (see Boswell's Johnson) to the grocer's counter and the gipsy-murderess's bacon!!!

What would he have said? what can any body say, save what Solomon said long before us? After all, it is but passing from one counter to another, from the bookseller's to the other tradesman's-grocer or pastry-cook.[14]

Where Eliot and Gladstone link the circulation of books to readers' mortality, Byron compares it to authors': the fate of books allows a poet to denigrate his trade. The "leaf of Pamela wrapt round the bacon" prefigures our own metaphor that names worthless information by analogy with canned meat.

In fact, Byron's elegiac tone has more in common with twenty-first-century complaints of information overload than it does with the rags-to-pages triumphalism elaborated a century earlier by one of the periodicals most frequently named in Mayhew's inventories of waste-dealers' stock (e.g., 1:293). The *Spectator* for 1 May 1712 begins by observing, "When I trace in my Mind a bundle of Rags to a Quire of *Spectators*, I find so many Hands employ'd in every Step they take thro' their whole Progress, that while I am writing a *Spectator*, I fancy my self providing Bread for a Multitude"; but he goes on to confess that

> I have lighted my Pipe with my own Works for this Twelve-month past; my Landlady often sends up her little Daughter to desire some of my old *Spectators*, and has told me, more than once, the paper they are printed on is the best in the World to wrap Spice in. They likewise make a good Foundation for a Mutton pye.

The shift from metaphorical food ("I fancy my self providing Bread") to its literal counterpart ("They . . . make a good Foundation for a Mutton pye") echoes the demotion of "paper" from a count noun (a "Spectator paper") to a mass noun ("the paper they are printed on").

The essay counterbalances that fall, however, by a celebration of the benefits that papermaking diffuses among classes and even nations. "It is pleasant enough," Addison adds (the word here connotes pleasure as well as the older sense of "comic"),

> to consider the Changes that a Linnen-fragment undergoes, by passing through the several Hands above-mentioned. The finest pieces of Holland, when worn to tatters, assume a new Whiteness more beautiful than their first, and often return in the shape of Letters to their Native Country. A Lady's Shift may be metamorphosed into Billet-doux, and come into her Possession a second time. A beau may peruse his Cravat after it is worn out, with greater Pleasure and Advantage than he ever did in a Glass. In a word, a piece of Cloath, after having officiated for some Years as a Towel or a Napkin, may by this means be raised

from a Dung-hill; and become the most valuable piece of Furniture in a Prince's Cabinet.[15]

Where Mr. Spectator's writing descended from study to kitchen, here cloth rises from the hands of a servant ("a Towel or a Napkin") to a prince's study. Literally handed down from the *piano nobile*, paper can rise again, phoenix-like, through writing.

If Byron invokes the obituary, Addison channels the it-narrative. Like it-narrators, papers establish transitive relations among the rich and the poor through whose hands they pass in succession. Tracing the book's origins exalts; predicting the book's fate degrades: a similar symmetry links *Sartor Resartus*'s reflection on the mortality of books—"is it not beautiful to see five million quintals of Rags picked annually from the Laystall; and annually, after being macerated, hot-pressed, printed on, and sold,—returned thither; filling so many hungry mouths by the way?"—with Carlyle's boast that a letter from John Sterling attacking *Sartor Resartus* would be "made into matches" (Carlyle, *Sartor Resartus* 35, 233). (The material quoted within the novel itself is presented as having reached the narrator's hands via "one of those Book-packages, which the *Stillschwegen'she Buchhandlung* is in the habit of importing from England [. . . with] as is usual, various waste printed-sheets by way of interior wrappage; into these the Clothes-philosopher, with a certain Mohamedan reverence even for waste paper, where curious knowledge will sometimes hover, disdains not to cast his eye" [211].)

In an equally self-referential object narrative published a few years before *London Labour*, Charles Knight traces the pages that we are reading to exchanges among nations—but also among a descending series of social classes. "The material of which this book is formed existed a few months ago, perhaps in the shape of a tattered frock, whose shreds, exposed to the sun and wind, covered the sturdy loins of the shepherd watching his sheep on the plains of Hungary"; or

> it might have been swept, new and unworn, out of the vast collection of the shreds and patches, the fustian and buckram, of a London tailor; or it might have accompanied every revolution of a fashionable coat in the shape of lining—having travelled from St. James's to St. Giles's, from Bond Street to Monmouth Street, from Rag Fair to the Dublin Liberty, till man disowned the vesture, and the kennel-sweeper claimed its miserable remains. In each or all of these forms, and in hundreds more which it would be useless to describe, this sheet of paper a short time since might have existed. No matter, now, what the colour of the rag—how lily the cotton—what filth it has gathered and harboured through all its transmutation—the scientific paper-maker can produce

out of these filthy materials one of the most beautiful productions of manufacture." (*The Old Printer and the Modern Press* 256–57)

The cloth itself trickles down the social ladder from St. James to the secondhand markets of Monmouth Street, but the next phase of its life reverses that process, turning "filthy" raw materials into the medium of pure thought.

HANDMAIDS OF LITERATURE

Scatological jokes age as badly as bookish puns. The idea of paper falling into the hands of a servant—or tradesman, or Irishman, or woman—had a better run than most, leading a long and happy life in the most recyclable of genres, the anecdote (Donaldson 4). Between 1791 and 1823, Isaac D'Israeli's six-volume *Curiosities of Literature* cataloged a mind-numbingly comprehensive series of variations on this theme. "We are," D'Israeli concludes, "indebted to the patriotic endeavours of our grocers and trunkmakers, alchemists of literature! they annihilate the gross bodies without injuring the finer spirits" (D'Israeli 3:311). One antiquarian "left [old manuscripts] carelessly in a corner, and during his absence his cook used them for culinary purposes" (D'Israeli 3:54)—rather as Percy found the seventeenth-century manuscripts that would form the basis of his *Reliques of Ancient English Poetry* "lying dirty on the floor under a bureau . . . being used by the maids to light the fire."[16]

Gender could easily enough be replaced by class, as when another antiquarian "one day at his tailor's discovered that the man was holding in his hand, ready to cut up for measures—an original Magna Charta" (D'Israeli 3:54), or by race, as when Allan Quatermain complains at the beginning of a tale that "a Kaffir boy found my manuscript convenient for lighting kitchen fire" (Haggard 11). Alternatively, gender could trump class: in a third anecdote retailed by D'Israeli, a scholar's "niggardly niece, although repeatedly entreated to permit them to be published, preferred to use [his] learned epistles occasionally to light her fires!" (D'Israeli 3:54).[17] Servants recycle what masters once read; women discard what men once created. Like Gladstone's comparison of librarian to warehouseman, Thackeray's "littery footman" makes manual laborers a figure for the degeneration of the literary at once into the literal and into litter. When Harriet Taylor's maid kindled a fire with the first draft of Carlyle's *French Revolution*, she walked straight into a literary convention.[18]

In the era of wood pulp and Tupperware, with fireplaces and therefore servants on the wane, the trope should have died. Instead, it adapted, by substituting a figurative for a literal maidservant, and a preserver for a destroyer. As early as 1881, Jon Klancher points out, Andrew Lang declared that "the book-collector may regard his taste as a kind of hand-maid of critical science" (Lang 23); and as late as 1939, the *Cambridge History of English Literature* pronounced that "bibliography . . . is only the handmaid of literature [and] . . . it is only because of their loyal ser-vices to letters that [bibliographers] can claim a place in these pages."[19] David Greetham riffs on this tradition when he declares that "bibliogra-phers are the good housekeepers of the world of books" (Eliot and Rose 9). Where the figure of the maid who reads her master's books survives even into the age of the public library (as we saw in the previous chapter), so the maid survives within scholarly discourse as a figure for literal-ism. Without disappearing, the maid subsides into metaphor—with her historical belatedness making her an apt image for those bibliographers whose relation to the text has never made the leap into New Critical idealism and abstraction.

The fall from gentlemen's minds to servants' bodies pivots on a fall from figurative language. The O'Connell anecdote hinged on competing senses of the preposition "in": does "I've seen something good in your book" refer to the content of the text or to the contents of the object? "Support" performs a similar function in Fielding's punning description of "writings, which were calculated to support the glorious cause of dis-affection or infidelity, humbled to the ignoble purpose of supporting a tart or a custard!" (Gigante, *The Great Age of the English Essay* 172). Isak Dinesen's *Out of Africa* opens with a Kikuyu servant criticizing the manuscript's fragmentary quality, explaining that, in contrast, the au-thor's copy of the *Odyssey* "is a good book. It hangs together from the one end to the other. Even if you hold it up and shake it strongly, it does not come to pieces" (44). In making the African servant literalize the European mistress's fear that her collection of vignettes lacks formal coherence, the anecdote substitutes race for gender as the factor that dif-ferentiates naive materialists from those who reduce phrases like "hang together" to their metaphorical sense.

To be a gentleman is to be all mind. When the playwright in *La Bo-hème* sacrifices his manuscript to heat an icy garret, he describes its con-tents in a series of equally broad double entendres, as "scintillating" and "ardent" (Puccini and Murger 26).[20] The ambiguous class position of the bohemian is announced, in this opening scene, by the fact that we watch him burning papers before he is ever shown marking them. We know that we stand outside of the class system, in Murger's imagined Latin Quarter, when we see a single person combining composition (the mental labor of a gentleman) with destruction (the proper work of a maid).

The costermonger who chooses the largest newspaper, in anticipation of the moment when he will have "got the reading out of it," similarly collapses into a single person the functions traditionally parceled out between gentleman-scholar and female servant. Yet Mayhew breaks down the distinction between writers and wrappers in part by placing such puns in the mouth not of his middle-class narrator—as Murger and others do—but of working-class informants themselves. The narrator's indifference to literary genres ("it is no matter what kind") finds its foil in a flypaper manufacturer's comically elaborate taxonomy of newspapers:

> I use the very best "Times" paper for my "catch-em-alives." I gets them kept for me at stationers' shops and liberaries, and such-like. I pays threepence a-pound, or twenty-eight shillings the hundred weight. That's a long price, but you must have good paper if you want to make a good article. I could get paper at twopence-a-pound, but then it's only the cheap Sunday papers, and they're too slight. (3:32)

An "article," of course, usually refers to a news item, just as a "paper" designates a periodical. By the same token, "cheap" and "slight" figure prominently among the insults dear to Victorian literary critics. By stripping each term of its metaphoric or metonymic charge, Mayhew shifts our attention from texts to objects. Not for nothing does his informant call the paste that gums the paper a "composition."[21]

"Composition" in the usual sense remains absent from Mayhew's pages. Where Addison starts from a legible object and goes on to trace its afterlife, Mayhew's descriptions of the resale trade open in medias res: third- or fourthhand paper forms their starting point, not their punch line. By exploring from the inside the uses to which the formerly legible can be put, Mayhew forbids his readers to dismiss recycling as a D'Israelian "curiosity," much less an Addisonian joke or a Carlylean disaster. To take paper seriously is to resist at once the generic logic that reduces it to fodder for anecdote and the social logic that ridicules those who notice it.

By extension, Mayhew refuses to align the life cycle of paper with the social class of its users. The anecdotes that I quoted a moment ago delegated to servants the dirty work of noticing paper's material attributes: its impermeability, its inflammability, its absorbency. Here, in contrast, the middle-class implied reader (or should that be implied handler?) is never allowed to forget his or her own body. *London Labour* constantly reminds us that we're wearing out its pages, if only because the weekly numbers published in 1851–52 were protected by wrappers printed with letters from readers and answers from Mayhew. Even as the wording of the wrappers incorporated readers' writing, their material form kept dirty hands at a distance.

A similar paradox governs the fate of the wrappers themselves. Generically, the "answers to correspondents" look like the most explicitly theoretical section of the work, substituting finished economic principles for ethnographic raw material. In physical terms, however, they've proven the most disposable, missing from most library copies today because they were tossed aside when the consecutively numbered pamphlets were bound into volumes. This would have come as no surprise to anyone involved in the production of the pamphlets: Mayhew himself termed the wrappers "waste"—as if pamphlets could be packaged like pies—and his publisher announced that "the outer pages of this periodical will, in future, be used as a wrapper, intended to be cut off in binding" (Mayhew, *Essential Mayhew* 87).

In anticipating its own disposal, *London Labour* drags its readers down to the level of grocers. Where most Victorian reformist genres, from the political speech to the industrial novel, leveled up—asking middle-class readers to endow working-class characters with an interiority that mirrored their own—Mayhew levels down, reducing the page in front of us to tomorrow's fish-and-chip paper. Texts from *Oliver Twist* to *Ranthorpe* to *David Copperfield* withdraw the book from the marketplace; Mayhew instead reminds us of the resale value of the page before our eyes, and the paper in our hands.

The corollary is that handling replaces reading as the locus of interactivity—of the end user's ability and even responsibility to customize and repurpose books. More specifically, the material attributes of paper, which had once been imagined to constrain appropriation and adaptation, now become a prompt for both. And as in the it-narrative, the handling of paper also replaces the reading of text as the activity that unites different social classes. As it weaves among successive users and uses, each page remains itself (although the same can rarely be said of a bound book). In that sense, Mayhew's model reverses the logic that we tend to take for granted in the history of the Victorian *novel*, where a single text cuts across different media as its plot migrates fluidly from monthly numbers to library-issue triple-decker to cheap reprint to an equally varied series of theatrical adaptations, falling (or occasionally rising) in price at each remove. Here, in contrast, the medium remains stable—at the level of the page, if not of the book—but its uses change over time.

BREAD ON THE WATERS

Like any act of recycling, then, Mayhew's invocation of used paper is at once old and new. Old, because his interest in the economics of wastepaper plays on the worn-out trope of the pastry-cook and the trunk-maker; but also new, because where the tradition on which he draws overlays the difference between book and text onto a social hierarchy, Mayhew himself maps that distinction onto a life cycle.

Even that second distinction breaks down, however, as Mayhew subjects temporal priority to the same questioning that undermined social precedence. As perversely as he revealed handlers of food as readers of books, he now scrambles the order in which one follows the other. One "fancy-cabinet-maker," now unemployed thanks to the underselling of slop-masters,

> enjoyed no reading, when I saw him last autumn, beyond the book-leaves in which he received his quarter of cheese, his small piece of bacon or fresh meat, or his saveloys; and his wife schemed to go to the shops who 'wrapped their things from books,' in order that he might have something to read after his day's work. (2:114)

Where D'Israeli represents women confusing texts with pie plates, here it's the wife's job to spot literary value in cheese packaging. However similar Mayhew's sociology of recycling might appear to Addison's or D'Israeli's, the urge to reanimate waste that the wife shares with *London Labour* reverses the destruction wrought by landladies and maidservants. In the costermonger's language, the cabinetmaker's wife doesn't get the reading out so much as put it back in. The contrast between shrewish landlady and devoted wife thus personifies the tension between one model that casts women as the weak link in chains of textual transmission, and another that credits female librarians and editors with copying, carrying, and cataloging text, whether for love or for money.

In practice, we owe the survival of many manuscripts to the efforts of wives and daughters—not to mention stenographers, typists, and other female professionals. To compare the bibliographer to a "handmaid," in this context, is not simply to reduce bibliography to a service industry; it's also to remember how close the archive lies to the pantry, the oven, and the toilet. Every scholar knows how hard it is to untangle transmission from destruction—whether for material objects (think of printer's waste) or verbal texts (think of the relations between editing and expurgation). It-narratives admit as much when they struggle to disentangle the indifference that leaves the pages uncut from the respect that protects them from dirty fingers—to distinguish the Pocket Bible's "shelf of banishment" from its "place of repose." Untouched by the middle-class male reading public, it-narrators are all too greasily touched by women and servants. Yet far from maids ignoring the value of the papers that they dust, there it's the infidel master who leaves the Bible to gather dust and the simple maid who puts it to use. In one case, the lower orders leach books of legibility; in the other, they render readable what the upper classes have reduced to objects of display. In whichever direction, the relation of reading to handling is mapped onto class. In satire and reviews as in the marketplace described by Mayhew and the homes represented in religious tracts, to articulate the relation of textual to nontextual uses of books is to theorize a social order.

Compare a contemporaneous review that insults a novel by imagining it falling into the hands of a female Irish street vendor: "'Hawkstone'—if it has not gone to the butter-shop, and enlightened Irish barrow-women before that time—'Hawkstone,' if surviving, will teach [the reader] how important it was once thought to furnish a model-protestant hero with a rosary" ("The Progress of Fiction as an Art" 361). It's not entirely clear where to situate the irony in "enlighten": whether the reviewer is implying that the book will go from cradle to grave, from press to barrow, without ever finding a reader, if only because the hypothetical barrow women aren't literate; or whether, on the contrary, the point is that to read a page of *Hawkstone*—even for a barrow woman—would be the reverse of "enlightening."

This article has now been attributed to Marian Evans, at that time the editor in all but name of the *Westminster Review*. Two decades later, with the luxury of shunning magazine serialization, let alone the labor of reviewing novels more fit for the pastry-cook than for her audience, George Eliot returned to the metaphor. Chapter 41 of *Middlemarch* opens by comparing two forms of evidence, stone and paper:

> As the stone which has been kicked by generations of clowns may come by curious little links of effect under the eyes of a scholar, through whose labors it may at last fix the date of invasions and unlock religions, so a bit of ink and paper which has long been an innocent wrapping or stop-gap may at last be laid open under the one pair of eyes which have knowledge enough to turn it into the opening of a catastrophe. (412)

Remember how effectively the plot of *Middlemarch* is jump-started by the it-narrative that Lydgate reads, *Chrysal, or the Adventures of a Guinea*, a classic whose characteristically eighteenth-century interest in numismatic "circulation" is eventually displaced by Lydgate's more modern interest in the circulation of blood. At the other end of *Middlemarch*, however, "a bit of ink and paper" comes back into circulation to provide the same impetus that paper does in nineteenth-century children's books organized around the travels of a bible or a hymnbook. At the same time, the opening of chapter 41 glances back at the question that Casaubon spent the early books of the novel dodging: whether biblical scholarship should rely on philological or archaeological evidence—on words or things, on metaphorical digging through sources or literal digging through dirt, on the tradition of the library or the transmission of the potsherd.

Butter in the review of *Hawkstone*, sandwiches in the *Dublin University Magazine*: traditionally, to couple books with food is to strip them of their textual value. In the resolutely materialist landscape of *London Labour*, however, the specificity of "bacon or fresh meat, or . . . saveloys" comes across as more than simply comic. Another vendor, after complaining that "we should both be tired" if he were to inquire too closely into the contents of his wastepapers, adds: "Very many were religious, more's the pity. I've heard of a page round a quarter of cheese, though, touching a man's heart" (2:114). For the *Monthly Review,* nothing was lower than cheese: even butter would be less smelly. For Mayhew, in contrast, wrapping doesn't preclude reading. The paper may touch the cheese, but the page still touches the heart. In that sense, *London Labour* has less in common with the scatological humor of satire than with the providential logic invoked in *The Romance of Tract Distribution* (1934), which describes a tract used to wrap groceries converting the shopper (N. Watts 16). If book distribution can be narrated as "romance," it's less for its entertainment value than because its providential logic shares the circularity of a foundling's wanderings. It-narrative simply makes explicit the assurance that no matter how many times it's given away, lost, stolen, pawned, shipwrecked, and shot at, a book will always find its way home (Yeames).

Even the form of *London Labour* endorses the cabinetmaker's wife's sense that wastepapers remain continuous with legible texts. It's true that a volume break quarantines those vendors who resell objects for a further iteration of their original use (including secondhand booksellers) from those who bill them as raw material for a new purpose (including wastepaper dealers).[22] The first sell by the piece, the second by weight (2:107). But the distinction that works for tools or crockery applies less neatly to paper. As soon as printed matter enters *London Labor*, the aging process begins to run in reverse, if only because where scrap metal is counted, wastepaper is quoted:

> The letters which I saw in another waste-dealer's possession were 45 in number, a small collection, I was told; for the most part they were very dull and common-place. Among them, however, was the following, in an elegant, and I presume a female hand . . . The letter is evidently old, the address is of West-End gentility, but I leave out name and other particularities:
>
> 'Mrs—[it is not easy to judge whether the flourished letters are 'Mrs.' or 'Miss,' but certainly more like 'Mrs.'] Mrs.—(Zoological Artist) presents her compliments to Mr.—, and being commissioned to communicate with a gentleman of the name, recently arrived at Charing-cross, and presumed by description to be himself, in a matter of delicacy and confidence, indispensably verbal; begs to say, that if interested in the ecclaircissement and necessary to the same, she may

be found in attendance, any afternoon of the current week, from 3 to
6 o'clock, and no other hours. (2:114)

The "however" that singles out the quotable exception from the illegible
mass also substitutes reproduction for description. Once the *roman à
clef* breaks in on an economic tract, wastepaper becomes hard to distinguish from used books. As always with *London Labour*, it would be
easy to tabulate the stylistic tricks that make this passage so novelistic:
the ostentatious substitution of blanks for names; the silver-fork gesture
toward a West End address; the hints of "delicacy" and "confidence";
and, of course, the speculation over the marital status of the owner of
the "elegant, and I presume female" hand. But what makes this passage
novelistic is ultimately less a stylistic or even thematic toolbox than the
faith that a legible narrative can be assembled from a pile of "dull and
common-place" letters. As an article on wastepaper declares in an Evangelical magazine, "Of waste paper it may truly be said 'Resurgam'; it undergoes processes of apparent destruction only to rise Phoenix-like from
its ashes—or rather pulp" ("Waste Paper" 419). In the miraculous logic
that makes Mayhew's pages touch the heart of any literary critic, entropy
can always be reversed.[23]

In their own perverse way, even Mayhew's infidel informants draw
on the parable of the sower. When he cites a pickpocket as authority
for the report that tracts are good to light pipes with, he borrows the
logic that classical satire bequeaths to the book reviews where a social
inferior is pictured disposing of competing publications.[24] Yet Mayhew's
mention of pipe lighting draws as much on Evangelical literature as on
the classical tradition. One RTS account compares handing out tracts to
"casting pearls before swine," explaining that "what grieved me was, to
see these tracts torn in pieces before my eyes, to light pipes with" (Jones
169). Another Evangelical magazine acknowledges that "cautious min-
isters have shaken their heads, and told stories of the burning of Bibles,
and Testaments, and tracts; and how the priest got them, and sold them
for waste paper; and even that they have received their butter from the
shops wrapped in pages of the Bible. The soldiers, they admit, receive
them freely, but it is because the size of the paper exactly suits for making
cigarettes"; yet these misuses disseminate good books among precisely
those populations that would not seek them out to read (Manning). A
third tract recounts that "an enterprising grocer had picked up and ap-
propriated to himself the Bible which the priest had thrown away, and
had thought to turn to good account the large leaves of the book, and
so he had wrapped up his wares—his soap and cheese and candles—in
the leaves which he tore out"; as customers unwrapped their groceries,
they read and were converted (Borrett White 126). As the American Tract

Society explains, "If made waste paper of, as some of them are and will be, even in that state the scattered leaves of the Bible, or of other religious books have been made, and will continue to be made, the means of exciting serious and godly thoughts" (*Annual Report of the American Tract Society* 130). One 1885 sermon praises "the enterprise of Andrew Fuller and some others along ago, who printed hymns upon papers which were to be used in the sale of cottons and other small wares . . . I knew a friend who in purchasing his tobacco found it done up in a passage of the Word of God, and, by the perusal of that portion, became a converted man" (R. Stewart). Dispersal didn't need to mean disposal: in some cases, it could become the most effective means for the providential spread of the Word.

Administrators of free libraries after 1850 translated the tension between preservation and circulation into a secular, civic register. Over a decade after the Public Libraries Act, when Edward Edwards set out to attack fellow librarians who put the needs of books above the interests of borrowers, his language became at once mock-antiquarian and mock-blasphemous: a library that reduces the "public stock of learning" into an "exclusive" closed-stack collection was "a talent digged in the ground," "an idol to be respected and worshipped for a raritie by an implicite faith, without anie benefit to those who did esteam it far off."[25] Love of the text can shade into idolatry toward the book, or reverence be coded as coolness.

Sowing or digging into the ground: the broadcast metaphor remains ambiguous. Even disasters far worse than finding bible pages wrapped around Indian food can be a blessing in disguise: one missionary shipwrecked in 1814 off the African coast consoles himself for the natives' seizure of his goods by reflecting that "my having been cast away, may perhaps be the saving of many of those into whose hands these Bibles have fallen, or shall fall in the future" (Howsam 150).[26] The *Wesleyan Magazine* observes that

> in three days 2000 Chinese Gospels and more than a hundredweight of the Scriptures in Mongolian were disposed of to Chinamen, Mongols, and Mohammedans, thus securing their distribution over vast tracts of country. Even where the book sold fails to interest the purchaser, it by no means follows that it is lost. Among the men who have come to crave further teaching concerning "the way of life," one was questioned as to how he had received his Christian books; he replied that he had bought them from an old woman who was selling them as waste paper. Some men buy every book that is published, and study them all. One such bought ninety books and tracts from Dr. Edkins, of the London Mission at Peking, and by the time he had got through

them all, he was so thoroughly convinced of the folly of idol-worship that he pronounced sentence of death on the whole regiment of his domestic idols, numbering nearly a hundred, and representing a ton weight of copper! ("Pioneer Work in China" 902)

The trope of wastepaper miraculously turning back into legible text is undercut, however, by the parallel linking "a hundredweight of the Scriptures" with the comparison to copper. Measurement by weight restores the very materiality that the anecdote tries to leach out of wastepaper. Like Edwards, Mayhew translates the trope of broadcasting from a religious to an economic register. Where the it-narrative once shifted its voice box from coins to bibles, exalting the commercial providentialism of the hidden hand into an Evangelical defense of circulation, Mayhew now makes circulation in the marketplace the source of paper's mysterious power.

Evangelicals' faith in self-propelling paper could just as easily be borrowed by infidel radicals, like William Hone, who traced his political beliefs to a scrap of a description of Lilburne happened upon at the cheesemonger's. The attorney general explained at the trial of *The Rights of Man* that there had been no reason to prosecute Part I, whose audience was limited, but that Part II was "thrust into the hands of subjects of all descriptions, even children's sweetmeats being wrapped in it" (Nelson 227). In *Memoirs of Modern Philosophers*—itself framed as a found manuscript—Bridgetina is corrupted by a proof sheet of Godwin's *Political Justice* wrapped around her mother's snuff: "Notwithstanding the frequent fits of sneezing it occasioned, from the quantity of snuff contained in every fold, I greedily devoured its contents. I read and sneezed, and sneezed and read" (Hamilton and Grogan 176). In a novel that exemplifies the battle between everyday truths and pernicious abstractions by comparing the mother's knowledge of cookery with the daughter's of books, the association of pages with groceries reminds us of their radical authors' social inferiority.

London Labour itself lumps tracts together with every other form of free—and therefore worthless—print: the difference between bibles and advertisements disappears when "sham indecent" packets are described as being stuffed with "a religious tract, or a slop-tailor's puff" (1:241). Political tracts prove as hard to get rid of as religious tracts: "the anti-Corn-Law League paper, called the *Bread Basket*, could only be got off by being done up in a sealed packet, and sold by patterers as a pretended improper work" (1:241). Free tracts, junk mail, bill sticking: Mayhew testifies to a moment when the auditory overload of early modern cities was giving way to the curse of cheap paper.

WELL-WORN PAGES

That shift forms part of a larger economic story. By bracketing books with humbler objects, *London Labour* also places the falling price of paper in the context of a general decline in the life span of manufactured goods. Mayhew registers the extent to which, for books as for other commodities, disposability was beginning to replace secondhandedness as the sign of cheapness. What made an object cheap was no longer that it had been repurposed in the past (think of "the unmistakable look of much of the attire of his class, that it was not made for the wearer"), but that it *lacked* the potential to be recycled in the future.

Such disposability characterizes "slop-goods," whose flimsy fabrics (in the case of clothes) or rosewood veneers (in furniture) place them opposite Mayhew's own "history of a people, from the lips of the people themselves . . . in their own 'unvarnished' language." The more immediate their appeal to buyers, the more probable that their first owner will be their last: one used-clothes vendor insists that a secondhand cloak "always bangs a slop . . . because it was good to begin with" (2:41). Mayhew gives the best lines to secondhand sellers and buyers, distancing himself from "clerks and shopmen . . . often tempted by the price, I was told, to buy some wretched new slop thing rather than a superior coat second-hand" (2:29). When Mayhew pits mended goods against manufactured "slops," an older culture in which clothes provided what Peter Stallybrass and Ann Rosalind Jones call the "materials of memory" meets planned obsolescence.[27]

In 1854, Charles Knight compared the shortsightedness of the previous century's booksellers to the shortsightedness of contemporaneous fishmongers who destroyed their surplus at the end of each day rather than sell it at half-price: "The dealers in fish had not recognized the existence of a class who would buy for their suppers what the rich had not taken for their dinners . . . The fishmongers had not discovered that the price charged to the evening customers had no effect of lowering that of the morning. Nor had the booksellers discovered that there were essentially two, if not more, classes of customers for books—those who would have the dearest and the newest, and those who were content to wait till the gloss of novelty had passed off, and good works became accessible to them, either in cheaper reprints, or in 'remainders' reduced in price" (*The Old Printer and the Modern Press* 225–26). William St Clair has shown how strongly the age of texts determined the price of books: texts published in the nineteenth century remained inaccessible to early nineteenth-century readers (St Clair). His point about new *texts* could be extended to new *books*—and a fortiori to new periodicals. This is a question not just of copyright, however, but also of fashion: when a commentator in 1871 remarks that "thirty-shilling novels are sold for three shillings" once circulating libraries experience no more demand for them, the explanation has nothing to do with intellectual property (Friswell 522). Newspapers decline most steeply in value, of course: as Richard Altick explains, newspapers rented out as many as seventy times on the day of their publication could then be sent to country subscribers who paid threepence for a copy mailed on the day of publication but only two for one mailed the following day. In a town described by Altick in 1799, the London *Courier* passed from the surgeon, to a French émigré, to the Congregational minister, to the druggist, to a schoolmaster, to a sergemaker, and so on (R. Altick 323; see also Colclough 266). Yet these transactions were not unique to political news: like clothes, magazines, too, were handed from mistress to maid (Beetham, "In Search of the Historical Reader" 98): both lost value according to a cycle that had less to do with their material durability than with the short shelf life of fashion.

The life cycle that fashion plates share with fashionable clothes reminds us that what holds for textiles applies to texts as well. The nineteenth-century book trade formed a key site in the struggle between an economy that paired high prices with a succession of multiple users, and an economy that produced cheap single-use goods. Until the 1890s, as we know, the distribution mechanisms of fiction reflected a trade-off between short- and long-term popularity. On both sides of the Atlantic, publishers balanced their backlist of steady sellers (reference books, schoolbooks, bibles) against novels whose sales spiked in the year following publication. In the United States, this meant large print runs produced so cheaply as to be disposable; in Britain, it meant circulating libraries that saved subscribers from being stuck with stockpiles of last season's best sellers.

Natalka Freeland has shown that the second strategy was what allowed fiction to exemplify an emerging culture of built-in obsolescence: the novel in general, and the detective novel in particular, are defined by resistance to rereading.[28] Books in general, and novels and periodicals in particular, come to epitomize a trickle-down economy. In fact, Simon Eliot has shown that books descended the social scale not only from individual to individual, but also from institution to institution. In the public sector, bigger metropolitan libraries off-loaded their old best sellers to smaller provincial libraries; in the private sector, Mudie's resold books and magazines that had been lent out, sometimes as little as a month old ("Circulating Libraries in the Victorian Age and After" 131, 139). Like the protagonists of eighteenth-century it-narratives, nineteenth-century books moved across—or rather down—the social scale. In the process, they mirrored the less concrete movement of texts from library-issue triple-decker to cheap reprint.

Slops are to castoffs as best sellers to chapbooks. That analogy between clothes and pages requires no ingenuity on my part: it was a well-worn one (as *Sartor Resartus* still testifies) for anyone writing or reading before wood pulp.[29] Paper came from cloth and borrowed the language of cloth; both are metonymically referred to as "prints." Conversely, Dickens brings the manufacturing process full circle by comparing old clothes to parchment: in *A Tale of Two Cities*, the shoemaker's "poor tatters of clothes, had, in a long seclusion from direct light and air, faded down to such a dull uniformity of parchment-yellow," and in "Meditations in Monmouth-Street" "there was the man's whole life written as legibly on those clothes, as if we had his autobiography engrossed on parchment before us" (*A Tale of Two Cities* 43; *Sketches by Boz* 78). The analogy gains weight from the geography that Mayhew describes: Holywell Street, off the Strand, was the center both for the sale of dirty books and for Jewish rag-traders (Nead 178).

One slum that Mayhew visits contains clotheslines "on which hung . . . handkerchiefs looking like soiled and torn paper" (2:89). The flypaper maker's room, too, reminds him of "a washerwoman's back-yard, with some thousands of red pocket-handkerchiefs suspended in the air . . . I had to duck my head down, and creep under the forest of paper strips rustling above us . . . A pile of entire newspapers was here brought out, and all of them coloured red on one side, like the leaves of the books in which gold-leaf is kept" (3:31). In a different study of London, he accuses readers of "dig[ging] their scissors into [my] results, taking care to do with them the same as is done with the stolen handkerchiefs in Petticoat Lane—viz., pick out the name of the owner."[30] Such analogies weave the author's stylistic signature (as needlework or as watermark) into what might otherwise look like others' words. In the process, they equate the repetition of words with the recycling of fibers. Although the section devoted to sellers and buyers of secondhand goods begins, "In commencing a new volume . . . " (2:1), the description of rag-buyers that follows will remind us that no volume is entirely new, either in its material form or in its verbal content.

The rise of extensive wearing signals the fall of intensive reading. Mayhew couldn't yet have predicted the technological changes that papermaking would undergo in the decades following, but what he did sense was a cultural and economic shift toward limiting goods to a single owner. Mayhew's liveliest informants tend to share his nostalgia for that older dispensation; thus one complains that customers who disdain the secondhand trade are

> often green, and is had by 'vertisements, and bills, and them books about fashions which is all over both country and town. Do you know,

sir, why them there books is always made so small? The leaves is about four inches square. That's to prevent their being any use as waste paper. I'll back a coat such as is sometimes sold by a gentleman's servant to wear out two new slops. (2:29)

Remember Chadwick campaigning to scale down blue books to an octavo less negotiable on the resale market, or missionaries in Madras printing tiny books to avoid tempting recipients to put them to base uses. The imaginary fight on which this vendor bets isn't just between one kind of clothes and another; it's also between two kinds of book about clothes. On one side, artificial safeguards against resale unite fashion-books with the goods that they advertise; on the other, Mayhew himself invokes an old play to explain how cloth can be renewed. The dealer's attack on four-inch books is prefaced by an antiquarian digression:

> In the last century, I may here observe, . . . when woollen cloth was much dearer, much more substantial, and therefore much more durable, it was common for economists to have a good coat "turned." It was taken to pieces by the tailor and re-made, the inner part becoming the outer. This mode prevailed alike in France and England: for Molière makes his miser, *Harpagon,* magnanimously resolve to incur the cost of his many-years'-old coat being "turned," for the celebration of his expected marriage with a young and wealthy bride. (2:29)

The quotation marks around "turned" draw our attention to Mayhew's own reuse of an old text—the same one mentioned ten sheets of paper earlier among the uncut copies of "Shakespeare, Molière, [and] Bibles . . . found in the insatiate bag of the waste collector." Mayhew's Molière has been cut—and not for tailor's measures, either. Although the pages that hawk slop-goods can't even be torn off for wrapping, the texts that describe mended clothes can still be reproduced a century and a half later.

FOUND OBJECTS

I'd like to conclude by asking how economic history (in this case, the price of paper both new and used) relates to literary history (that is, to representations on—and of—paper). On the one hand, the raw material of books cheapened even over the period that this monograph covers, thanks to factors both political (the 1861 repeal of the paper tax) and technological (the spread of mechanization in the first decades of the nineteenth century and of rag substitutes in the last).[31] Begun in serial form a year before wood pulp was commercially produced for the first time, *London Labour* appeared in full more than a decade later, once the new technology had established its viability. The wastepaper trade that Mayhew represents was on the verge of being destroyed by the increasing cheapness of new paper ("Traffic in Waste Paper" 135): the halving of its cost between 1840 and 1910 hit the waste trade harder than the trade in secondhand books (Weedon 67). As the price of its raw materials dropped, the book's life span came to coincide with the text's.

On the other hand, the late nineteenth-century drop in wastepaper's market value is anticipated, within literature, by an early nineteenth-century decline in the power of fiction to imbue wastepaper with *narrative* value. If Mayhew's scenes of miraculous cheese packaging draw on the language of contemporary religious tracts, those tracts borrow their fascination with wastepaper in turn from the fiction of a century earlier. Old paper is tracked through, or stars in the backstory of, almost every eighteenth-century subgenre: not just object narratives like *Adventures of a Quire of Paper* (1779), but epistolary novels, pseudomemoirs, and most of all gothic romance.

The trope receives its death blow only once Catherine Morland, coming across a manuscript in an old chest, is crestfallen to discover that "an inventory of linen, in coarse and modern characters, seemed all that was before her" (Austen, *Northanger Abbey* 134). So far, so quixotic: pages are to romance what paper is to the real. *Don Quixote* itself, readers will remember, traces its own pages to "waste papers" found by the narrator, not at a bookseller's but in a silk merchant's shop. In this universe, books may be burned but waste remains recuperable. In contrast, Austen's refusal to quote the roll verbatim denies new life to old papers. Once Catherine convinces herself that no amount of framing can turn used paper back into usable text, illegibility becomes a guarantor of realism.

In narrowly literary-historical terms, Catherine's awakening puts an end to framed narrative—the conceit that until that moment undergirded the novel in forms as various as interpolated tales and inset correspondences. Ian Duncan has incisively described the turn of the nineteenth century as a moment when "as 'literature' becomes a commercial object (a printed book for sale) it seeks to recuperate its 'contexts,' a lost world of organic relations of production, by representing itself as a relic of precapitalist origins" (Duncan, *Scott's Shadow* 274) What Duncan shows for the shift from oral to written culture holds equally true within written culture for the shift from manuscript to print (better, the competition between the two)—or, more to the point for Mayhew, from recycled found objects to single-use manufactured books. Austen's rejection of the gothic pivots less on the improbability of wife murder than on the outdatedness of narrative hand-me-downs. No longer can oral tales or handwritten letters be translated into print; no longer do readers situate themselves within a food chain. To the found objects of older fiction, *Northanger Abbey* opposes the manufactured goods of the circulating library; to dusty papers, the "ten or twelve" new novels that Catherine and Isabella Thorpe borrow before throwing aside. From Cervantes's narrator onward, the figure who treasures wastepaper is a naïf; to be disillusioned is to accept that paper is mortal.

If realism is the fictional mode that grants finality to waste, *London Labour* would be "novelistic" not in the sense that's usually supposed—that its reality effect depends on formal conventions borrowed from contemporary fiction—but, on the contrary, in that it shares the novel's constitutive oscillation between the real and the romantic. The former finds its thematic corollary in a narrative of mortality, whether of humans or of papers; the latter, in the hope that characters can return from the grave, or papers from the grocery.

Of Mayhew's series in the *Morning Chronicle*, Thackeray claimed that "readers of romance own they never read anything like it" ([W. M. Thackeray]). The title of the RTS's internal history, *The Romance of Tract Distribution*, too, inscribed tracts within the plot of the discarded manuscript that finds readers as providentially as orphans find parents. The found manuscript trope is recycled by Mayhew's brother Horace in *Letters Left at the Pastrycook's: Being the Clandestine Correspondence between Kitty Clover at School and her "Dear, Dear Friend" in Town* (1853), whose contents are presented as a set of letters about to be sold for wastepaper. Framed by a reminiscence couched in terms of more durable media ("I had carefully noted down in the porcelain tablets of my recollection . . . "), the letters themselves are punctuated by metaphors of paper put to nontextual uses: one character is described as having "hair the colour of blotting-paper"; another "felt I was going to be turned inside out, like a paper bag" (1, 29, 80, 48).

Expelled from the central tradition of British fiction as thoroughly as the wastepaper trope from book reviewing, the found manuscript migrates in two directions. One is the providentialism of religious tracts; the other is the closed economy of the detective novel. Natalka Freeland shows that the "conservation of information" governing mid-nineteenth-century detective fiction means that any document that a character tries to destroy will come back to haunt him: a pile of ashes will always turn out to contain fragments of a will; the contents of any wastebasket can be pieced together to form a letter (Freeland). That convention of trashy fiction finds its Evangelical counterpart in the "story of a man who tore up a tract in the face of a distributor on board a ship. Two fragments, however, were blown by the wind into the folds of his coat, which he was surprised to see on the cabin floor the next morning. On one fragment was the word 'God,' on the other 'Eternity' " (N. Watts 9). Where D'Israeli tells us that destruction is easy and preservation is difficult, detective novel and tract alike suggest just the reverse.

A mechanistic model of the relation between literary history and economic history might lead us to expect a four-part correlation: the found manuscript trope should decline in tandem with the price of paper, and the wastepaper trope should disappear from book reviewing once paper comes to be discarded rather than repurposed.[32] Certainly the *practice* of reading old papers seems to have declined over the course of the nineteenth century. David Vincent has argued that "the spread of literature and cheap literacy in the nineteenth century represented a shift for the bulk of the population from rereading to reading. It became acceptable for the first time to throw print away before every word had been perused" (*The Rise of Mass Literacy* 103). What's more crucial for our purposes is that it became acceptable for the first time to throw print away *after* every word had been perused: that is, that reading material was no longer designed with "after-uses" in mind.

If we look at the *representation* of wastepaper, however, no such story emerges. Neither found manuscripts and legible grocery wrapping, nor the pastry-cook and the trunk-maker, disappear. Instead, both tropes adapt: the latter, as I suggested, by subsiding into the metaphor of the bibliographer as servant, and the former, I want to argue now, by migrating into stories of childhood. On the one hand, stories *for* children are now framed as found manuscripts: thus manuscripts conspire with oral tradition to eclipse the printed form of Hans Christian Andersen's tales, introduced as something fished out of a barrel ("Auntie Toothache") or unwrapped from a piece of cheese ("The Goblin and the Huckster") (Andersen 1096, 457). On the other, stories *about* children rely on the serendipitous or even providential discovery of books or papers discarded by adults. Remember the antiprovidential language in Anne Mozley's assertion that any child "will surely find that the book thus influential came to him by a sort of chance, through no act of authority or intention" (195). Such "chances" shade into something more like a miracle, however, when Edmund Gosse associates found manuscripts with the "wonder" of childhood, describing the skin-trunk and hat-box that he came across in (cue to *Copperfield*) the lumber room:

> The hat-box puzzled me extremely, till one day, asking my Father what it was, I got a distracted answer which led me to believe that it was itself a sort of hat, and I made a laborious but repeated effort to wear it. The skin-trunk was absolutely empty, but the inside of the lid of it was lined with sheets of what I now know to have been a sensational

novel. It was, of course, a fragment, but I read it, kneeling on the bare floor, with indescribable rapture . . . This ridiculous fragment filled me with delicious fears; I fancied that my Mother, who was out so much, might be threatened by dangers of the same sort; and the fact that the narrative came abruptly to an end, in the middle of one of its most thrilling sentences, wound me up to almost a disorder of wonder and romance. (51)

What unites the trunk with the box is the child's confusion of insides with outsides. In both cases, he treats container as content. (A leather book is traditionally more legible than a skin-trunk.) Gosse implies that one would need to be a child—in fact, a child of a generation ago, brought up by religious fundamentalists—to find "wonder and romance" in an empty trunk.

For Gosse as for Dickens, romance isn't just a textual mode: it's also a distribution method (accident), a setting (the lumber room), and a medium (waste). Its prelapsarian model of reading is premised on a providential model of circulation: if the pages are cut out, so is the middleman. Or indeed middlewomen like Gosse's own mother, whose tract writing and tract distributing are described a page earlier with the caveat that "I would not for a moment let it be supposed that I regard her as a Mrs. Jellyby": the author of "a tract, called 'The Guardsman of the Alma,' of which I believe that more than half a million copies were circulated," reappears as a character in either a gothic romance (the genre with which tracts compete) or a modern novel in which tracts are ridiculed.

The parallelism between box wearing and trunk reading produces another, subtler, effect. The comic mode of the former—the deadpan "laborious but repeated" juxtaposed with the image of a boy with a box on his head—can't help rubbing off on the latter. "What I now know" marks a distance between naive child and world-weary narrator that bears more resemblance to the gap separating narrator from heroine in *Northanger Abbey* than child from adult in *David Copperfield*. Where Dickens pictures the child reading less into books than would an adult—"whatever harm was in some of them was not there for me"—Gosse and Austen show the child reading something into them that the adult narrator knows was never there.

With one crucial difference: while *Northanger Abbey* reveals a laundry list where we expected to find a story, Gosse gives us a story in place of the clothing that we expect to find in a trunk. In early modern satire as in book reviewing, found manuscripts signify realism: the humbling truths of embodiment and mortality, whether of persons or of papers. In a world where wastepaper has lost its economic value, however, found manuscripts can also, on the contrary, mark "wonder" and "romance."

As accident gives way to miracle, Catherine Morland's naïveté is replaced by David Copperfield's innocence. More specifically, where Catherine is excessively enmeshed in genre (in this case, the gothic), the children in the later narratives stand blissfully outside of it. The young Gosse doesn't even recognize "what I now know to have been a sensational novel," any more than the young David can identify "a cheap series of reprints then in course of publication." It's not enough to share the ignorance that characters *within* romance model (not to know, for example, that innkeepers get paid); closer to home, romance can't withstand the knowledge that books themselves are for sale. Yet Austen forbids us even to feel nostalgia for an age where found manuscripts crowded out bought books. Even the progression from a "roll . . . of trifling size" to a tripledecker's telltale compression of pages is undercut, in turn, by Catherine's disappointment that what looked like a scroll is really more like a codex: "she now plainly saw that she must not expect a *manuscript* of equal length with the generality of what she had shuddered over in *books*, for the *roll*, seeming to consist entirely of small disjointed *sheets*, was . . . much less than she had supposed it to be" (137; my emphasis).

WRAPPING UP

If some materials are more renewable than others, some genres are more rereadable. Artificial safeguards are needed to prevent fashion-books from being recycled, but literary texts remain quotable; the ephemerality of laundry lists provides a foil for the timelessness of the novel that frames them.

> [Catherine] seized another sheet, and saw the same articles with little variation; a third, a fourth, and a fifth presented nothing new. Shirts, stockings, cravats and waistcoats faced her in each. Two others, penned by the same hand, marked an expenditure scarcely more interesting, in letters, hair-powder, shoe-string and breeches-ball. (Austen, *Northanger Abbey* 134)

Where Catherine Morland fails to turn laundry list into story, Austen's critics have succeeded—through stories about the reality effect, about money, about women's domestic oppression. Equivalent operations could be performed upon *London Labour*: a reader could notice, for example, that the sheet of paper contains a list of sheets, or that for Austen's narrator as for Mayhew's flypaper manufacturer, the word "article" doesn't refer to an essay.[33] In that McLuhanesque reasoning, Austen's reference to "a roll of paper" would tip us off to the fact that the "inventory of linen" is literally *of* linen—manufactured from recycled rags, as the *Spectator* reminds us, themselves produced by the washing described. If Catherine finds "nothing new" on the page, the explanation may be that its material is secondhand.

The language that conveys Catherine's disappointment in the found manuscript mirrors the language that earlier described critics' disapproval of novels—their "talk in threadbare strains of the trash with which the press now groans" (Austen, *Northanger Abbey* 21). If that image is to be believed, novels no sooner emerge from the "press" than they become "threadbare": the moment between production and disposal dwindles to a vanishing point. *Northanger Abbey* reverses that time line and rehabilitates the "threadbare," making the exposure of the textile's grid an image for the laying bare of textual devices. Like the cabinetmaker's wife turning wrapping back into reading, Austen fishes the cliché of the found manuscript out of the dustbin of literary history.

Even as fashions change (whether in sartorial style or literary conventions), materials endure. Such, at least, is the principle linking the laundry list that appears at the end of the novel to the dialogue with which the novel begins.

> "And pray, sir, what do you think of Miss Morland's gown?"
>
> "It is very pretty, madam," said he, gravely examining it; "but I do not think it will wash well; I am afraid it will fray . . . But then you know, madam, muslin always turns to some account or other; Miss Morland will get enough out of it for a handkerchief, or a cap, or a cloak. Muslin can never be said to be wasted." (14)

In this analysis, the end of the novel would force us to cycle back to the beginning, as if to test whether its text is as disposable as a laundry list or as reusable as a length of muslin. Once the question of how muslin will wear gives way to the question of whether the clock can be turned back on the telltale compression of pages, the novel's structure mimics the passage from clothing to books that defines the life cycle of linen. What Mayhew says of Harpagon's coat holds true for the novel: in a mirror image of Gosse's "laborious . . . effort" to wear a hatbox, "the outer part becomes the inner."

Yet to notice the movement from cloth to paper, or from content to form, would be to ignore the lesson that the scene so painfully impresses on Catherine. The critical urge to transmute empty surfaces into hidden depths—beginning with the two-dimensional plane of the page—replicates Catherine's faith that the most throwaway document will repay interpretation. Perhaps the more throwaway the better: a dusty manuscript "bangs" a circulating-library novel just as rusty knives sell best (according to one of Mayhew's informants) because "folks like to clean up a thing theirselves, and it's as if it was something made from their own cleverness" (2:11). In Mayhew's scrap heap, found objects outlast manufactured goods. Once interviews become "more novel, curious and interesting" than fiction, then recycling begins to look like a name for reading.

Rags to riches, or at least rags to text: the fantasy that nothing lies outside the realm of interpretation was as crucial to semiotics in the twentieth century as to the gothic in the eighteenth. No cultural artifact too nonverbal to be metaphorically "read"; no butter wrapping too smelly to be reinvested with legibility. Unfortunately, most scholars' experience approximates Catherine Morland's or the costermonger's more closely than a gothic heroine's or the cabinetmaker's. The laundry lists that we find in the archive refuse to yield stories. Nor do volumes in rare-book libraries tell as much about previous handlers as do the heroes of it-narrative. Seeking marginalia, we see uncut pages; craving ink marks, we find drink stains.

Blackstone's *Commentaries* famously claimed that "the identity of a literary composition consists intirely in the *sentiment* and the *language*," while "the paper and print are merely accidents" by which "that composition [is 'conveyed'] to the ear or eye of another." In Mayhew's world, texts reach readers thanks not to intentions but to side effects. Instead of food stains rendering the page illegible, the paper's usability as fish wrap ensures the transmission of text. As a result, Mayhew's characters don't just have the ears and eyes that Blackstone mentions; he also endows them with noses.[34] Although the mention of "a once perfumed love-note" reminds us that papers lose some of their sensory attributes as they age, the persistence of excremental humor points in the opposite direction, toward a materiality that intensifies over time. As the items listed in a "West-End Perfumer's" account books lose their fragrance, the books themselves accrue smells.

A mystery story published a few decades after *London Labour* reminds us that the smell of paper can provide as much information as its textual content. After sniffing a letter, the detective tells us, "it became clear to me that I was wasting my time on the handwriting and the contents. I needed to concentrate on outward appearances and use them as my starting-point" (Groller 239). Where literary critics ignore outsides, criminal forensics ignores insides. The historian Paul Duguid dramatizes the clash between those two models in an anecdote about his own archival work: suffering from asthma, he holds his nose in the reading room; in the next seat, however, he notices another researcher passing letter after letter under his nostrils without even glancing at their contents. The researcher turns out to be sniffing for vinegar, to determine whether the port from which the letter was sent was under quarantine. Duguid's point is that digitizing a paper document can cause relevant information to be lost or changed, but the shock value of his anecdote comes as his own persona shifts from brainy researcher to embodied asthmatic; he chokes, quite literally, on the realization that the nose can convey more information than the eye (Brown and Duguid 173). A different mind-body contrast structures another early twenty-first century argument against digitization, Sean Latham and Robert Scholes's "The Rise of Periodical Studies," which faults an editor for failing to reproduce the advertisements from a 1711 *Spectator*. The example given is an advertisement for a "Tincture to restore the Sense of Smelling, tho' lost for many Years" (Latham and Scholes 526). For these literary critics as for the business historian, to think of the book as something more than words is to recover a long-lost body. For them, too, to use the nose along with the eyes is to identify oneself as a handmaiden of literature.[35]

The "accidental" can refer either to mistakes like the burning of a manuscript of *The French Revolution*, or to punctuation, spacing, and other features that have been excluded from the historically shifting boundaries of what we call "the text itself."[36] In neither case does an absence of authorized intention imply an absence of legible meaning. The dog-eared page, the uncut page, the faded page, even the page that smells of the cheese it once wrapped or the vinegar that once disinfected it— each of these can tell us something about what users have done to their books and books have done to their users. What they can't tell us about is the way in which a text was read, much less in forms that a literary-historical or intellectual-historical training renders interesting.

The cabinetmaker's wife teaches us to see the page at once as the lining of a barrow and the lines of a poem. Looking backward to the satirical tradition, Mayhew also looks forward to the question that faces historians of the book: how to disentangle reading from handling. *London Labour* forbids us to parse that difference as hierarchy: to position the text as prior to the book (the book as a residue left over once the text has been used up); or as superior to the book (the book as the province first of illiterate grocers, then of theoretically illiterate bibliographers); or as purer than the book (the book as manual, the text as digital; the book as the dusty residue clogging up our libraries, the text as abstract thought streaming into an ethereal future).

Conclusion

FOR GISSING AT century's end, membership in an audience tainted the aesthetic: "My pleasure in the finest music would be greatly spoilt by having to sit amid a crowd, with some idiot audible on the right hand or left" (*The Private Papers of Henry Ryecroft* 125).[1] In *New Grub Street* the shallow and quick writer's wife ceases to be a sympathetic character at the moment when she stops noticing defects "to which the common reader would be totally insensible" and starts commenting instead "on the features of the work which had made it popular": when, in other words, she positions herself within a public rather than against it. In the process, texts give way to "persons":

> [Amy Reardon's] interests were becoming more personal; she liked to hear details of the success of popular authors—about their wives or husbands, as the case might be, their arrangements with publishers, their methods of work ... She talked of questions such as international copyright, was anxious to get an insight into the practical conduct of journals and magazines, liked to know who 'read' for the publishing houses. (99)

"Questions such as international copyright," "practical conduct of journals and magazines," "who 'read' for the publishing houses": Amy's new concerns read like a parody of the book-historical scholarship that would emerge half a century later. The quotation marks around "read" reduce the agent to a focus group, responsible no longer for describing her own response but for predicting the response of other potential buyers. Like the midcentury tract-lady, the late-century businesswoman cares more about others' reading than about her own; both show more interest in distributing books than in reading texts. Once a frivolous dreamer, the female inscribed reader is now too hardheaded: instead of a passive consumer, an overactive middlewoman. In this overcrowded landscape, the book no longer serves as block.

The book, or more specifically the novel: *New Grub Street*'s sensitivity to reading by proxy may stem from its own genre's susceptibility to such secondhand reading. From the beginning, its reviewers frame their arguments not around their own response, but around speculations about the response of a collective, commercialized readership: thus a 1809 review of Edgeworth's tales begins, "If the importance of a literary work is to be estimated by the number of readers which it attracts, . . . a novel or a tale cannot be deemed a trifling production" (quoted in Ferris, *The Achievement of Literary Authority* 32), and James Fitzjames Stephen's 1855 review of Dickens opens by establishing the popularity of novels and adding almost apologetically that "measured by these standards, their importance must be considered very great" (J. F. Stephen 148).

Tracts worry less, we've seen, about reading the wrong texts than about touching a book that had passed through the wrong hands or the wrong number of handlers. Gissing's satire identifies an even more diffuse threat: that the book might become a vector for the social entanglements from which it's supposed to provide an escape. The reader's relation with the book (or the author) is pure, his relation with other readers (or handlers) tainted. If the happiest books have no history, the happiest reader is also the one who can imagine himself to be their first, their only, their implied, their intended.[2]

Where fantasized romances with literary characters crowd out the consciousness of one's own husband, wife, or stepparent, so fantasized intimacy with authors upstages any sense of commonality with other readers. As Woolf observed, "it is precisely because we hate and we love that our relation with the poets and novelists is so intimate that we find the presence of another person intolerable" ("How Should One Read a Book?" 268). The fin-de-siècle cult of the uncut page forms the mirror image of the book fumigator. In Emerson's analogy, "a man's library is a sort of harem, and tender readers have a great pudency in showing their books to a stranger" (Emerson and Lubbock 21).

Common appreciation of literary texts had long been imagined as a social cement. When Diderot writes that "I have never met anyone who shared my enthusiasm [for Richardson's characters] without being tempted to hug him," or when the preface to *Pamela* quotes a master's praises of the serving-boy who cries upon overhearing the novel read aloud, they invest the book with the power to bridge distances of space and rank (Diderot 1099; S. Richardson, *Pamela* 19). By the nineteenth century, however, the emotions generated by shared reading were coded less positively. Fellow readers can avoid one another (like those public library users who wanted their books disinfected) or even compete for space—sometimes quite literally for room in the margin. Where the virgin page vehicles a meeting of minds between author and reader, an already-annotated book forces each reader to recognize himself as only one of a series. H. J. Jackson asks, "do second readers respond to the first? Identify with the first? Seek to differentiate themselves? Ignore the other's notes altogether?" (*Romantic Readers* 263) Those twenty-first-century scholars who debate whether to speak of "the reader" in the abstract or of empirical, plural readers echo nineteenth-century writers' recourse to reading as one of the arenas in which to explore the relation of the individual to the mass.[3]

Today, a gulf separates any literary critic's description of his own reading of a particular text—whose interest lies in its atypicality, even its perverseness—from a scholar's description of readings that are removed from his own world (whether in time or in habitus: an academic sociologist describing a middlebrow book club, for example) and whose agent is imagined as either collective or representative (Long). In recent memory, that gulf has mapped on to a division of labor between two disciplines, literary criticism and cultural history. It's true that the latter doesn't always focus on collectives; on the contrary, microhistory has produced a spate of exemplary figures whose liveliness depends precisely on their eccentricity. Yet in grammatical terms, literary critics continue to speak of "the reader" (or "this reader"), where historians more easily speak of "readers." The contrast between the novel of manners described in chapter 2 and the romantic bildungsroman analyzed in chapter 3 prefigures the disciplinary tension between reception studies and literary-critical introspection.

My argument could be taken as one very long footnote to Natalie Davis's long-ago reminder that the book constitutes "not merely a source for ideas . . . but a carrier of relationships" (N. Davis 192). Reciprocally, relationships are carriers of texts: human interactions transmit books, change books, and imbue books with meaning. Those relationships can be serial (think of the marginalia contributed by successive readers of a single copy of the *Imitation of Christ*) or synchronic (think of the simultaneous reading of multiple copies of the newspaper). Benedict Anderson's famous account of the second scenario has focused scholars' attention on the public sphere of the "subway" and the "barbershop," but the homes described in religious tracts, and the parishes in which tracts themselves circulated, form more fraught—as well as more feminine—venues for book sharing (Anderson 35). I contrasted the newspaper's power to build Andersonian bridges among same-sex strangers with its power to drive Trollopian wedges between opposite-sex intimates. In place of the "community in anonymity" that Anderson describes, tracts both create and represent face-to-face relationships. What links members of this community is something more concrete than the newspaper as conceptualized by Anderson: a book that cannot be in two places at once (though it can be put to two successive uses), and whose different copies are far from interchangeable. The point is not simply that a model of print culture that made tracts its exemplar would look less heroic than one that took the newspaper as representative. More strikingly, even Anderson's own example presumes a presentist model of how (and when) the newspaper circulates. Taking for granted what he calls "the obsolescence of the newspaper on the morrow of its printing," Anderson adds that "we know that particular morning and evening editions will overwhelmingly be consumed between this hour and that, only on this day, not that. (Contrast sugar, the use of which proceeds in an unclocked, continuous flow; it may go bad, but it does not go out of date)" (35).

We've seen, however, what a small and unrepresentative fraction of any newspaper's life cycle was formed by its first day. One of the many objects handed along—or in social terms, handed down—the paper subsided by easy stages from rich readers to poor readers and finally from readers to the grocery, the kitchen, and the privy. Oliphant's historical novel *Kirsteen* (published in 1890 but set at the other end of the century) describes "a Glasgow paper, posted by its first reader the day after publication to a gentleman on Loch Long, then forwarded by him to Inveralton, thence to Drumcarro. Mr Pyper at the Manse got it at fourth hand. It would be difficult to trace its wanderings after that" (40). In this narrative, the newspaper marks not the functional equality of men within the public sphere, but on the contrary the differences of rank among readers (the later the lower), and even more sharply between readers and those end users (wrappers? wipers?) who populate *London Labour* but which Oliphant's more genteel final clause declines to name.

By assuming that "the daily newspaper was made to be perishable, purchased to be thrown away," *Imagined Communities* confuses the real simultaneity of the "news" with the putative synchronicity of the "paper" (Terdiman 120).[4] And in the absence of synchronicity, virtuality disappears as well. No longer dematerialized, papers become paper; no longer disembodied, its users not only read, but eat and defecate. As virtuality goes, so goes equality: if words knit individuals into a nation, paper splinters them into masters and servants, men and women, stepparents and orphans.[5] About those relations, books bear tales out of which we can never get the reading.

Notes

1. Letter to C.W.W. Wynn, 25 June 1805, in Robert Southey, "State and Prospects of the Country," *The Emergence of Victorian Consciousness, the Spirit of the Age*, ed. George Lewis Levine (New York: Free Press, 1967), 239; he elsewhere reiterated that "the main demand for contemporary literature comes from [circulating] libraries, books being now so inordinately expensive that they are chiefly purchased as furniture by the rich. It is not a mere antithesis to say that they who buy books do not read them, and that they who read them do not buy them. I have heard of one gentleman who gave a bookseller the dimensions of his shelves, to fit up his library." Robert Southey, *Letters from England*, The Cresset Library (London: Cresset Press, 1951), 349. Compare Walter Benjamin's declaration that "the non-reading of books . . . should be characteristic of all collectors," and Italo Calvino's list of "Books Made for Purposes Other than Reading." Walter Benjamin, "Unpacking My Library," *Illuminations*, ed. Hannah Arendt (New York: Harcourt Brace, 1985), 62; Italo Calvino, *If on a Winter's Night a Traveler*, 1st ed. (New York: Harcourt Brace Jovanovich, 1981), 5. On Victorian anxiety about the materiality of the book, see Kevin Dettmar, "Bookcases, Slipcases, Uncut Leaves: The Anxiety of the Gentleman's Library," *Novel* 39 (2005 [i.e., 2006]); and, for a longer history, Jeffrey Todd Knight, "'Furnished' for Action: Renaissance Books as Furniture," *Book History* 12 (2009). On book historians' inconsistent usages of "materiality," see David Ayers, "Materialism and the Book," *Poetics Today* 24.4 (2003). My subject draws more generally on Carla Mazzio and Bradin Cormack's elegant description of what they call "book use" and "book theory." Bradin Cormack and Carla Mazzio, *Book Use, Book Theory: 1500–1700* (Chicago: University of Chicago Library, 2005).

2. Cp. James Kearney, "The Book and the Fetish: The Materiality of Prospero's Text," *Journal of Medieval and Early Modern Studies* 32.3 (2002): 449.

3. In Robin Bernstein's alternative taxonomy, the book functions at once as a "text" that encloses meaning, as a "script" that instructs past and future performances, and as a "prop" that carries evidence of past uses; and lest terms like "script" and "performance" should make one think of an actor reading aloud, it's worth adding that those performances don't need to take the verbal contents of the book as their prompt at all. Robin Bernstein, "Dances with Things: Material Culture and the Performance of Race," *Social Text* 101 (2009): 40–41.

4. Even that would have done nothing to address the related problem that the English language has an umbrella term for different genres of writing—"text"— but no equally wieldy term to encompass different kinds of inscribed object. To avoid mouthfuls like "pieces of printed matter," I group newspapers and magazines

among "book-objects"—even though much iconography contrasted the two (as we'll see in chapter 2), and, in a period rife with reprinting, the same text was read and used very differently depending on whether it appeared in the pages of a periodical or a freestanding volume. For precedent, I can plead the most authoritative recent reference book's declaration that "the use of the term 'book' in our title does not exclude newspapers, prints, sheet music, maps, or manuscripts." Michael Felix Suarez and H. R. Woudhuysen, "Introduction," *The Oxford Companion to the Book*, ed. Michael Felix Suarez and H. R. Woudhuysen (Oxford: Oxford University Press, 2010), x; or, further back, Lamartine's much-quoted 1831 slogan: "the only book possible from today is a Newspaper." See also Thomas Tanselle's remark that "symptomatic of the confusion is the use of 'book' to mean both intangible work and physical object—more often the former, necessitating the use of such phrases as 'the book as a physical object' when speaking of the latter. The expression 'a good book' does not normally refer to a well-designed book." G. Thomas Tanselle, "Libraries, Museums, and Reading," *Literature and Artifacts* (Charlottesville: Bibliographical Society of the University of Virginia, 1998), 10.

5. Paul Duguid, "Thought for Food," quoting W.J.T. Mitchell and Bill Hill: http://www.icdlbooks.org/meetings/duguid.html.

6. http://g-ecx.images-amazon.com/images/G/01/digital/fiona/general/Jeff_letter_narrow._V5047014_.png.

7. Conrad of Hirsau explains that "a book is the name given to parchment with marks on it. This name originated from the bark of a tree on which men used to write before the use of animal skin . . . 'Book' (*liber*) is so called from the verb 'to free', because the man who spends his time reading often releases his mind from the anxieties and chains of the world." A. J. Minnis, A. Brian Scott, and David Wallace, *Medieval Literary Theory and Criticism c.1100–c.1375: The Commentary Tradition*, rev. ed. (Oxford: Clarendon Press, 1991), 42.

8. Mitch traces popular literacy to the growth of a national sport network made possible in turn by the growth of railways and telegraphs. Betting spurred interest in sporting news and sales of newspapers: "Many a man made the breakthrough to literacy by studying the pages of the *One O' Clock*" (quoted in David Mitch, *The Rise of Popular Literacy in Victorian England* [Philadelphia: University of Pennsylvania Press, 1992], 60–61).

CHAPTER 1
READER'S BLOCK

1. See, e.g., Robert Darnton, "Readers Respond to Rousseau," *The Great Cat Massacre and Other Episodes in French Cultural History* (London: Allen Lane, 1984); for a suggestive analogy in the visual arts, see James Elkins, *Pictures & Tears: A History of People Who Have Cried in Front of Paintings* (New York: Routledge, 2001).

2. As Ina Ferris observes, "the history of the book remains primarily oriented toward a history of publishing, so that books function mainly as physical units available to empirical study and description (verbal, statistical, graphic, etc.),

while its model of history (like that of most historicisms) privileges the parameters of production." Ina Ferris, "Introduction," *Romantic Libraries* (College Park: University of Maryland, 2004).

3. See, e.g., Meredith McGill, *American Literature and the Culture of Reprinting, 1834–1853* (Philadelphia: University of Pennsylvania Press, 2003), and William H. Sherman, *Used Books: Marking Readers in Renaissance England*, Material Texts (Philadelphia: University of Pennsylvania Press, 2008); for a survey of the field, Leah Price, "Reading: The State of the Discipline," *Book History* 7 (2004).

4. Other models to which mine is indebted include what Mark McGurl describes (with a keen sense of its ironies) as "philistine literary criticism," and the approach to media studies that Lisa Gitelman (borrowing from Alfred Gell) calls "methodological philistinism": Lisa Gitelman, *Always Already New: Media, History, and the Data of Culture* (Cambridge, Mass.: MIT Press, 2006); Mark McGurl, *The Novel Art: Elevations of American Fiction after Henry James* (Princeton, N.J.: Princeton University Press, 2001), 19–20.

5. On the competing metaphors, see G. N. Cantor, *Science in the Nineteenth-Century Periodical: Reading the Magazine of Nature*, Cambridge Studies in Nineteenth-Century Literature and Culture, 45 (Cambridge: Cambridge University Press, 2004), 1.

6. See also Roger Chartier's distinction between "heteronomous but nonetheless interconnected forms of logic—the ones that organize utterances and the ones that command action and behavior." Roger Chartier, ed., *On the Edge of the Cliff: History, Language and Practices* (Baltimore, Md.: Johns Hopkins University Press, 1997), 1. Thomas Richards points out that it's in this period that the meaning of "reading" expands from the interpretation of texts to the indication of graduated instruments. Thomas Richards, *The Imperial Archive: Knowledge and the Fantasy of Empire* (London: Verso, 1993), 18.

7. Charles Dickens, *Our Mutual Friend*, ed. Stephen Gill (Harmondsworth: Penguin, 1985), 605, 263, 81, 636. On this passage, see also Garrett Stewart, *Dear Reader: The Conscripted Audience in Nineteenth-Century British Fiction* (Baltimore, Md.: Johns Hopkins University Press, 1996), 232.

8. The article dates to 1991; I owe the reference to Richard Biernacki, "Method and Metaphor after the New Cultural History," *Beyond the Cultural Turn: New Directions in the Study of Society and Culture*, ed. Victoria E. Bonnell and Lynn Avery Hunt (Berkeley: University of California Press, 1999), 74. See also Mark A. Schneider, "Culture-as-Text in the Work of Clifford Geertz," *Theory and Society* 16.6 (1987) on the conflation of "culture" with "text," or Ian Hunter's critique of "the imperative to expand the esthetic concept of culture . . . to all social activities and relations," Ian Hunter, "Setting Limits to Culture," *New Formations* 2 (Spring 1988): 115, or Bourdieu's claim that "to read a ritual—which is something like a dance—as if it were discursive and could be expressed in mathematical terms is to transform it profoundly." Todd W. Reeser and Steven D. Spalding, "Reading Literature/Culture: A Translation of 'Reading as a Cultural Practice,'" *Style* 36.4 (2002): 664. Hayden White charges that "the interpretation of cultural phenomena is regarded merely as a special case of the act of reading, in which the manipulation and exchange of signs is carried out most self-consciously, the act

of reading literary texts." Hayden White, *Tropics of Discourse: Essays in Cultural Criticism* (Baltimore, Md.: Johns Hopkins University Press, 1978), 262. The emphasis might usefully be placed instead on "literary *texts*."

9. On the place of smell in book history, see Sean Latham and Robert Scholes, "The Rise of Periodical Studies," *PMLA* 121 (2006): 526.

10. I borrow the term "deflation" from Bruno Latour, "Drawing Things Together," *Representation in Scientific Practice* (Cambridge, Mass.: MIT Press, 1990). See also Mary Poovey's argument that "our modern interpretational habits typically track at too high a level of abstraction to give [material vessels] their due"; and David Greetham's argument that if mid-twentieth-century "textual activity could be called Platonist, then the postmodern descriptive mode might be seen as Aristotelian." Mary Poovey, "The Limits of the Universal Knowledge Project: British India and the East Indiamen," *Critical Inquiry* 31.1 (2004): 183–202; 202. David Greetham, "What Is Textual Scholarship?" *A Companion to the History of the Book*, ed. Simon Eliot and Jonathan Rose (Malden, Mass.: Blackwell Pub., 2007), 29.

11. Matthew Brown argues that the one category of object that scholars of material culture shy away from is the book. Even their successors, "thing theorists" like Bill Brown, ask how books represent things, not how books are things (Harvard Humanities Center talk, 2007). See, however, Bill Brown, "Introduction: Textual Materialism," *PMLA* 125.1 (2010).

12. See, e.g. (among many possible examples), Friedrich A. Kittler, *Discourse Networks*, trans. Michael Metteer and Chris Cullens (Stanford, Calif.: Stanford University Press, 1990); Gitelman, *Always Already New: Media, History, and the Data of Culture*; Dora Thornton, *The Scholar in His Study: Ownership and Experience in Renaissance Italy* (New Haven, Conn.: Yale University Press, 1997); Patricia Crain, *The Story of A: The Alphabetization of America from the New England Primer to the Scarlet Letter* (Stanford, Calif.: Stanford University Press, 2002); Pamela Thurschwell, "Henry James and Theodora Bosanquet: On the Typewriter, in the Cage, at the Ouija Board," *Textual Practice* 13 (1999); Lawrence S. Rainey, *Institutions of Modernism: Literary Elites and Public Culture* (New Haven, Conn.: Yale University Press, 1998); Robert J. Mayhew, "Materialist Hermeneutics, Textuality and the History of Geography: Print Spaces in British Geography, c. 1500–1900," *Journal of Historical Geography* 33 (2007). Kate Flint performs such a move in the Victorian period itself, asking what happens to our metaphor of texts "transporting" readers when those readers (or, one might add, those texts) are being very literally transported—on a train, for example. Kate Flint, *The Feeling of Reading: Affective Experience and Victorian Literature*, ed. Rachel Ablow (Ann Arbor: University of Michigan Press, 2010), 28–29.

13. See, e.g., Edgar Johnson, *Charles Dickens: His Tragedy and Triumph* (New York: Viking, 1977), 389. For other examples of nineteenth-century dummy books, see H. J. Jackson, *Marginalia: Readers Writing in Books* (New Haven, Conn.: Yale University Press, 2001), 134–35.

14. See also Charlotte Allen, "Indecent Disposal: Where Academic Books Go When They Die," *Lingua Franca* 5.4 (1995).

15. See John Plotz's argument that the book is defined by the tension between its "metonymic" and "metaphoric" powers—functions that correspond to the two

poles that I call "material form" and "verbal content." John Plotz, "Out of Circulation: For and Against Book Collecting," *Southwest Review* 84.4 (1999): 472.

16. The corresponding scene in the parallel subplot comes when Rebecca interrupts her singing to burn a letter from *her* lover. Asked why the music has stopped, she answers in a voice that echoes the narrator's, or rather the captioner's: "it's a false note." William Makepeace Thackeray, *Vanity Fair*, ed. Peter L. Shillingsburg (New York: Norton, 1994), 113. At the other end of the narrative, once financial exchanges have replaced amorous ones, the roles are reversed: Rawdon himself hopes to "wrap a ball in the note"—this time, a banknote— "and kill Steyne with it." Thackeray, *Vanity Fair*, 539.

17. William Makepeace Thackeray, "The Memoirs of Mr. C. J. Yellowplush," *Christmas Books; Snobs and Ballads* (New York: Metropolitan Publishing Company), 292, 288. For a very different argument about Victorian literalization, see Margaret Homans, *Bearing the Word: Language and Female Experience in Nineteenth-Century Women's Writing* (Chicago: University of Chicago Press, 1986), 29–30.

18. Thanks to Nick Dames for sharpening my thinking on this point.

19. In *The Struggles of Brown, Jones and Robinson* (1862), Trollope returns to the theme: the narrator is rudely told, "You've been making out all these long stories about things that never existed, but what's the world the better for it;— that's what I want to know. When a man makes a pair of shoes—" Anthony Trollope, *The Struggles of Brown, Jones and Robinson* (Harmondsworth: Penguin, 1993), 214.

20. Marian Evans, too, opens a slashing review of Lord Brougham's *Lives of Men of Letters* by asking readers what they would think of a wealthy amateur pretending to make boots. George Eliot, "J. A. Froude's *The Nemesis of Faith*," *Selected Essays, Poems, and Other Writings*, ed. Nicholas Warren (London: Penguin, 1990), 265. For the relationship of shoes to intellectuals, see also Jacques Rancière, *The Philosopher and His Poor*, trans. Andrew Parker (Durham, N.C.: Duke University Press, 2004), 67–69, and "Political Shoemakers," in E. J. Hobsbawm, *Workers: Worlds of Labor*, 1st American ed. (New York: Pantheon Books, 1984), esp. 107.

21. James Lackington, *Memoirs of the First Forty-Five Years of the Life of James Lackington, Bookseller* (1791; 1792), reprinted in Paul Keen, *Revolutions in Romantic Literature: An Anthology of Print Culture, 1780–1832* (Peterborough, Ont.: Broadview, 2004), 48.

22. John Ruskin, *The Political Economy of Art* (London: Dent, 1968), 39.

23. George Gissing, *The Private Papers of Henry Ryecroft* (New York: Modern Library, n.d.), 43, 125; see also Janice Radway, "Reading Is Not Eating: Mass-Produced Literature and the Theoretical, Methodological, and Political Consequences of a Metaphor," *Book Research Quarterly* 2 (1986). On food as metaphor in this period more generally, see Denise Gigante, *Taste: A Literary History* (New Haven, Conn.: Yale University Press, 2005).

24. Cp. Mary Jean Corbett's argument that "masculine authorship requires women's domestic labor." Mary Jean Corbett, *Representing Femininity: Middle-Class Subjectivity in Victorian and Edwardian Women's Autobiographies* (New York: Oxford University Press, 1992), 63.

25. I am indebted here to Regenia Gagnier's account of an "abstract individualism" that emerged in the nineteenth century in opposition to "material conditions and social environment." Regenia Gagnier, *Subjectivities: A History of Self-Representation in Britain, 1832–1920* (New York: Oxford University Press, 1991), 39.

26. Discussing nineteenth-century thinkers' tendency to take reading as an exemplar of automaticity, Deidre Lynch cites Erasmus Darwin's *Zoomania* (1796), where the words "PRINTING PRESS" are followed by a passage asking readers whether they noticed the shape and size of the thirteen letters or simply conjured up a mental image of the "most useful of modern inventions." Deidre Lynch, "Canon's Clockwork: Novels for Everyday Use," *At Home in English: A Cultural History of the Love of Literature* (Chicago: University of Chicago Press, forthcoming).

27. My discussion exemplifies—for better and worse—what Ian Hunter has called a "European university metaphysics" characterized by the tension between "an infinite, atemporal, self-active, world-creating intellect, and a finite, 'duplex' (intellectual-corporeal) worldly being." Ian Hunter, "The History of Theory," *Critical Inquiry* 33.1 (2006): 98. I am indebted, too, to Ina Ferris's critique of "a long-standing European tradition of looking through actual books to an immaterial 'wisdom' they make available to readers, usually understood (as here) primarily as 'intellectual beings.'" Ferris, "Bibliographic Romance: Bibliophilia and the Book-Object," *Romantic Libraries*.

28. In Catherine Gallagher's pithy formulation, "print paradoxically gave material evidence for a text surpassing all copies." Catherine Gallagher, *Nobody's Story: The Vanishing Acts of Women Writers in the Marketplace, 1670–1820* (Berkeley: University of California Press, 1994), 65.

29. Compare Garrett Stewart's argument that "reading dematerializes signs on the way to its imagined scene." Garrett Stewart, *The Look of Reading: Book, Painting, Text* (Chicago: University of Chicago Press, 2006), 118.

30. For a fuller discussion, see Price, "Reading: The State of the Discipline."

31. Victorianists became interested in reading before scholars in many other fields, and the result is a rich body of secondary literature, stretching from the period itself up to the present, to which this book is everywhere indebted. See especially (in addition to studies cited elsewhere in this book) Robert K. Webb, *The British Working Class Reader, 1790–1848: Literacy and Social Tension* (London: Allen & Unwin, 1955); Guinevere Griest, *Mudie's Circulating Library and the Victorian Novel* (Bloomington: Indiana University Press, 1970); John Sutherland, *Victorian Novelists and Publishers* (Chicago: University of Chicago Press, 1976); Robert Patten, *Charles Dickens and His Publishers* (Oxford: Clarendon Press, 1978); Scott Bennett, "Revolutions in Thought: Serial Publication and the Mass Market in Reading," *The Victorian Periodical Press: Samplings and Soundings*, ed. Joanne Shattock and Michael Wolff (Leicester: Leicester University Press, 1982); Raymond Williams, *Culture and Society, 1780–1950* (New York: Columbia University Press, 1983); Laurel Brake, "Literary Criticism in Victorian Periodicals," *Yearbook of English Studies* 16 (1986); Jon P. Klancher, *The Making of English Reading Audiences, 1790–1832* (Madison: University of Wisconsin Press, 1987); Margaret Diane Stetz, "Life's 'Half-Profits': Writers and Their Read-

ers in Fiction of the 1890s," *Nineteenth-Century Lives*, ed. Laurence Lockridge et al. (Cambridge: Cambridge University Press, 1989); Ina Ferris, *The Achievement of Literary Authority: Gender, History, and the Waverley Novels* (Ithaca, N.Y.: Cornell University Press, 1991); Linda K. Hughes and Michael Lund, *The Victorian Serial* (Charlottesville: University of Virginia Press, 1991); Kate Flint, *The Woman Reader, 1837–1914* (Oxford: Clarendon Press, 1993); Alan Richardson, *Literature, Education, and Romanticism: Reading as Social Practice, 1780–1832* (Cambridge: Cambridge University Press, 1994); Jonathan Rose, "How to Do Things with Book History," *Victorian Studies* 37.3 (1994): 461–71; Margaret Beetham, *A Magazine of Her Own? Domesticity and Desire in the Woman's Magazine, 1800–1914* (London: Routledge, 1996); John Jordan and Robert Patten, *Literature in the Marketplace: Nineteenth-Century British Publishing and the Circulation of Books* (Cambridge: Cambridge University Press, 1995); Helen Small, "A Pulse of 124: Charles Dickens and a Pathology of the Mid-Victorian Reading Public," *The Practice and Representation of Reading in England*, ed. James Raven, Helen Small, and Naomi Tadmor (Cambridge: Cambridge University Press, 1996); Stewart, *Dear Reader: The Conscripted Audience in Nineteenth-Century British Fiction*; Peter D. McDonald, *British Literary Culture and Publishing Practice, 1880–1914* (Cambridge: Cambridge University Press, 1997); Richard Altick, *The English Common Reader: A Social History of the Mass Reading Public, 1800–1900* (Chicago: University of Chicago Press, 1998); Patrick Brantlinger, *The Reading Lesson: The Threat of Mass Literacy in Nineteenth-Century Britain* (Bloomington: University of Indiana Press, 1998); Bill Bell, "Bound for Australia: Shipboard Reading in the Nineteenth Century," *Journeys through the Market: Travel, Travellers, and the Book Trade*, ed. Robin Myers and Michael Harris (New Castle, Del.: Oak Knoll Press, 1999); Martyn Lyons, "New Readers in the Nineteenth Century: Women, Children, Workers," *A History of Reading in the West*, ed. Guglielmo Cavallo and Roger Chartier (Cambridge: Polity Press, 1999); James A. Secord, *Victorian Sensation: The Extraordinary Publication, Reception, and Secret Authorship of Vestiges of the Natural History of Creation* (Chicago: University of Chicago Press, 2000); Philip Connell, "Bibliomania: Book Collecting, Cultural Politics, and the Rise of Literary Heritage in Romantic Britain," *Representations* 71 (2000); Graham Law, *Serializing Fiction in the Victorian Press* (Houndsmills: Palgrave, 2000); Jonathan Rose, *The Intellectual Life of the British Working Classes* (New Haven, Conn.: Yale University Press, 2001); William St Clair, *The Reading Nation in the Romantic Period* (Cambridge: Cambridge University Press, 2004); Daniel Hack, *The Material Interests of the Victorian Novel*, Victorian Literature and Culture Series (Charlottesville: University of Virginia Press, 2005); H. J. Jackson, *Romantic Readers: The Evidence of Marginalia* (New Haven, Conn.: Yale University Press, 2005); Mary Hammond, *Reading, Publishing, and the Formation of Literary Taste in England 1880–1914* (Aldershot: Ashgate, 2006); Nicholas Dames, *The Physiology of the Novel: Reading, Neural Science, and the Form of Victorian Fiction* (Oxford: Oxford University Press, 2007); David McKitterick, *The Cambridge History of the Book in Britain, 1830–1914* (Cambridge: Cambridge University Press, 2009).

32. On Protestantism as a religion of the book in this period, see Jon P. Klancher, "The Bibliographer's Tale, or the Rise and Fall of Book History in Britain

(1797–1825)" (paper presented at the American Society for Eighteenth-Century Studies, Montreal, 31 March 2006); on the language of the "fetish," Peter Melville Logan, *Victorian Fetishism: Intellectuals and Primitives*, SUNY Series, Studies in the Long Nineteenth Century (Albany: State University of New York Press, 2009), 81.

33. Kearney, "The Book and the Fetish: The Materiality of Prospero's Text," 436; William Tyndale, *The Obedience of a Christian Man*, ed. David Daniell, Penguin Classics (London: Penguin, 2000), 161.

34. Edwards, manuscript scrapbook, Book E (1860–1876), Manchester Public Library Archives, p. 189, quoted in Alistair Black, *A New History of the English Public Library: Social and Intellectual Contexts, 1850–1914* (London: Leicester University Press, 1996), 93.

35. "The Role of the Black West Indian Missionary in West Africa, 1840–1890" (PhD diss., Temple University, 1972), 121, quoted in Isabel Hofmeyr, "Metaphorical Books," *Current Writing* 13.2 (2001): 105.

CHAPTER 2
ANTHONY TROLLOPE AND THE REPELLENT BOOK

1. See also Nicholas Dames's observation that "the very straightforward gerund 'reading' is almost invisible in Thackeray who prefers the use of slightly imperfect synonyms which reflect discontinuity, such as 'subsiding into' or 'simpering over' a book; to 'turn over the leaves,' 'dip into,' or 'muse over' a volume; or the virtually constant use of 'peruse' or 'perusal' to stand in for any more continuous 'reading'." Dames, *The Physiology of the Novel: Reading, Neural Science, and the Form of Victorian Fiction*, 108.

2. By Eliot's time, the joke is already a hackneyed one: when Lydia Languish asks, "What are those books by the glass?" her maid answers, "The great one is only The Whole Duty of Man—where I press a few blondes, ma'am." Richard Sheridan, *The Rivals*, 1.2.

3. On this passage, see also Kevin McLaughlin, *Paperwork: Fiction and Mass Mediacy in the Paper Age* (Philadelphia: University of Pennsylvania Press, 2005), 111–12.

4. On the relation between novel and newspaper in Victorian culture, see Richard D. Altick, *The Presence of the Present: Topics of the Day in the Victorian Novel* (Columbus: Ohio State University Press, 1991), and Matthew Rubery, *The Novelty of Newspapers: Victorian Fiction after the Invention of the News* (Oxford: Oxford University Press, 2009). See also one commentator's mock lament that "during the morning meal, [the newspaper] interferes seriously, and amid protest, with the attention [the householder] ought to pay to the remarks of his wife." James David Symon, *The Press and Its Story; an Account of the Birth and Development of Journalism up to the Present Day, with the History of All the Leading Newspapers: Daily, Weekly, or Monthly, Secular and Religious, Past and Present; Also the Story of Their Production from Wood-Pulp to the Printed Sheet* (London: Seeley Service & Co., 1914), 1.

5. Cp. "in the leisured, stylish world of [James's] late novels in particular, books are a part of the whole material of social exchange, and James's interest in them is almost anthropological." Tessa Hadley, "Seated Alone with a Book . . . " *Henry James Review* 26 (2005): 231.

6. Benedict Anderson, *Imagined Communities: Reflections on the Origin and Spread of Nationalism* (London: Verso, 1991), 35. Goffman's study of the 1954 newspaper strike argues that its main impact was not on public access to information, but rather on subway commuters' use of space: Erving Goffman, *Behavior in Public Places: Notes on the Social Organization of Gatherings* (New York: Free Press, 1963), 52. For an eloquent recent argument for Goffman's importance to literary interpretation, see Jeff Nunokawa, *Tame Passions of Wilde* (Princeton, N.J.: Princeton University Press, 2003), 58–59. My understanding of the paradoxes of silent reading in public draws on Georg Simmel, *The Sociology of Georg Simmel*, ed. Kurt Wolff (New York: Free Press, 1950), 337, as well as on David Henkin's argument that reading in antebellum New York constituted a public performance, even—or especially—when silent. David Henkin, *City Reading: Written Words and Public Spaces in Antebellum New York* (New York: Columbia University Press, 1998), 110. David Vincent, too, associates privacy with "control" rather than with "isolation": David Vincent, *The Rise of Mass Literacy: Reading and Writing in Modern Europe* (Cambridge: Polity, 2000), 103.

7. See, e.g., http://www.nytimes.com/2010/06/10/garden/10childtech.html?ref=garden.

8. Wharton's "The Line of Least Resistance" contrasts the boudoir where the real work of letter writing is done with the library where nothing happens: "Mr. Mindon has never quite known what the library was for; it was like one of those mysterious ruins over which archaeology endlessly disputes. It could not have been intended for reading, since no one in the house ever read, except an underhousemaid charged with having set fire to her bed in her surreptitious zeal for fiction." Edith Wharton, *The Collected Short Stories of Edith Wharton*, ed. R.W.B. Lewis (New York: Scribner, 1968), 219.

9. For a similar face-off between one spouse absorbed in reading and another who reduces the book to a projectile, see the Brontë-esque Helen Mathers, *Comin' Thro' the Rye: A Novel* (New York: A. L. Burt, 1876), 219.

10. Leslie Kaufman, "Making Their Own Limits in a Spiritual Partnership," *New York Times*, 15 May 2008.

11. As Mikita Brottman asked more recently, "how many nights do you have to spend with a new lover before it's acceptable to reach out and take a book from the nightstand?" Mikita Brottman, *The Solitary Vice: Against Reading* (Berkeley, Calif.: Counterpoint, 2008), 71. On the honeymoon as a Victorian institution more generally, see Helena Michie, *Victorian Honeymoons: Journeys to the Conjugal*, Cambridge Studies in Nineteenth-Century Literature and Culture, 53 (Cambridge: Cambridge University Press, 2006).

12. Mary Elizabeth Braddon, *The Doctor's Wife*, ed. Pykett Lynn (Oxford: Oxford University Press, 1998), 108; see also Barbara Leckie, *Culture and Adultery: The Novel, the Newspaper, and the Law, 1857–1914* (Philadephia: University of Pennsylvania Press, 1999), 147.

13. Unisar announces that its silent "'TV ears' will save your marriage"—and the gender of that "you" is made clear when the Web site adds, "No more arguments about watching TV in bed. And no more waking up your wife or baby because the TV is too loud." One satisfied customer on Amazon.com substitutes Unisar's product for the newspapers represented in Trollope, testifying that the "Marriage Saver" means that "he can watch TV in bed, while I can read in peace and quiet!" (http://www.amazon.com/gp/pdp/profile/A2YJORXMEWOXX5).

14. On the problem of attention in the nineteenth century, see Jonathan Crary, *Suspensions of Perception: Attention, Spectacle, and Modern Culture* (Cambridge, Mass.: MIT Press, 1999).

15. For a suggestive analogy from cinema, see Paul Willemen's analysis of "the look at the viewer": Paul Willemen and British Film Institute, *Looks and Frictions: Essays in Cultural Studies and Film Theory* (Bloomington: Indiana University Press; London: British Film Institute, 1994), 107.

16. *Sylph* no. 5, 6 October 1795, quoted in John Tinnon Taylor, *Early Opposition to the English Novel: The Popular Reaction from 1760 to 1830* (New York: King's Crown Press, 1943), 53.

17. Denis Diderot, "Eloge de Richardson," *Oeuvres complètes*, ed. Herbert Dieckmann, Jean Fabre, and Jacques Proust (Paris: Hermann, 1975). 1066 (my translation).

18. Robert Fraser quotes a play called *The Blinkards* (1915) by the Fante lawyer Kobina Sekyi, in which a lady intellectual scolds her maid for disturbing the petals pressed between the leaves of a book, explaining, "Haven't I told you that, in England, leaves are placed in the books to dry, the books when the leaves are dry being placed in drawing rooms?" Robert Fraser, *Book History through Postcolonial Eyes* (London: Routledge, 2008), 90.

CHAPTER 3
DAVID COPPERFIELD AND THE ABSORBENT BOOK

1. See also Charles Bernstein's analysis of "'absorption' and its obverses—impermeability, imperviousness, ejection, repellence." Charles Bernstein, *A Poetics* (Cambridge, Mass.: Harvard University Press, 1992), 20, and Michael Fried, *Realism, Writing, Disfiguration: On Thomas Eakins and Stephen Crane* (Chicago: University of Chicago Press, 1987), 120.

2. On the interchangeability of boots with books in a different Dickens text, *American Notes*, see Meredith McGill, "American Pickwick; or, The Artist in Boots" (unpublished paper, 2007).

3. On the distinction between online and offline processes in reading, see Andrew Elfenbein, "Cognitive Science and the History of Reading," *PMLA* 121 (2006).

4. Compare Garrett Stewart's argument that scenes of reading in modern easel painting (as opposed to, say, illuminated manuscript) require the represented book to be illegible: "the inner is guaranteed by its retreat from the seen." Stewart, *The Look of Reading: Book, Painting, Text*, 9.

5. When the late twentieth-century cyberpunk novelist Neal Stephenson has the heroine's stepfather throw a "Vicky" book at her head, he assumes that readers will remember the Victorian novel's tendency to reduce books to projectiles. Neal Stephenson, *The Diamond Age; or, a Young Lady's Illustrated Primer* (New York: Bantam, 1995), 94.

6. Grand is careful to point out the gender reversal at work here: the husband is disconcerted at this reaction because, although "it was inevitable that the man should tire . . . he would smile at pictures of the waning of the honeymoon, where the husband returns to his book and his dog, and the wife sits apart sad and neglected," but "that the wife should be the first to be bored was incredible." Sarah Grand, *The Beth Book* (New York: D. Appleton, 1897), 7, 372.

7. Even in *Vanity Fair* itself, Becky's book throwing is quickly enough redressed by a rather different school scene in which Dobbin, hiding from the rest of the school by dreaming over a book, is violently interrupted:

> [P]oor William Dobbin . . . was lying under a tree in the playground, spelling over a favourite copy of the Arabian Nights which he had apart from the rest of the school, who were pursuing their various sports . . . Well, William Dobbin had for once forgotten the world, and was away with Sindbad the Sailor in the Valley of Diamonds, or with Prince Ahmed and the Fairy Peribanou in that delightful cavern where the Prince found her, and whither we should all like to make a tour; when shrill cries, as of a little fellow weeping, woke up his pleasant reverie; and looking up, he saw Cuff before him, belabouring a little boy . . . Down came the stump with a great heavy thump on the child's hand. A moan followed. Dobbin looked up. The Fairy Peribanou had fled into the inmost cavern with Prince Ahmed: the Roc had whisked away Sindbad the Sailor out of the Valley of Diamonds out of sight, far into the clouds: and there was everyday life before honest William; and a big boy beating a little one without cause. (Thackeray, *Vanity Fair* 41)

Nothing further from Becky's performance than Dobbin's absorption: her worldly histrionics against the inwardness that allows the reader, "apart from the rest of the school," to "forget the world." The illustration places the book in the foreground, making its spine larger than Cuff's cane; the narrator's voice, too, pushes fairies and princes onto the same ontological plane as prosaic Dobbin. Such scenes confirm Deidre Lynch's description of the hero of late eighteenth-century biography as "endowed with a double life": "set in front of one book, his Latin grammar, for instance, he dreams of another." Lynch, "Canon's Clockwork: Novels for Everyday Use," 57.

8. See also Gayatri Spivak's argument that Jane's preference for image over text "can make the outside inside" and Antonia Losano on the contradiction that Brontë represents Jane choosing images over letterpress but herself refuses to have *Jane Eyre* illustrated. Gayatri Chakravorty Spivak, "Three Women's Texts and a Critique of Imperialism," *Critical Inquiry* 12.1 (1985): 246; Antonia Losano, "Reading Women/Reading Pictures: Textual and Visual Reading in Charlotte Brontë's Fiction and Nineteenth-Century Painting," *Reading Women: Literary and Cultural Icons from the Victorian Age to the Present*, ed. Janet Badia and Jennifer Phegley (Toronto: University of Toronto Press, 2005), 29.

9. On reverie in the novel, see Debra Gettelman, "Reverie, Reading, and the Victorian Novel" (PhD diss., Harvard University, 2005).

10. I am indebted here and throughout to Sharon Marcus's argument that "Jane experiences attacks on her body that often stem from her verbal outbursts or from an insufficiently abstract relation to reading and writing . . . The very text becomes a catastrophically material object." Sharon Marcus, "The Profession of the Author: Abstraction, Advertising, and *Jane Eyre*," *PMLA* 110.2 (1995): 209.

11. On this construction in *Jane Eyre*, see Stewart, *Dear Reader: The Conscripted Audience in Nineteenth-Century British Fiction*.

12. "Tolkien, the Book; Rereading Lord of the Rings," *Weekly Standard*, 31 December 2001, 43; Martha Nussbaum, *Love's Knowledge: Essays on Literature and Philosophy* (Oxford: Oxford University Press, 1992), 8; Charles M. Reed, *Reading As If for Life: Preparing Young Women for the Real World* (speech delivered at the Women's College at Brenau University, Gainesville, Georgia, 13 September 2001). Carla Peterson, *The Determined Reader: Gender and Culture in the Novel from Napoleon to Victoria* (New Brunswick, N.J.: Rutgers University Press, 1986), begins with an acknowledgment to the daughter whom she describes "curled up in an armchair with a book, 'reading as if for life' " (ix). See also Anne Fadiman, *Rereadings* (New York: Farrar, Straus and Giroux, 2005), xv. On this passage, see also Rosemary Lloyd, "Reading As If for Life," *Journal of European Studies* 22.3 (1992).

13. On reading aloud and library visits as compared to owning books, see Steven D. Levitt and Stephen J. Dubner, *Freakonomics: A Rogue Economist Explores the Hidden Side of Everything*, rev. and expanded ed. (New York: William Morrow, 2006), 172; as Maria Tatar nicely summarizes their claim, "parents (their socioeconomic status level of education and so on) matter [more than] parenting (reading to children, taking them to museums and so on." Maria Tatar, *Enchanted Hunters: The Power of Stories in Childhood*, 1st ed. (New York: W. W. Norton & Co., 2009), 33. For parents' educational level compared to presence of books in the home, National Endowment for the Arts *To Read or Not to Read*, 73. Similar results are cited in McQuillan, *The Literacy Crisis: False Claims and Real Solutions*, 1998, and Stephen D. Krashen, *The Power of Reading: Insights from the Research*, 2nd ed. (Westport, Conn.: Libraries Unlimited; Portsmouth, N.H.: Heinemann, 2004). In another counterintuitive study, "the amount of shared parent-child reading time did not matter, on average, for the reading skills of either group of kids [mothers who read an average amount and mothers who spent an unusually high amount of time reading]. What mattered instead . . . for the kids of the high-reading moms was how orderly the family's home was" (http://www.slate.com/id/2212318/).

14. As Andrew Miller shrewdly argues, "George de Barnwell's mistake—and, implicitly Bulwer's as well—is to conceive of books he reads as fundamentally superior to, and different in kind from, commodities, the currants and tea and cocoa sold in the shop . . . For Thackeray, however, books simply are objects, and are governed by the forces that govern the production of objects"—and, one could add, their consumption. Andrew Miller, *Novels behind Glass: Commodity Culture and Victorian Narrative* (Cambridge: Cambridge University Press, 1995), 45–46.

15. The trope of childhood reading as more authentic than adult reading persists in recent examples of the genre that Seth Lerer has dubbed "biblioautobiography" (i.e., nonacademic essays about the history of reading that contain a significant autobiographical component). Seth Lerer, "Falling Asleep over the History of the Book," *PMLA* 21 (2006): 230. See, e.g., Alberto Manguel, *A History of Reading* (London: HarperCollins, 1996), 10, and Sven Birkerts, *The Gutenberg Elegies: The Fate of Reading in an Electronic Age* (Winchester, Mass.: Faber and Faber, 1994), 37, 88–89. One of the few examples of such an essay that shows great sensitivity to the social structures that shape a child's imaginative engagement is Francis Spufford, *The Child That Books Built* (London: Faber and Faber, 2002)—but then the narrator's mother happens to be a historian of literacy.

16. On David's shift toward a more utilitarian model of literacy, see Leah Price, "Stenographic Masculinity," and Ivan Kreilkamp, "Speech on Paper: Charles Dickens, Victorian Phonography, and the Reform of Writing," both in *Literary Secretaries/Secretarial Culture*, ed. Leah Price and Pamela Thurschwell (Aldershot: Ashgate, 2005).

17. National Endowment for the Arts, *Survey of Public Participation in the Arts* (2002), and National Endowment for the Arts, *Reading at Risk: A Survey of Literary Reading in America.* (2004).

18. Mrs. Sherwood, *A Drive in the Coach through the Streets of London: A Story Founded on Fact* (London: F. Houlston and Son, 1824); see Katie Trumpener, "The Making of Child Readers," *The Cambridge History of English Romantic Literature*, ed. James Chandler, New Cambridge History of English Literature (Cambridge: Cambridge University Press, 2009), 564.

19. Thanks to Catherine Robson for this reference.

20. For analogous issues in Hardy, see Jonathan Wike, "The World as Text in Hardy's Fiction," *Nineteenth-Century Literature* 47.4 (1993).

21. Gaskell reports that when Patrick Brontë "asked Charlotte what was the best book in the world; she answered, 'The Bible.' And what was the next best; she answered, 'The Book of Nature.'" Gaskell herself described Miss Brontë's "careful examination of the shape of the clouds and the signs of the heavens, in which she read, as from a book, what the coming weather would be"; she also quotes a letter from Charlotte asserting that "if no new books had ever been written, some of these minds would themselves have remained blank pages: they only take an impression." Elizabeth Cleghorn Gaskell, *The Life of Charlotte Brontë*, ed. Angus Easson (Oxford: Oxford University Press, 1996), 48, 353, 85.

22. Sir Charles Russell, speech at a dinner of the First International Shorthand Congress, Tuesday, 27 September 1887, in *Proceedings of the First International Shorth and Congress* (London: Isaac Pitman, 1888), 159.

23. *Manual of Phonography* (1860), 11–13, quoted in Lisa Gitelman, *Scripts, Grooves, and Writing Machines: Representing Technology in the Edison Era* (Stanford, Calif.: Stanford University Press, 1999), 28.

24. Charles Dickens, *The Speeches of Charles Dickens*, ed. K. J. Fielding (Oxford: Clarendon Press, 1960), 347.

25. Kreilkamp, "Speech on Paper: Charles Dickens, Victorian Phonography, and the Reform of Writing."

26. On Dickens's eagerness to protect his study from acoustic interference, see John M. Picker, *Victorian Soundscapes* (New York: Oxford University Press, 2003), 52–65.

27. See James Buzard, "Home Ec. with Mrs. Beeton," *Raritan—A Quarterly Review* 17.2 (1997), and Chris Vanden Bossche, "Cookery, not Rookery: Family and Class in *David Copperfield,*" *David Copperfield and Hard Times, ed. John Peck* (London: St. Martin's, 1995), 31–57. On preprinted account books, see also Leonore Davidoff and Catherine Hall, *Family Fortunes: Men and Women of the English Middle Class, 1780–1850* (Chicago: University of Chicago Press, 1987), 384.

28. On the figure of Mr. Dick, see Alexander Welsh, *From Copyright to Copperfield: The Identity of Dickens* (Cambridge, Mass.: Harvard University Press, 1987), 117–18; on copying more generally, see Alexander Welsh, "Writing and Copying in the Age of Steam," *Victorian Literature and Society*, ed. James Kincaid and Albert Kuhn (Columbus: Ohio State University Press, 1984), 30–45.

29. On the body as a writing surface, see Sara Thornton, *Advertising, Subjectivity and the Nineteenth-Century Novel: Dickens, Balzac and the Language of the Walls*, Palgrave Studies in Nineteenth-Century Writing and Culture (Basingstoke: Palgrave Macmillan, 2009), 38–41.

30. On this passage, see also D. A. Miller, *The Novel and the Police* (Berkeley: University of California Press, 1988), 198. Compare Thackeray's parody, whose narrator complains, "The big boys keep me awake telling stories to 'em *all night*; and I know *ever so many*, and am always making *stories in my head*; and somehow I feel that I'm better than *many of the chaps*—only *I can't do anything.*" [William Thackeray], "Why Can't They Leave Us Alone in the Holidays?" *Punch,* 1851.

31. Walter J. Ong, *Rhetoric, Romance, and Technology: Studies in the Interaction of Expression and Culture* (Ithaca, N.Y.: Cornell University Press, 1971), 113–41.

32. For a subtle reading of this logic, see Rosemarie Bodenheimer, *Knowing Dickens* (Ithaca, N.Y.: Cornell University Press, 2007).

CHAPTER 4
IT-NARRATIVE AND THE BOOK AS AGENT

1. This is not to say that books were the *only* speaking objects: the genre was revived in large part by Douglas Jerrold's *Story of a Feather* (1843). Catherine Waters, *Commodity Culture in Dickens's Household Words: The Social Life of Goods* (Aldershot: Ashgate, 2008), 14.

2. "Cheap Books," *Christian Spectator,* July 1846, 146, quoted in Aileen Fyfe, *Science and Salvation: Evangelical Popular Science Publishing in Victorian Britain* (Chicago: University of Chicago Press, 2004), 159.

3. The metaphor is common enough: compare "Julia has free access to all her father's books, with the exception of one book-case; this has glazed doors, and the books are all turned with their titles inward, so we do not know their

names or contents, but Mr. Waldron says they are not suitable for us to read. Julia looks at them with a sort of awe; she says they are criminals locked up in their condemned cell." *She Would Be a Governess: A Tale* (London: Routledge, 1861), 25. The carceral metaphor sometimes gives way to a sexual one, as when one article compares "books behind their glass doors" to "Persian beauties behind their jalousies, to be looked at." Sarah G. S. Pratt, *First Homes* (1882), American Periodicals Series Online, ProQuest, 1 November 2009.

4. On the magazine, see Rosemary Scott, "The Sunday Periodical: Sunday at Home," *Victorian Periodicals* 25.4 (1992).

5. An exception is the *Story of a Pocket Bible*: according to Jacky Bratton, "an edition of the early 1860s offers evidence of the value which was placed upon it; inscriptions carefully record that it was bequeathed to a beloved daughter in 1875, who received it in 1884, and left it in turn to her dear daughter." J. S. Bratton, *The Impact of Victorian Children's Fiction* (London: Croom Helm, 1981), 65.

6. Thanks to Ellen Garvey for this reference.

7. The narrator figures itself even more insistently as a body in "Adventures of a Robinson Crusoe. Written by Itself. To the Editors of the Young Gentleman's and Lady's Magazine," where "I was clothed in a most elegant and substantial manner, in what might be called a military dress; for it was red laced with gold." The little girl who chooses the narrator over other editions, it adds, "was not the only young lady that has been captivated with scarlet and gold!" Vanity is followed by sexual consummation, itself naturally followed by abandonment: "Being carried home, she was for some time so fond of me, that she was constantly tumbling over my leaves by day, and even took me to bed with her at night." Once she becomes "mistress of my contents," however, "I became more estranged from my young mistress, or rather she from me." Eventually the book is passed on to her young son: "Hitherto my coat was not much the worse for the wear. The red was indeed a little faded, and the gold tarnished; a few wrinkles and spots deformed my substance; but still I would have passed for a middle-aged book, and among younger competitors would not have appeared to much disadvantage." Like any fallen woman, the book ages in proportion to how much it's been loved, or at least used: "The more I became a favourite, the harder was my usage. I was thumbed without mercy, my leaves began to curl, my binding to break . . . I had been used till novelty was no more." A gaudy dress, an initial physical closeness followed by boredom and estrangement once the owner "soon became mistress of" its insides: we don't need to wait for the book's later reference to being "in keeping" to understand that it tells its story by analogy to a prostitute's. Finally traded for half a dozen apples, the book blames its owner for never once reflecting how much "it must wound the feelings of a tender mother to part with me on such easy terms, and for such an ignoble end . . . he was sensible of his error when it was too late to recover me—when my destiny was sealed for ever." Finally, the narrator is traded for a copy of—the *Young Gentleman's and Lady's Magazine*. Abandoned in favor of "a little new flaunting pamphlet in a French gray patent wrapper, tricked out with painted flowers, and composed of more subjects than there are colours in the rainbow," the "old faithful servant" has nothing left to tell but that "my poor remains were gathered up, and I was sold to a chandler for a penny." "Adventures of a Robinson Crusoe. Written by

Itself. To the Editors of the Young Gentleman's and Lady's Magazine," *Young Gentleman's & Lady's Magazine, or Universal Repository of Knowledge, Instruction and Amusement* 1 (1799): 186–92.

8. When Ben Franklin composed an epitaph for himself comparing the body to "an old Book . . . stript of its Lettering and Gilding" but insisting that "the Work shall not be lost" but rather "Appear once more / In a new and elegant Edition / Corrected and improved / By the Author," he made clear that the hackneyed analogy (text is to book as soul is to body) gains its greatest anthropomorphic force at the moment of death (quoted in Robert Darnton, *First Steps toward a History of Reading* [New York: Norton, 1990], 186).

9. Compare "But here is the old family Bible. Not one with double clasp, gilt back, and full morocco binding, with colored marriage certificate, and two pages for photographs; not the kind which is enveloped in its own dignity, and seems to say to any who would venture to open its stiff clasps, I am only to be looked at. The Bible they read is in grandmother's room." Charles Le Roy Goodell, *My Mother's Bible: A Memorial Volume of Addresses for the Home* (Boston: Lee and Shepard, 1891), 24.

10. Jonathan Lamb, "Modern Metamorphoses and Disgraceful Tales," *Critical Inquiry* 28 (2001): 158; Lynn M. Festa, *Sentimental Figures of Empire in Eighteenth-Century Britain and France* (Baltimore, Md.: Johns Hopkins University Press, 2006); see also Paul Collins, "You and Your Dumb Friends," *Believer*, March 2004. On the relation of slave narrative to bildungsroman more generally, see Julia Sun-Joo Lee, *The American Slave Narrative and the Victorian Novel* (New York: Oxford University Press, 2010).

11. The same could be said of his namesake, David Grieve, who finds "an old calf-bound copy of Paradise Lost" in a meal chest; "all the morning he had been lying hidden in a corner of the sheepfold devouring it, the rolling verse imprinting itself on the boy's plastic memory by a sort of enchantment." Mary Arnold Ward, *David Grieve* (London: Macmillan, 1891), 88.

12. I owe the idea of bibliomorphizing to Mark Kauf (Tufts).

13. On the extent to which it-narrators are empowered to speak for themselves, see Mary Poovey's discussion of the shifting positions occupied by the narrator of Bridges's *Adventures of a Bank-note*, which sometimes represents its powers of speech as an idiosyncrasy and sometimes seems to assume that its fellow banknotes can speak as well, sometimes imagines itself as a speaker whose words are being recorded by a human "secretary" and sometimes uses phrases that imply its own power to write and even to transact business with booksellers. Mary Poovey, *Genres of the Credit Economy: Mediating Value in Eighteenth- and Nineteenth-Century Britain* (Chicago: University of Chicago Press, 2008), 149.

14. For the paradox of the weak hero in the nineteenth-century British novel, see Alexander Welsh, *The Hero of the Waverley Novels* (New Haven, Conn.: Yale University Press, 1963). Compare Michael Gamer's reading of *Waverley* as an object narrative, where the protagonist's passivity doesn't prevent him from providing the novel's formal center—with the difference, of course, that object narratives, like *Copperfield* but unlike *Waverley*, are conventionally first-person. Michael Gamer, "Waverley and the Object of (Literary) History," *Modern Language Quarterly* 70.4 (2009).

15. Cp. David Pearson, "What Can We Learn by Tracking Multiple Copies of Books?" *Books on the Move: Tracking Copies through Collections and the Book Trade*, ed. Robin Myers, Michael Harris, and Giles Mandelbrote (New Castle, Del.: Oak Knoll Press; London: British Library, 2007), 18.

16. On nonhuman agents, see also Bruno Latour, "On Technical Mediation—Philosophy, Sociology, Genealogy," *Common Knowledge* 3.2 (1994).

17. See also Aileen Fyfe's argument that the Evangelical press praised tracts for their ability to reach places where middle-class bodies could not safely venture. Fyfe, *Science and Salvation: Evangelical Popular Science Publishing in Victorian Britain*, 29.

18. See also Seth Lerer, *Inventing English: A Portable History of the Language* (New York: Columbia University Press, 2007), 21.

19. Robert Darnton, *The Business of Enlightenment: A Publishing History of the Encyclopédie, 1775–1800* (Cambridge, Mass.: Harvard University Press, 1979); Cathy Davidson, "The Life and Times of *Charlotte Temple*: The Biography of a Book," *Reading in America: Literature and Social History*, ed. Cathy Davidson (Baltimore, Md.: Johns Hopkins University Press, 1989). Not that this metaphor is confined to books in particular: witness Igor Kopytoff's essay "The Cultural Biography of Things." Igor Kopytoff, "The Cultural Biography of Things," *The Social Life of Things: Commodities in Cultural Perspective*, ed. Arjun Appadurai (Cambridge: Cambridge University Press, 1986). For the longer history of the "talking book," see Henry Louis Gates, *The Signifying Monkey: A Theory of Afro-American Literary Criticism* (New York: Oxford University Press, 1988), 152–69.

CHAPTER 5
THE BOOK AS BURDEN: JUNK MAIL AND RELIGIOUS TRACTS

1. Collier, "Of the Entertainment of Books," quoted in Holbrook Jackson, ed., *The Anatomy of Bibliomania* (Chicago: University of Illinois Press, 2001), 71; see also Collet Dobson Collet, *History of the Taxes on Knowledge; Their Origin and Repeal* (London: T. F. Unwin, 1899), 35. For the prehistory, see Ann Blair, *Too Much to Know: Managing Scholarly Information before the Modern Age* (New Haven, Conn.: Yale Unversity Press, 2010).

2. On the taxes on knowledge, see Collet, *History of the Taxes on Knowledge; Their Origin and Repeal*; Webb, *The British Working Class Reader, 1790–1848: Literacy and Social Tension*; Joel H. Wiener, *The War of the Unstamped: The Movement to Repeal the British Newspaper Tax, 1830–1836* (Ithaca, N.Y.: Cornell University Press, 1969); Alan Rauch, *Useful Knowledge: The Victorians, Morality, and the March of Intellect* (Durham, N.C.: Duke University Press, 2001); Simon Eliot, "The Business of Victorian Publishing," *The Cambridge Companion to the Victorian Novel*, ed. Deirdre David (Cambridge: Cambridge University Press, 2001).

3. On the Malthusian metaphor, see Lucy Newlyn, *Reading, Writing, and Romanticism: The Anxiety of Reception* (Oxford: Oxford University Press, 2000), 40. Thomas Hood invokes a similar logic when he prefaces his "National Tales"—

coming right after "My son and heir" (originally published in the Comic Annual for 1831), in which a father laments that "James is too big a boy, like book, / To leave upon the shelf unbound"—with an apology to "the learned Malthusians of our century" for "increasing my family in this kind; and by twin volumes, instead of the single octavos which have hitherto been my issue." Thomas Hood, "The Choice Works of Thomas Hood, in Prose and Verse, Including the Cream of the Comic Annuals. With Life of the Author, Portrait, and over Two Hundred Illustrations" (1883?), 655.

4. On junk mail in the nineteenth-century United States, see David M. Henkin, *The Postal Age: The Emergence of Modern Communications in Nineteenth-Century America* (Chicago: University of Chicago Press, 2006), 153, and Richard Menke, *Telegraphic Realism* (Stanford, Calif.: Stanford University Press, 2008), 31–67.

5. On postal scams, see Henkin, *The Postal Age,* 155–58.

6. On the history of the metaphor, see James Hamilton, "Unearthing Broadcasting in the Anglophone World," *Residual Media*, ed. Charles R. Acland (Minneapolis: University of Minnesota Press, 2007). Peters points out that the etymology of "parable" itself involves sowing. John Durham Peters, *Speaking into the Air: A History of the Idea of Communication* (Chicago: University of Chicago Press, 1999), 52.

7. On books as gifts between men, see also Margot Finn, "Men's Things: Masculine Possession in the Consumer Revolution," *Social History* 25.2 (2000): 133–55.

8. On the *Fraser*'s hoax (designed to trick Martineau and other celebrities into providing references for an imaginary ex-servant), see *Harriet Martineau's Autobiography*, ed. Maria Chapman, 2 vols. (Boston: Osgood, 1877), 1:320.

9. James Raven, *Free Print and Non-Commercial Publishing since 1700* (Aldershot: Ashgate, 2000), 4, 5; on free books in the Romantic period, see Jackson, *Romantic Readers: The Evidence of Marginalia,* 36–37. By its own partisan calculations, Metro PE dwarfs all Pan-European daily and weekly titles with a reach over four times higher than that of the *Financial Times* and nearly three times higher than the *Economist's* (http://hugin.info/132142/R/1496601/432324.pdf).

10. My model owes much to Elaine Hadley's analysis of Victorian liberalism: Elaine Hadley, *Living Liberalism: Practical Citizenship in Mid-Victorian Britain* (Chicago: University of Chicago Press, 2010).

11. See also Michael Ledger-Lomas, "Mass Markets: Religion," *The Cambridge History of the Book in Britain, 1830–1914*, ed. David McKitterick (Cambridge: Cambridge University Press, 2009), 326, and Boyd Hilton, *The Age of Atonement: The Influence of Evangelicalism on Social and Economic Thought, 1785–1865* (Oxford: Clarendon Press, 1991), 314.

12. Margot Finn dates to the 1820s and 1830s "the inroads made by increasingly individualistic and contractual ways of thinking upon traditions of consumer behavior which had earlier been entangled in a dense reticulation of personal ties and affective obligations." Finn, "Men's Things," 137.

13. See also Gilmartin on "the circular structure of a print economy of charitable provision." Kevin Gilmartin, "'Study to Be Quiet': Hannah More and the Invention of Conservative Culture in Britain," *ELH* 70.2 (2003): 513.

14. See Leslie Howsam, *Cheap Bibles: Nineteenth-Century Publishing and the British and Foreign Bible Society*, Cambridge Studies in Publishing and Printing History (Cambridge: Cambridge University Press, 1991), as well as, more generally, Raven, *Free Print and Non-Commercial Publishing since 1700*.

15. On the system of subsidized pricing, see Anne Stott, *Hannah More: The First Victorian* (Oxford: Oxford University Press, 2003), 175.

16. The narrator of a *Punch* sketch, "Bachelor Days," plays on the convention when he explains the accumulation of wastepaper in his cupboard by the speculation that his housekeeper must be overcompensating for having discarded a valuable manuscript in some previous job. "Bachelor Days. IV," *Punch*, 1907, 17.

17. *Missionary Register*, Church Missionary Society, May 1817, 186, quoted in Homi K. Bhabha, "Signs Taken for Wonders: Questions of Ambivalence and Authority under a Tree outside Delhi, May 1817," *Critical Inquiry* 12.1 (1985): 164.

18. See, e.g., Motoko Rich, "With Kindle, the Best Sellers Don't Need to Sell," *New York Times*, 22 January 2010.

19. Compare a review of the *Twenty-third Annual Report of the Bombay Tract and Book Society* in a local journal, which remarks upon "the millions of tracts which have been distributed, and have served for waste paper in India, China, and other heathen countries." "Review of the Twenty-Third Annual Report of the Bombay Tract and Book Society," *Bombay quarterly magazine and review* 7 (1852): 497. See also Graham Shaw's observation that "large single sheets containing the Ten Commandments were used by boys to make kites." Graham Shaw, "South Asia," *A Companion to the History of the Book*, ed. Simon Eliot and Jonathan Rose (Malden, Mass.: Blackwell Pub., 2007), 133.

20. For a rebuttal, see, e.g., http://illandancient.blogspot.com/2010/01/this-pensioners-burning-books-story.html.

21. Sherwood invokes the same verse in "The Red Book" and in "The Penny Tract." Mary Sherwood, *The Works of Mrs. Sherwood. Being the Only Uniform Edition Ever Published in the United States* (New York: Harper & Brothers, 1834), 13:146, 351.

22. "Answers to Correspondents," 47.5 (1 November 1851), reprinted in Henry Mayhew, *The Essential Mayhew: Representing and Communicating the Poor*, ed. Bertrand Taithe (London: Rivers Oram Press, 1996), 212.

23. L. N. R[anyard], *The Missing Link, or, Bible-Women in the Homes of the London Poor* (New York: Robert Carter, 1860), 116–17. On Mrs. Ranyard, see John Matthias Weylland, *These Fifty Years Being the Jubilee Volume of the London City Mission* (London: Partridge, 1884).

24. On charity visiting in *Bleak House*, see also Beth Fowkes Tobin, *Superintending the Poor: Charitable Ladies and Paternal Landlords in British Fiction, 1770–1860* (New Haven, Conn.: Yale University Press, 1993), 129–52.

25. See Dorothy J. Hale, "Fiction as Restriction: Self-Binding in New Ethical Theories of the Novel," *Narrative* 15 (2007).

26. M. O. Grenby argues that tracts could themselves address middle-class children as well as working-class adults: M. O. Grenby, "Chapbooks, Children, and Children's Literature," *The Library* 3 (2007): 290.

27. As Laura Green argues, "because its destination is always ultimately the self, literary identification . . . tends to consolidate rather than expand the subject's consciousness." Laura Green, "'I Recognized Myself in Her': Identifying with the Reader in George Eliot's *The Mill on the Floss* and Simone De Beauvoir's *Memoirs of a Dutiful Daughter*," *Tulsa Studies in Women's Literature* 24.1 (2005): 57.

28. As William McKelvy points out, *Felix Holt*, too, has a character asserting his identity by rejecting a father's archival bequest. William R. McKelvy, *The English Cult of Literature: Devoted Readers, 1774–1880*, Victorian Literature and Culture Series (Charlottesville: University of Virginia Press, 2007), 230.

29. For an example of a Victorian reader's annotations to the *Imitation of Christ*, see Florence Nightingale, *Florence Nightingale on Mysticism and Eastern Religions*, vol. 4 of *The Collected Works of Florence Nightingale*, ed. Gerard Vallee, 16 vols. (Waterloo, Ont.: Wilfrid Laurier University Press, 2001-), 81–104.

30. Though Eliot, as usual, expressed an anxious investment in the production values of her own novels: Leah Price, *The Anthology and the Rise of the Novel* (Cambridge: Cambridge University Press, 2000), 122.

31. On those other possessions, see Jeff Nunokawa, *The Afterlife of Property: Domestic Security and the Victorian Novel* (Princeton, N.J.: Princeton University Press, 1994).

32. Similarly, John Kucich argues that Maggie values objects as intensely as Tom does, though more diffusely: see John Kucich, "George Eliot and Objects: Meaning and Matter in the Mill on the Floss," *Dickens Studies Annual* 12 (1983).

33. On the "reified communication" that makes the value of words depend on their embodiment in an object, see Kucich, "George Eliot and Objects: Meaning and Matter in the Mill on the Floss," 326.

34. Compare Gaskell's *My Lady Ludlow*: where the bible "was not opened at any chapter, or consoling verse; but at the page whereon were registered the births of her nine children." Elizabeth Cleghorn Gaskell, *My Lady Ludlow and Other Stories*, ed. Edgar Wright (Oxford: Oxford University Press, 1989), 162.

35. In formal terms, however, the family bible bears more resemblance to the newspaper, as a source that narrators can quote to introduce new information into the plot: as late as 1901, *Buddenbrooks* cites its flyleaf as authority for three plot points. Thomas Mann, *Buddenbrooks: The Decline of a Family*, trans. John E. Woods, 1st American ed. (New York: Knopf, 1993), 155, 71, 510. Yonge cuts even straighter to the chase, introducing a near facsimile with the remark that "for the convenience of our readers we subjoin the first page of the family Bible." Charlotte Mary Yonge, *The Pillars of the House; or, under Wode, under Rode* (London: Macmillan and Co., 1874), 82. What the confidante performed in classical drama, the bible and the newspaper parcel out between them in the modern novel.

On the peculiarities of manuscript-print interaction in bibles, see Sherman, *Used Books: Marking Readers in Renaissance England*, 71–86. Sherman notes that the Bible less often provided a place for the family record than did the Book of Common Prayer and collections of sermons and homilies (60).

36. My thoughts on association copies owe much to the late Jay Fliegelman.

37. On the moral ambiguity of a model of sympathy in which "another's internal state becomes 'intimately present' only by losing its distinct quality of belonging to someone else," see Gallagher, *Nobody's Story: The Vanishing Acts of*

Women Writers in the Marketplace, 1670–1820, 170; similarly, Caroline Levine argues that identification in Eliot ends up obliterating the alterity of the Other. Caroline Levine, *The Serious Pleasures of Suspense: Victorian Realism and Narrative Doubt,* Victorian Literature and Culture Series (Charlottesville: University of Virginia Press, 2003), 108. For a counterargument, however, see Suzy Anger's critique of recent critics' symptomatic readings of Eliotic sympathy. Suzy Anger, *Victorian Interpretation* (Ithaca, N.Y.: Cornell University Press, 2005), 96.

CHAPTER 6
THE BOOK AS GO-BETWEEN: DOMESTIC SERVANTS AND FORCED READING

1. Compare Roger Chartier's argument that as the spread of literacy made the *fact* of reading, or even of reading a particular book, less significant, *ways* of reading became a surer marker of social position—and therefore that historians should pay special attention to divergent uses of shared texts. Roger Chartier, *The Order of Books* (Cambridge: Polity Press, 1994), 15–16.

2. Compare the 1868 cartoon in *Punch* showing a maid borrowing a lodger's sensation novel—with the difference that here the book crosses a gender divide. Flint, *The Woman Reader, 1837–1914,* 279.

3. On differences among formats, see also Fyfe, *Science and Salvation: Evangelical Popular Science Publishing in Victorian Britain,* 157.

4. On Dickens's method of "working the copyrights," see Patten, *Charles Dickens and His Publishers,* 236–65.

5. Dusting is also, as Carolyn Steedman points out, the *least* dirty of household activities, and one of the few that the lady of the house might condescend to perform herself: *Labours Lost: Domestic Service and the Making of Modern England* (Cambridge: Cambridge University Press, 2009), 347.

6. Andrew Lang remarks similarly that the amateur "loves to have his study, like Montaigne's, remote from the interruption of servants, wife, and children." Andrew Lang, *The Library* (London: Macmillan & Co., 1881), 34.

7. On the problem of master-servant privacy and secrecy, see David Vincent, *The Culture of Secrecy in Britain, 1832–1998* (Oxford: Oxford University Press, 1998), 71–73. In the opposite direction, locks may have been even more important for servants than for masters, since the former's privacy was more precarious, as were their belongings: see "Thresholds and Boundaries at Home" in Amanda Vickery, *Behind Closed Doors: At Home in Georgian England* (New Haven, Conn.: Yale University Press, 2009).

8. With exceptions, as in *Ward and Lock's Home Book,* which urges householders to banish their own worn-out volumes to the kitchen, because servants "should be allowed access to the books of the house, always, of course, under proper restrictions. In these auxiliary cases may be kept such books as those whose original cloth binding is dilapidated, but not yet sufficiently so as to consign them to the binder's hand." *Ward and Lock's Home Book,* (London: Ward, Lock, 1882), 281. On servants' libraries, see Felicity Stimpson, "Servants' Reading: An Examination of the Servants' Library at Cragside," *Library History* 19.1 (2003).

9. Quoted in Flint, *The Woman Reader, 1837–1914,* 234; for a rich account of servants' reading more generally, see 232–34.

10. In Worboise's *Thornycroft Hall*, the character who corresponds most closely to Mrs. Reed complains that "the dust was so thick that I could have written my name on every article of furniture." "And did you write it, my dear?" asks her husband. "No, Mr. Ward, I did not; but I wrote 'SLUT' in great capital letters on the looking glass, and on both the tables." Emma Worboise, *Thornycroft Hall: Its Owners and Its Heirs* (London: J. Clarke, 1886), 21.

11. On breakage of china, see, e.g., *Domestic Management, Or the Art of Conducting a Family; with Instructions to Servants in General. Addressed to Young Housekeepers* (London: printed for H. D. Symonds at the Literary Press, No. 62 Wardour-Street, Soho, 1800), 87.

12. "My mistress's bonnet," *Godey's Lady's Book* 37 (September 1848): 119, reproduced in Anna Vemer Andrzejewski, *Building Power: Architecture and Surveillance in Victorian America*, 1st ed. (Knoxville: University of Tennessee Press, 2008), 129.

13. *Pawnbroker's Gazette* 865 (1855): 194–95; many thanks to Beth Womack for this reference.

14. On the politics of literacy in More, see Brantlinger, *The Reading Lesson: The Threat of Mass Literacy in Nineteenth-Century Britain,* 6.

15. In the following chapter, Jane asks of John: "Master! How is he my master? Am I a servant?" To which a maid replies, "No; you are less than a servant, for you do nothing for your keep." Charlotte Brontë, *Jane Eyre*, ed. Michael Mason (Harmondsworth: Penguin, 2003), 19. See also Leah Price, "The Life of Charlotte Brontë and the Death of Miss Eyre," *SEL: Studies in English Literature, 1500–1900* 35.4 (1995).

16. Cp. one American decorating manual's claim that with a locked bookcase "one does not soil one's hands, which is inevitable where books stand uncovered and undusted for weeks; and he is no lover of his books who will allow the housemaid to include them in her daily duties, for she is usually far more dangerous than the corrupting moth and dust." Maria Oakey Dewing, *Beauty in the Household* (New York: Harper, 1882), 84.

17. On the relation between finger and bookmark, see Peter Stallybrass, "Books and Scrolls: Navigating the Bible," *Books and Readers in Early Modern England*, ed. Jennifer Andersen and Elizabeth Sauer (Philadelphia: University of Pennsylvania Press, 2002).

18. Similarly, a novel cowritten by (among others) Charlotte Yonge has a clergyman's wife complain that her husband "never can remember to leave the right tracts; the drunkards don't get the intemperate ones, and that elect woman Mrs. Scroggs was so offended because he gave her 'Are you Converted?' " Frances Awdry, *The Miz Maze; or, the Winkworth Puzzle. A Story in Letters* (London: Macmillan, 1883), 139.

19. Compare Joseph Slaughter's argument that the bildungsroman characteristically depicts "the passage from orality to writing: Joseph R. Slaughter, *Human Rights, Inc.: The World Novel, Narrative Form, and International Law*, 1st ed. (New York: Fordham University Press, 2007), 284.

20. Remember again Vincent's argument that "the self-educated reader was as much of a [Victorian] myth as the self-made millionaire." David Vincent, *Literacy and Popular Culture: England 1750–1914* (Cambridge: Cambridge University Press, 1989), 259.

21. Thus Samuel F. Pickering, Jr., "The Old Curiosity Shop—a Religious Tract?" *Illinois Quarterly* 36.1 (1973), compares the plots of *The Old Curiosity Shop* and *The Dairyman's Daughter*. Tracts may, however, overstate the appeal of novels to readers like Caroline Cox: Jan Fergus and Carolyn Steedman both argue that, in the eighteenth century at least, verse was more accessible to them. Jan S. Fergus, "Provincial Servants' Reading in the Late 18th Century," *The Practice and Representation of Reading in England*, ed. James Raven, Helen Small, and Naomi Tadmor (Cambridge: Cambridge University Press, 1996); Steedman, *Labours Lost: Domestic Service and the Making of Modern England*.

22. The one exception marks Becky Sharp's nadir: when reduced to living off pious ladies, "she not only took tracts, but she read them." Thackeray, *Vanity Fair,* 642.

23. On the difficulty of getting the poor to buy tracts, see Webb, *The British Working Class Reader, 1790–1848: Literacy and Social Tension,* 56.

24. If, as John Plotz argues, the moonstone exemplifies "reverse portability" (from imperial periphery to center), it also shares with the religious tract a kind of reverse desirability. John Plotz, *Portable Property: Victorian Culture on the Move* (Princeton, N.J.: Princeton University Press, 2008), 40–44.

25. Penny Fielding argues that "the distinctions between private collection and lending library are broken down" in "The Tractate Middoth" and "Casting the Runes": Penny Fielding, "Reading Rooms: M. R. James and the Library of Modernity," *Modern Fiction Studies* 46.3 (2000): 764. One might add that the library is dangerous precisely because it puts one reader into relation with others: speculating about the reviewer's confidentiality, the secretary concludes that "the only danger is that Karswell might find out, if he was to ask the British Museum people who was in the habit of consulting alchemical manuscripts" (132). In that sense, the relation among scholarly colleagues looks less cozy than the cross-class collaboration of the passenger and conductor examining a streetcar ad together.

26. On reading aloud in this period, see Philip Arthur William Collins, *Reading Aloud; a Victorian Métier* (Lincoln: Tennyson Research Centre, 1972); Ivan Kreilkamp, *Voice and the Victorian Storyteller*, Cambridge Studies in Nineteenth-Century Literature and Culture, 49 (Cambridge: Cambridge University Press, 2005); and on Nightingale's opposition to it, John Plotz, *Semi-detached: The Aesthetics of Partial Absorption* (forthcoming).

27. Pearson, too, observes that "in didactic novels characters are often judged by their willingness to read to entertain others." Jacqueline Pearson, *Women's Reading in Britain, 1750–1835* (Cambridge: Cambridge University Press, 1999), 171. But she locates women's resistance to being read to at the level of content (a woman more concerned with a broken needle than with the death of Nelson is resisting patriarchal culture) rather than, like Nightingale, at the level of form (174).

28. Compare the gender politics of another Charlotte Adams tract:

It is too common to see men and boys engrossed in the selfish indulgence of reading to themselves books which are highly interesting, while a sister or a wife sits by labouring with her needle for the supply of indispensable wants to the family, without any share in the enjoyment of her companion. Some say they cannot read aloud—it hurts their chest, or it tires them—they never could do so in their lives. Let these tender gentlemen, who will shout for hours in some masculine amusement, try if they cannot acquire the power. (Charlotte Adams, *Boys at Home* [New York, 1854], 167)

29. See also Monica Lewis, "Anthony Trollope among the Moderns: Reading Aloud in Britain 1850–1960" (PhD diss., Harvard University, 2006).

30. In the early Victorian debates over postage, print was strategically conflated with manuscript. Postal reformers overwhelmingly couched their arguments in terms of personal correspondence, often invoking the *Pamela*-esque scenario of a young worker's virtue saved by letters from home. Mary Favret, *Romantic Correspondence: Women, Politics, and the Fiction of Letters* (Cambridge: Cambridge University Press, 1993), 204. Thus Rowland Hill quoted employers testifying to the "vice and profligacy . . . among the young persons in our establishment . . . resulting from want of communication with their parents by letter." Rowland Hill and George Birkbeck Norman Hill, *The Life of Sir Rowland Hill . . . And the History of Penny Postage*, 2 vols. (London: Thos. De La Rue, 1880), 1:308. Likewise, a member of the Board of Trade attributed "vicious courses" to young workers arriving in London and unable to afford a letter home. Sir Rowland Hill, *Post Office Reform, Its Importance and Practicability* (London: C. Knight and Co., 1837), 93. But the sharpest jump in postal traffic after 1840 did not involve letters from one individual to another: the bulk of objects that passed through the postal system were "printed circulars, prospectuses, catalogues, and prices current." *Administration of the Post Office: From the Introduction of Mr. Rowland Hill's Plan of Penny Postage up to the Present Time* (London: J. Hatchard and Son, 1844), 55. Conservatives could counter Hill's logic by pointing out that "where the poor man receives, say eight letters, from his sailor-son, or his daughter in service in the capital, or in some distant town, and thus gains a shilling in the year by cheap postage, let any one consider how much is gained and saved by this penny postage in such houses as Lloyd, Jones, and Co; Baring Brothers & Co." *Administration of the Post Office: From the Introduction of Mr. Rowland Hill's Plan of Penny Postage up to the Present Time*, 196. Like Miss Clack copying tracts out by hand, or like tract societies stuffing pamphlets into bottles, postal reformers borrowed the prestige of manuscript to ease the distribution of print.

CHAPTER 7
THE BOOK AS WASTE: HENRY MAYHEW AND THE FALL OF PAPER RECYCLING

1. Polastron observes that books buried in tombs are at least protected from wear and tear: "it is certainly possible to view as a fairly honest conservation system this egotistical practice that wipes them from the face of the earth" Lucien X. Polastron, *Books on Fire: The Destruction of Libraries Throughout History*,

1st U.S. ed. (Rochester, Vt.: Inner Traditions, 2007) 10. On the other hand, the salvage value of the carrier can sometimes hinder the survival of its contents, as when silent films were melted down for silver.

2. See also Jonathan Bloom, *Paper before Print: The History and Impact of Paper in the Islamic World* (New Haven: Yale University Press, 2001).

3. On Victorian attitudes toward waste, see Natalka Freeland, "Trash Fiction: The Victorian Novel and the Rise of Disposable Culture"; on wastepaper more specifically, Talia Schaffer, "Craft, Authorial Anxiety, and 'the Cranford Papers,'" *Victorian Periodicals Review* 38.2 (2005). On paper more generally, see McLaughlin, *Paperwork: Fiction and Mass Mediacy in the Paper Age* (which focuses more on nineteenth-century representations *of* paper than on representations *on* paper); Lee Erikson, *The Economy of Literary Form* (Baltimore, Md.: Johns Hopkins University Press, 1996) (whose analysis of the relation of paper to nineteenth-century literature is weakened by its emphasis on technology at the expense of changing taxation regimes); Christina Lupton, "Theorizing Surfaces and Depths: Gaskell's Cranford," *Criticism* 50.2 (2008); Andrea Pellegram, "The Message in Paper," *Material Cultures: Why Some Things Matter*, ed. Daniel Miller (Chicago: University of Chicago Press, 1998); and Joshua Calhoun, "The Word Made Flax: Cheap Bibles, Textual Corruption, and the Poetics of Paper," *PMLA* 126.2 (2011).

4. Henry Mayhew, *London Labour and the London Poor; Cyclopaedia of the Condition and Earnings of Those That Will Work, Those That Cannot Work, and Those That Will Not Work* (London: Griffin Bohn, 1861), 1:289–90, 3:33, 1:40. On the sham indecent trade in Mayhew, see also Lynda Nead, *Victorian Babylon: People, Streets and Images in Nineteenth-Century London* (New Haven, Conn.: Yale University Press, 2000), 157. On the dustman in Victorian culture, see Brian Maidment, *Dusty Bob: A Cultural History of Dustmen, 1780–1870* (Manchester: Manchester University Press, 2007).

5. For informants reading earlier installments of *London Labour*, see, e.g., 3:214. For subtle analyses of Mayhew's characterization of his informants' literacy, see Brantlinger, *The Reading Lesson: The Threat of Mass Literacy in Nineteenth-Century Britain*, 85–91, and Victor Neuburg, ed., *The Invention of the Streets*, 2. vols. (London: Routledge & Kegan Paul, 1973). On the other hand, one wastepaper dealer who declares that "the people as sells 'waste' to me is not such as can read" adds that "I don't understand much about books"; the point is proven a moment later when he describes a customer asking, "Have you any black lead?"— which Mayhew's readers must themselves gloss as "black letter" (2:110). Another dustman adds, "I niver vos at a school in all my life; I don't know what it's good for. It may be wery well for the likes o' you, but I doesn't know it 'u'd do a dustie any good. You see, ven I'm not out with the cart, I digs here all day; and p'raps I'm up all night, and digs away agen the next day. Vot does I care for reading, or anythink of that there kind, ven I gets home arter my vork? I tell you vot I likes, though! vhy, I jist likes two or three pipes o' baccer, and a pot or two of good heavy and a song" (2:178). The ballad locution "What cares I" asserts that orality does for him what literacy does for "the likes o'" the researcher.

6. Brantlinger situates Mayhew within a tradition of both romantic and realist fiction that contrasts "oral vitality" with "literate domestication." Brantlinger, *The Reading Lesson: The Threat of Mass Literacy in Nineteenth-Century Britain*, 91.

7. Nineteenth-century English commentators alternately praise and blame their own nation for recycling less than others. After remarking that old account-book covers can be made into soles of shoes that leak on rainy days, one writer in 1896 remarks, "It is only wicked foreigners who are said to do such things." "Government Waste-Paper," *Chambers's Journal* 13 (1896): 749; see also "Waste Paper," *Leisure Hour* 30 (1881).

8. On papermaking, see Dard Hunter, *Papermaking: The History and Technique of an Ancient Craft*, 2nd ed. (New York: A. A. Knopf, 1947), 555, and Adrian Johns, "Changes in the World of Publishing," *The Cambridge History of English Romantic Literature*, ed. James Chandler (Cambridge: Cambridge University Press, 2009), 392.

9. See also Johanna Drucker, *The Visible Word: Experimental Typography and Modern Art, 1909–1923* (Chicago: University of Chicago Press, 1994), 45–46.

10. Thanks to Natalka Freeland for pointing out the logical lapse to me. Andrew Miller suggests, more generally, that "no object can be owned which does not suggest to [Thackeray's] imagination the ruin and death of those who own it" (*Novels behind Glass*, 18).

11. On the paper bag, see Henry Petroski, *Small Things Considered: Why There Is No Perfect Design* (New York: Alfred A. Knopf, 2003), 99; becoming popular thanks to the cotton shortages during the Civil War, it gave birth to new metaphors, as when Barrie compared padded-out novels to "paper bags blown out with wind." *Two of Them* (1893), quoted in P. J. Waller, *Writers, Readers, and Reputations: Literary Life in Britain, 1870–1918* (Oxford: Oxford University Press, 2006), 33. For the place of waste in eighteenth-century culture, see Sophie Gee, *Making Waste: Leftovers and the Eighteenth-Century Imagination* (Princeton, N.J.: Princeton University Press, 2010).

12. "The Republican Refuted; in a Series of Biographical, Critical, and Political Strictures on Thomas Paine's Rights of Man," *Monthly Review* 7 (1792): 84; thanks to Paul Keen for suggesting this example to me.

13. On the phrase "not worth the paper," see Jacques Derrida, *Paper Machine*, trans. Rachel Bowlby, Cultural Memory in the Present (Stanford, Calif.: Stanford University Press, 2005), 44.

14. George Gordon Lord Byron, *Byron's Letters and Journals: The Complete and Unexpurgated Text of All the Letters Available in Manuscript and the Full Printed Version of All Others*, ed. Leslie Alexis Marchand 12 vols. (Cambridge, Mass.: Belknap Press of Harvard University Press, 1973), 8:11–12; thanks to Susan Wolfson and Betty Schellenburg for this reference.

15. *Spectator* 367 (1 May 1712), reprinted in Joseph Addison and Richard Steele, *The Spectator*, ed. Donald Frederic Bond (Oxford: Clarendon Press, 1965), 380–81.

16. Thomas Percy, *Bishop Percy's Folio Ms. Ballads and Romances*, ed. Hales and Furnivall (1868), 1:xii, quoted in Susan Stewart, *Crimes of Writing: Problems in the Containment of Representation* (Oxford: Oxford University Press, 1991), 119.

17. On bibliophilic misogyny, see Willa Silverman, "The Enemies of Books? Women and the Male Bibliophilic Imagination in *Fin-de-Siècle* France," *Contem-*

porary French Civilization 30.1 (Winter 2005/Spring 2006): 47–74, and Jackson, *The Anatomy of Bibliomania,* 137–68.

18. Carolyn Steedman offers a different analysis of what she calls the "servant joke": Steedman, *Labours Lost: Domestic Service and the Making of Modern England*, 222.

19. A. R. Waller and Adolphus William Ward, *The Cambridge History of English Literature* (Cambridge: Cambridge University Press, 1932), 12:362; see also W. W. Greg's description of bibliography as "the handmaid of literature": W. W. Greg, "What Is Bibliography?" *Transactions of the Bibliographical Society* 12 (1914): 47. On the bibliographer as service worker, see Jon Klancher, "Bibliographia Literaria: Thomas Dibdin and the Origins of Book History in Britain, 1800–1825" (unpublished paper), and, on the feminization of the literal, Homans, *Bearing the Word: Language and Female Experience in Nineteenth-Century Women's Writing.*

20. The libretto copies this scene roughly from chapter 9 of Henri Murger's eponymous novel: there, the "dénouement ne fit que flamber et s'éteindre." Giacomo Puccini and Henri Murger, *La Bohème* ([Paris]: Calmann-Lévy, Erato, 1988), 293.

21. The term puns on writing, but also on printing: the "secret composition" that makes the flypaper sticky bears some resemblance to the mixture of glue and treacle used to ink the device known as a "composition roller." See Annie Carey, *The History of a Book* (London: Cassell, Petter and Galpin, [1873]) 105.

22. Even the second volume itself distinguishes the tradesman who resells objects "to be disposed of as old metal or waste-paper" from "his brother tradesman [who] buys them to be resold and remanufactured for the purposes for which they were originally intended" (2:108).

23. See Christopher Herbert, *Culture and Anomie* (Chicago: University of Chicago Press, 1991), 221, and Michal Peled Ginsburg, "The Case against Plot in *Bleak House* and *Our Mutual Friend*," *ELH* 59.1 (1992): 179.

24. Altick, *The English Common Reader: A Social History of the Mass Reading Public, 1800–1900.* 108.

25. Edwards, manuscript scrapbook, Book E (1860–1876), Manchester Public Library Archives, p. 189, quoted in Black, *A New History of the English Public Library: Social and Intellectual Contexts, 1850–1914,* 93.

26. Similarly, a missionary in Jerusalem named Mr. Whiting explains that he gives tracts to illiterate Arab women who claim their sons can read "in the hope that, even if their object be to sell them, they will fall into the hands of some one who will derive benefit from them." William Jones, *The Jubilee Memorial of the Religious Tract Society Containing a Record of Its Origin, Proceedings, and Results, A.D. 1799 to A.D. 1849* (London: Religious Tract Society, 1850), 389.

27. See Peter Stallybrass and Ann Rosalind Jones, *Renaissance Clothing and the Materials of Memory* (Cambridge: Cambridge University Press, 2000). On clothing in the context of preindustrial recycling more generally, see Donald Woodward, "Swords into Ploughshares: Recycling in Pre-Industrial England," *Economic History Review* 38 (1985): 177–79.

28. Sutherland, *Victorian Novelists and Publishers,* 70; Freeland, "Trash Fiction," 11. Walter Siti, too, argues that the novel bears a special relationship to repetition, citing Vauvenargues's observation that "you never reread a novel." Franco Moretti, *The Novel,* 2 vols. (Princeton, N.J.: Princeton University Press, 2006), 1:99.

29. On the history of the text/textile metaphor, see Roger Chartier, *Inscription and Erasure: Literature and Written Culture from the Eleventh to the Eighteenth Century,* trans. Arthur Goldhammer (Philadelphia: University of Pennsylvania Press, 2007), 95–97.

30. Henry Mayhew and John Binny, *The Criminal Prisons of London and Scenes of Prison Life* (London: Frank Cass, 1968), 37; on this passage, see Anne Humpherys, *Travels into the Poor Man's Country: The Work of Henry Mayhew* (Athens: University of Georgia Press, 1977), 148.

31. The paper duty imposed in 1711 was reduced in 1836 and removed by Gladstone in 1860, the same year in which esparto grass was first used. George Richardson Porter, *The Progress of the Nation in Its Various Social and Economic Relations from the Beginning of the Nineteenth Century* (London: Methuen & Co., 1912), 405.

32. For an analogous argument that attributes the rise of the novel to the fall of paper prices, see Erikson, *The Economy of Literary Form.*

33. For a theory of such shifts from figurative to literal and metaphor to metonymy, see Elaine Freedgood, *The Ideas in Things: Fugitive Meaning in the Victorian Novel* (Chicago: University of Chicago Press, 2006), 20.

34. Cp. Bill Brown's shrewd observation that "the immaterial/material distinction often asserts itself as the difference between the visible and the tangible," as if sight were not just as physically embodied as touch. Bill Brown, "Materiality," *Critical Terms for Media Studies* (Chicago: University of Chicago Press, 2010), 51.

35. It's telling in this respect that an earlier study like Hans J. Rindisbacher, *The Smell of Books: A Cultural-Historical Study of Olfactory Perception in Literature* (Ann Arbor: University of Michigan Press, 1992), could analyze the literary representation of smell without ever mentioning the literal "smell of books" from which it took its title.

36. It's no accident (as literary critics used to say) that a single term designates both a maid's intrusion into the decisions about which version of a text survives—decisions that should remain the purview of gentlemanly editors—and those features of the text that are entrusted to a lesser functionary than the author (punctuation, for example, began as the publisher's responsibility and only gradually became part of the author's remit). David C. Greetham, *Theories of the Text* (Oxford: Oxford University Press, 1999), 142; Allan C. Dooley, *Author and Printer in Victorian England*, Victorian Literature and Culture Series (Charlottesville: University of Virginia Press, 1992).

CONCLUSION

1. Compare "Depuis que je savais que—contrairement à ce que m'avaient si longtemps représenté mes imaginations enfantines,—il n'y avait qu'une scène pour tout le monde, je pensais qu'on devait être empêché de bien voir par les autres spectateurs comme on l'est au milieu d'une foule; or je me rendis compte qu'au contraire, grâce à une disposition qui est comme le symbole de toute perception, chacun se sent le centre du théâtre" Marcel Proust, *A l'ombre des jeunes filles en fleurs*, 3 vols. (Paris: Gallimard, 1919), 1:26.

2. On "the fantasy of an ideal listener"—or rather, in what I think is an important distinction, ideal reader—see Carla Kaplan, "Girl Talk: *Jane Eyre* and the Romance of Women's Narration," *Novel: A Forum on Fiction* 30.1 (1996): 25.

3. As Jon Klancher argues, "the intense cultural politics of the Romantic period obliged writers not only to distinguish among conflicting audiences, but to do so by elaborating new relations between the individual reader and the collective audience." Klancher, *The Making of English Reading Audiences, 1790–1832* 11. Ian Duncan shows more specifically that "the Waverley novels, soliciting a 'universal' reading public, definitively establish the 'popular' form of an expanding national literacy, at the same time as they mark off a class boundary in economic terms." Ian Duncan, *Modern Romance and Transformations of the Novel: The Gothic, Scott, Dickens* (Cambridge: Cambridge University Press, 1992), 179.

4. Even Anderson's critics accept this premise unquestioningly, as when Culler's counterargument begins, "Since newspapers are read on the day of publication and thrown away." Jonathan Culler, "Anderson and the Novel," *Diacritics* 29.4 (1999): 27; for a more nuanced sense of Romantic-era newspaper circulation in Britain, see Kevin Gilmartin, *Print Politics: The Press and Radical Opposition in Early Nineteenth-Century England* (Cambridge: Cambridge University Press, 1996), 65–114.

5. Similarly, Elizabeth Eisenstein's claim that "printed materials encouraged silent adherence to causes whose advocates could not be located in any one parish and who addressed an invisible public from afar" does little to address the role of parishes in ensuring the circulation of printed matter: Elizabeth Eisenstein, "Some Conjectures about the Impact of Printing on Western Society and Thought: A Preliminary Report," *Journal of Modern History* 40.1 (1968): 42.

Works Cited

Ackland, Joseph. "Elementary Education and the Decay of Literature." *The Nineteenth Century* (1894): 412–23.

Adams, Charlotte. *Boys at Home*. New York, 1854.

———. *John Hartley, and How He Got on in Life*. London: Routledge, 1867.

———. *Little Servant Maids*. London: S.P.C.K., 1848.

———. *The Useful Little Girl*. London: S.P.C.K., 1865.

Adams, Thomas R., and Nicolas Barker. "A New Model for the Study of the Book." *A Potencie of Life: Books in Society*. Ed. Nicolas Barker. London: British Library, 1993. 5–43.

Addison, Joseph, and Richard Steele. *The Spectator*. Ed. Donald Frederic Bond. 5 vols. Oxford: Clarendon Press, 1965.

Administration of the Post Office: From the Introduction of Mr. Rowland Hill's Plan of Penny Postage up to the Present Time. London: J. Hatchard and Son, 1844.

Adventures of a Bible: Or, the Advantages of Early Piety. London: Dean and Munday, 1825.

"Adventures of a Quire of Paper." *London Magazine*, August–October 1779.

"Adventures of a Robinson Crusoe. Written by Itself. To the Editors of the Young Gentleman's and Lady's Magazine." *Young Gentleman's & Lady's Magazine, or Universal Repository of Knowledge, Instruction and Amusement* 1 (1799).

Allen, Charlotte. "Indecent Disposal: Where Academic Books Go When They Die." *Lingua Franca* 5.4 (1995): 44–53.

Altick, Richard. *The English Common Reader: A Social History of the Mass Reading Public, 1800–1900*. 1957. With a foreword by Jonathan Rose. Chicago: University of Chicago Press, 1998.

Altick, Richard D. *The Presence of the Present: Topics of the Day in the Victorian Novel*. Columbus: Ohio State University Press, 1991.

Amory, Hugh. "The Trout and the Milk: An Ethnobibliographical Talk." *Harvard Library Bulletin*, n.s., 7.1 (1996): 50–65.

Andersen, Hans Christian. *The Complete Fairy Tales*. Ware, Hertfordshire: Wordsworth Editions, 1997.

Anderson, Benedict. *Imagined Communities: Reflections on the Origin and Spread of Nationalism*. London: Verso, 1991.

Andrzejewski, Anna Vemer. *Building Power: Architecture and Surveillance in Victorian America*. Knoxville: University of Tennessee Press, 2008.

Anger, Suzy. *Victorian Interpretation*. Ithaca, N.Y.: Cornell University Press, 2005.

Annual Report of the American Tract Society. Boston, 1839.

Appadurai, Arjun. "Introduction: Commodities and the Politics of Value." *The Social Life of Things: Commodities in Cultural Perspective*. Ed. Arjun Appadurai. Cambridge: Cambridge University Press, 1986. 3–63.

Armstrong, Nancy. *How Novels Think: The Limits of Individualism from 1719–1900*. New York: Columbia University Press, 2005.

Austen, Jane. *Northanger Abbey*. Ed. John Davie. Oxford: Oxford University Press, 1990.

Austen, Jane. *Pride and Prejudice*. Ed. Donald Gray. New York: Norton, 2001.

[Austin, Alfred]. "Our Novels: The Sensational School." *Temple Bar*, July 1870, 410–24.

Awdry, Frances, et al. *The Miz Maze; or, the Winkworth Puzzle. A Story in Letters*. London: Macmillan, 1883.

Ayers, David. "Materialism and the Book." *Poetics Today* 24.4 (2003): 759–80.

Babbage, Charles. *The Ninth Bridgewater Treatise: A Fragment*. London: W. Pickering, 1989.

"Bachelor Days. IV." *Punch*, 3 July 1907, 16–17.

Bagehot, Walter. "Charles Dickens." *National Review,* October 1858, 458–86.

Barrie, J. M. *Two of Them*. New York: Lovell Coryell, 1893.

Battles, Matthew. *Library: An Unquiet History*. New York: Norton, 2003.

Beetham, Margaret. "In Search of the Historical Reader; the Woman Reader, the Magazine and the Correspondence Column." *Siegener Periodicum zur Internationalen Empirischen Literaturwissenschaft [SPIEL]* 19.1 (2000): 89–104.

———. *A Magazine of Her Own? Domesticity and Desire in the Woman's Magazine, 1800–1914*. London: Routledge, 1996.

Beeton, Isabella. *Mrs Beeton's Book of Household Management*. London: S. O. Beeton, 1863.

Bell, Bill. "Bound for Australia: Shipboard Reading in the Nineteenth Century." *Journeys through the Market: Travel, Travellers, and the Book Trade*. Ed. Robin Myers and Michael Harris. New Castle, Del.: Oak Knoll Press, 1999.

Bell, Florence Eveleen Eleanore Olliffe. *At the Works: A Study of a Manufacturing Town*. London: Nelson, 1911.

Benjamin, Walter. "Unpacking My Library." *Illuminations*. Ed. Hannah Arendt. New York: Harcourt Brace, 1985.

Bennett, Arnold. *Books and Persons; Being Comments on a Past Epoch, 1908–1911*. London: Chatto & Windus, 1917.

Bennett, Scott. "Revolutions in Thought: Serial Publication and the Mass Market in Reading." *The Victorian Periodical Press: Samplings and Soundings*. Ed. Joanne Shattock and Michael Wolff. Leicester: Leicester University Press, 1982. 225–57.

Bernstein, Charles. *A Poetics*. Cambridge, Mass.: Harvard University Press, 1992.

Bernstein, Robin. "Dances with Things: Material Culture and the Performance of Race." *Social Text* 101 (2009).

Best, Mrs. *The History of a Family Bible. A Tale of the American War*. London: John Farquhar Shaw, 1851.

Bhabha, Homi K. "Signs Taken for Wonders: Questions of Ambivalence and Authority under a Tree Outside Delhi, May 1817." *Critical Inquiry* 12.1 (1985): 144–65.

Biernacki, Richard. "Method and Metaphor after the New Cultural History." *Beyond the Cultural Turn: New Directions in the Study of Society and Culture*. Ed. Victoria E. Bonnell and Lynn Avery Hunt. Berkeley: University of California Press, 1999.

Binkley, Robert. "New Tools for Men of Letters." *Yale Review,* 1935, 519–37.

Birkerts, Sven. *The Gutenberg Elegies: The Fate of Reading in an Electronic Age*. Winchester, MA: Faber and Faber, 1994.

Birrell, Augustine. "Book-Buying." *Obiter Dicta*. New York: Charles Scribner's Sons, 1891. 284–91.

Black, Alistair. "The Library as Clinic: A Foucauldian Interpretation of British Public Library Attitudes to Social and Physical Disease, ca. 1850–1950." *Libraries & Culture* 40.3 (2005): 416–34.

———. *A New History of the English Public Library: Social and Intellectual Contexts, 1850–1914*. London: Leicester University Press, 1996.

Blades, William. *The Enemies of Books*. London: E. Stock, 1896.

Blair, Ann. "Note Taking as an Art of Transmission." *Critical Inquiry* 31.1 (2004): 85–107.

———. *Too Much to Know: Managing Scholarly Information before the Modern Age*. New Haven, Conn.: Yale Unversity Press, 2010.

Blair, Ann, and Peter Stallybrass. "Mediating Information, 1450–1800." *This Is Enlightenment*. Ed. Clifford Siskin and William Warner. Chicago: University of Chicago Press, 2010. 139–63.

Bloom, Jonathan. *Paper before Print: The History and Impact of Paper in the Islamic World*. New Haven, Conn.: Yale University Press, 2001.

Bodenheimer, Rosemarie. *Knowing Dickens*. Ithaca, N.Y.: Cornell University Press, 2007.

Bogue, David. *An Address to Christians on the Distribution of Religious Tracts*. London: G. Brimmer, 1799.

Borrett White, Lewis. *The Religious Condition of Christendom: Described in a Series of Papers Presented to the Eighth General Conference of the Evangelical Alliance in Copenhagen, 1884*. London: Office of the Evangelical Alliance, 1885.

Bosanquet, Helen. "Cheap Literature." *Contemporary Review* 79 (1901): 677–81.

Boston Society for the Religious and Moral Improvement of Seamen. *The Adventures of a Bible*. Boston: John Eliot, 1813.

Braddon, M. E. *Lady Audley's Secret*. Oxford World's Classics. Ed. David Skilton. Oxford: Oxford University Press, 2008.

Braddon, Mary Elizabeth. *The Doctor's Wife*. Ed. Pykett Lynn. Oxford: Oxford University Press, 1998.

Brake, Laurel. "Literary Criticism in Victorian Periodicals." *Yearbook of English Studies* 16 (1986): 92–116.

Brantlinger, Patrick. *The Reading Lesson: The Threat of Mass Literacy in Nineteenth-Century Britain*. Bloomington: University of Indiana Press, 1998.

Bratton, J. S. *The Impact of Victorian Children's Fiction*. London: Croom Helm, 1981.

"Brevities." *Figaro in London* 1 (1831): 7.

Brodhead, Augustus. *Conference on Urdu and Hindi Christian Literature, Held at Allahabad*. Madras: Christian Vernacular Education Society, 1875.

Brontë, Anne. *The Tenant of Wildfell Hall*. London: Dent, 1914.

Brontë, Charlotte. *Jane Eyre*. Ed. Michael Mason. Harmondsworth: Penguin, 2003.

———. *Shirley*. Ed. Andrew Hook and Judith Hook. London: Penguin Classics, 1985.

Brottman, Mikita. *The Solitary Vice: Against Reading*. Berkeley, Calif.: Counterpoint, 2008.

Broughton, Rhoda. *A Beginner; a Novel*. New York: D. Appleton and Company, 1894.

———. *Second Thoughts*. New York: Universal Publishing, n.d.

Brown, Bill. "Introduction: Textual Materialism." *PMLA* 125.1 (2010).

———. "Materiality." *Critical Terms for Media Studies*. Chicago: University of Chicago Press, 2010.

Brown, John Seely, and Paul Duguid. *The Social Life of Information*. Boston: Harvard Business School, 2002.

Bulwer-Lytton, Edward. *England and the English*. Ed. Standish Meacham. Chicago: University of Chicago Press, 1970.

———. "On Certain Principles of Art in Works of Imagination (1863)." *The Early and Mid-Victorian Novel*. Ed. David Skilton. London: Routledge, 1993. 174–75.

"The Bunch of Keys." *The Youth's Magazine or Evangelical Miscellany* 7 (1867).

Burnett, Frances Hodgson. *A Little Princess*. Ed. U. C. Knoepflmacher. Harmsondsworth: Penguin, 2002.

Butler, E. H. *The Story of British Shorthand*. London: Isaac Pitman, 1951.

Butor, Michel. *Les mots dans la peinture*. [Geneva]: A Skira, 1969.

Butterworth, C. H. "Overfeeding." *Victoria Magazine* 14 (1869–70): 500–504.

Buzard, James. "Home Ec. with Mrs. Beeton." *Raritan* 17.2 (1997): 121–35.

Byron, George Gordon Lord. *Byron's Letters and Journals*. Ed. Leslie Alexis Marchand. Vol. 8. 12 vols. Cambridge, Mass.: Belknap Press of Harvard University Press, 1973.

Calhoun, Joshua. "The Word Made Flax: Cheap Bibles, Textual Corruption, and the Poetics of Paper." *PMLA* 126.2 (2011): 327.

Calinescu, Matei. *Rereading*. New Haven, Conn.: Yale University Press, 1993.

Calvino, Italo. *If on a Winter's Night a Traveler*. New York: Harcourt Brace Jovanovich, 1981.

Canetti, Elias. *Auto-da-Fé*. Trans. D. V. Wedgewood. New York: Farrar, Straus and Giroux, 1984.

Cantor, G. N. *Science in the Nineteenth-Century Periodical: Reading the Magazine of Nature*. Cambridge: Cambridge University Press, 2004.

Carey, Annie. *The History of a Book*. London: Cassell, Petter and Galpin, [1873].

Carlyle, Thomas. *On Heroes, Hero-Worship, & the Heroic in History*. Berkeley: University of California Press, 1993.

———. *Sartor Resartus*. Ed. Kerry McSweeney and Peter Sabor. Oxford: Oxford University Press, 1987.

Certeau, Michel de. *The Practice of Everyday Life*. Trans. Steven F. Rendell. Berkeley: University of California Press, 1984.

"Charles Dickens and David Copperfield." *Fraser's Magazine,* December 1850, 698–710.

Charlesworth, Maria Louisa. *A Book for the Cottage: Or, the History of Mary and Her Family.* London: Seeley, Burnside and Seeley, 1848.

———. *The Female Visitor to the Poor; or, Records of Female Parochial Visiting.* London: Seeley, Burnside and Seeley, 1856.

———. *Ministering Children: A Tale Dedicated to Childhood.* New York: Robert Carter, 1867.

———. *The Old Looking-Glass; or, Mrs. Dorothy Cope's Recollections of Service.* London: Seeley, Jackson & Halliday, 1878.

Chartier, Roger. *Inscription and Erasure: Literature and Written Culture from the Eleventh to the Eighteenth Century.* Trans. Arthur Goldhammer. Philadelphia: University of Pennsylvania Press, 2007.

Chartier, Roger, ed. *On the Edge of the Cliff: History, Language and Practices.* Baltimore, Md.: Johns Hopkins University Press, 1997.

Chartier, Roger. *The Order of Books.* Cambridge: Polity Press, 1994.

Chesterfield, Philip Dormer Stanhope. *The Letters of Philip Dormer Stanhope, 4th Earl of Chesterfield.* Ed. Bonamy Dobrée. New York: AMS Press, 1968.

Chestnutt, Charles. "Baxter's Procrustes." *Atlantic Monthly* 93 (1904): 823–30.

"Circulation of the Scriptures." *Baptist Magazine* 1874: 308–11.

Cohen, Jessica, and Pascaline Dupas. "Free Distribution or Cost-Sharing? Evidence from a Randomized Malaria Prevention Experiment." *Quarterly Journal of Economics* 125.1 (February 2010): 1–45.

Colclough, Stephen. "Distribution." *The Cambridge History of the Book in Britain, 1830–1914.* Ed. David McKitterick. Cambridge: Cambridge University Press, 2009. 238–80.

Colclough, Stephen, and David Vincent. "Reading." *The Cambridge History of the Book in Britain, 1830–1914.* Ed. David McKitterick. Cambridge: Cambridge University Press, 2009. 281–323.

Coleridge, Samuel Taylor. *Biographia Literaria.* Ed. James Engell, and W. Jackson Bate. 2 vols. Princeton, N.J.: Princeton University Press, 1983.

Collet, Collet Dobson. *History of the Taxes on Knowledge; Their Origin and Repeal.* London: T. F. Unwin, 1899.

Collier, Jane. *An Essay on the Art of Ingeniously Tormenting.* Ed. Audrey Bilger. Peterborough, Ont.: Broadview Press, 2003.

Collins, Charles Alliston. "Our Audience." *Macmillan's Magazine,* June 1863, 161–66.

Collins, Paul. "You and Your Dumb Friends." *Believer,* March 2004.

Collins, Philip Arthur William. *Reading Aloud; a Victorian Métier.* Lincoln: Tennyson Research Centre, 1972.

Collins, Wilkie. *Basil: A Novel.* London: Sampson, Low, 1862.

———. *The Moonstone.* Ed. John Sutherland. Oxford: Oxford University Press, 1999.

———. *Poor Miss Finch.* Ed. Catherine Peters. Oxford: Oxford World's Classics, 1995.

Connell, Philip. "Bibliomania: Book Collecting, Cultural Politics, and the Rise of Literary Heritage in Romantic Britain." *Representations* 71 (2000): 24–47.

Cooke, Maud C. *Social Etiquette, Or: Manners and Customs of Polite Society.* Boston: Geo. Smith, 1893.

Cope, E. A. "Phonographic Associations: How to Organize and Conduct Them." *Phonetic Journal* (1885).

Corbett, Mary Jean. *Representing Femininity: Middle-Class Subjectivity in Victorian and Edwardian Women's Autobiographies.* New York: Oxford University Press, 1992.

Corelli, Marie. *Free Opinions Freely Expressed on Certain Phases of Modern Social Life and Conduct.* London: Archibald Constable, 1905.

Cormack, Bradin, and Carla Mazzio. *Book Use, Book Theory: 1500–1700.* Chicago: University of Chicago Library, 2005.

Coutts, Henry T. *Library Jokes and Jottings: A Collection of Stories Partly Wise but Mostly Otherwise.* London: Grafton & Co., 1914.

Craik, D. M. *A Life for a Life.* N.p.: Kessinger, n.d.

Craik, George L. *The Pursuit of Knowledge under Difficulties.* New ed. London: J. Murray, 1858.

Crain, Patricia. *The Story of A: The Alphabetization of America from the New England Primer to the Scarlet Letter.* Stanford, Calif.: Stanford University Press, 2002.

Crary, Jonathan. *Suspensions of Perception: Attention, Spectacle, and Modern Culture.* Cambridge, Mass.: MIT Press, 1999.

Cressy, David. "Books as Totems in Seventeenth-Century England and New England." *Journal of Library History* 21.1 (1986): 92–106.

Crowe, Catherine. *Susan Hopley or the Adventures of a Maid Servant.* London: Routledge, 1852.

Culler, Jonathan. "Anderson and the Novel." *Diacritics* 29.4 (1999): 20–39.

Curtius, Ernst Robert. "The Book as Symbol." Trans. Willard R. Trask. *European Literature and the Latin Middle Ages.* New York: Pantheon Books, 1953. 302–47.

D'Arcy, Ella. "Irremediable." *Victorian Short Stories 2: The Trials of Love.* 1899. London: Everyman, 1990. 94–115.

D'Israeli, Isaac. *Curiosities of Literature.* 3 vols. London: Frederick Warne, 1881.

Dames, Nicholas. *The Physiology of the Novel: Reading, Neural Science, and the Form of Victorian Fiction.* Oxford: Oxford University Press, 2007.

Dardier, J. P. "Evangelization by Book-Post." *Christian World* 35 (1884): 317–19.

Darnton, Robert. *The Business of Enlightenment: A Publishing History of the Encyclopédie, 1775–1800.* Cambridge, Mass.: Harvard University Press, 1979.

———. "First Steps toward a History of Reading." *The Kiss of Lamourette: Reflections in Cultural History.* New York: Norton, 1990.

———. "Readers Respond to Rousseau." *The Great Cat Massacre and Other Episodes in French Cultural History.* London: Allen Lane, 1984. 215–56.

Darwin, Charles. "'This Is the Question Marry Not Marry' [Memorandum on Marriage]." 1838. http://www.darwinproject.ac.uk/darwins-notes-on-marriage.

Daston, Lorraine. "Taking Note(s)." *ISIS: Journal of the History of Science in Society* 95.3 (2004): 443–48.

Davidoff, Leonore, and Catherine Hall. *Family Fortunes: Men and Women of the English Middle Class, 1780–1850.* Chicago: University of Chicago Press, 1987.

Davidson, Cathy. "The Life and Times of *Charlotte Temple*: The Biography of a Book." *Reading in America: Literature and Social History.* Ed. Cathy Davidson. Baltimore, Md.: Johns Hopkins University Press, 1989. 157–79.

Davies, Tony. "Transports of Pleasure: Fiction and Its Audiences in the Later Nineteenth Century." *Formations of Pleasure.* London: Routledge & Kegan Paul, 1983.

Davis, Natalie. *Society and Culture in Early Modern France.* Stanford, Calif.: Stanford University Press, 1975.

Davis, Natalie Zemon. "Beyond the Market: Books as Gifts in Sixteenth-Century France." *Transactions of the Royal Historical Society* 33 (1983): 69–88.

Davis, Nuel Pharr. *The Life of Wilkie Collins.* Urbana: University of Illinois Press, 1956.

Deakin, Mary H. *The Early Life of George Eliot.* Manchester: University of Manchester Press, 1913.

Derrida, Jacques. *Paper Machine.* Trans. Rachel Bowlby. Cultural Memory in the Present. Stanford, Calif.: Stanford University Press, 2005.

Dettmar, Kevin. "Bookcases, Slipcases, Uncut Leaves: The Anxiety of the Gentleman's Library." *Novel* 39 (2005 [i.e., 2006]): 1–24.

Dewing, Maria Oakey. *Beauty in the Household.* New York: Harper, 1882.

Dibdin, Thomas Frognall. *The Bibliomania: Or, Book-Madness; Containing Some Account of the History, Symptoms and Cure of This Fatal Disease, in an Epistle Addressed to Richard Heber, Esq.* London: Printed for Longman Hurst Rees and Orme by W. Savage, 1809.

Dickens, Charles. *American Notes and Reprinted Pieces.* London: Chapman and Hall, n.d.

———. *Bleak House.* Ed. Nicola Bradbury. New York: Penguin, 1996.

———. *David Copperfield.* Ed. Jeremy Tambling. New York: Penguin, 1996.

———. *Great Expectations.* New York: Bantam, 1981.

———. *Hard Times.* Ed. Paul Schlicke. Oxford: Oxford University Press, 1998.

———. *Nicholas Nickleby.* New York: Bantam, 1983.

———. *Oliver Twist.* Ed. Philip Horne. Harmondsworth: Penguin, 2003.

———. *Our Mutual Friend.* Ed. Stephen Gill. Harmondsworth: Penguin, 1985.

———. *Prospectus for the Cheap Edition of the Works of Mr. Charles Dickens.* London: Chapman & Hall, 1847.

———. *Sketches by Boz.* Ed. Michael Slater. New York: Phoenix, 1996.

———. *The Speeches of Charles Dickens.* Ed. K. J. Fielding. Oxford: Clarendon Press, 1960.

———. *A Tale of Two Cities.* Ed. Richard Maxwell. Harmondsworth: Penguin, 1994.

———. *Uncommercial Traveller.* Cambridge: Printed at the Riverside Press, 1869.

Diderot, Denis. "Eloge de Richardson." *Oeuvres complètes.* Ed. Herbert Dieckmann, Jean Fabre, and Jacques Proust. Paris: Hermann, 1975.

Dinesen, Isak. *Out of Africa.* London: Century Publishing, 1985.

Domestic Management, Or the Art of Conducting a Family; with Instructions to Servants in General. Addressed to Young Housekeepers. London: Printed for H. D. Symonds at the Literary Press, No. 62 Wardour-Street, Soho, 1800.

Donaldson, Ian. "The Destruction of the Book." *Book History* 1.1 (1998): 1–10.

Dooley, Allan C. *Author and Printer in Victorian England.* Victorian Literature and Culture Series. Charlottesville: University of Virginia Press, 1992.

Doveton, Frederick Brickdale. *Sketches in Prose and Verse.* London: Sampson Low, Marston, Searle, & Rivington, 1886.

Doyle, Arthur Conan. *Through the Magic Door.* New York: McClure, 1908.

Drucker, Johanna. *The Visible Word: Experimental Typography and Modern Art, 1909–1923.* Chicago: University of Chicago Press, 1994.

Drummond, William, et al. "Bibliotheca Edinburgena Lectori." *The Works of William Drummond, of Hawthornden: Consisting of Those Which Were Formerly Printed, and Those Which Were Design'd for the Press: Now Published from the Author's Original Copies.* Edinburgh: Printed by James Watson in Craig's-Close, 1711. 222.

Duguid, Paul. "Material Matters: The Past and Futurology of the Book." *The Future of the Book.* Ed. Geoffrey Nunberg. Berkeley: University of California Press, 1996. 63–102.

———. "The Quality of Information: High-Tech Supply and Low-Tech Command," 2007 PARC Forum, 22 February 2007.

Duncan, Ian. *Modern Romance and Transformations of the Novel: The Gothic, Scott, Dickens.* Cambridge: Cambridge University Press, 1992.

———. *Scott's Shadow: The Novel in Romantic Edinburgh.* Literature in History. Princeton, N.J.: Princeton University Press, 2007.

Dyer, S. "Bible Distribution in China—Its Methods and Results." *Records of the General Conference of the Protestant Missionaries of China* (1890): 116–35.

Edgerton, David. *The Shock of the Old: Technology and Global History since 1900.* London: Profile Books, 2006.

Edgeworth, Maria. *Patronage.* Vol. 6 of *The Novels and Selected Works of Maria Edgeworth.* Ed. Conor Carville and Marilyn Butler. 12 vols. London: Pickering & Chatto, 1999.

Edwards, Amelia Ann Blanford. *Barbara's History: A Novel.* New York: Harper, 1864.

Eisenstein, Elizabeth. "Some Conjectures about the Impact of Printing on Western Society and Thought: A Preliminary Report." *Journal of Modern History* 40.1 (1968): 1–56.

Elfenbein, Andrew. "Cognitive Science and the History of Reading." *PMLA* 121 (2006): 484–502.

Eliot, George. "J. A. Froude's *The Nemesis of Faith.*" *Selected Essays, Poems, and Other Writings.* Ed. Nicholas Warren. London: Penguin, 1990. 265–67.

———. *Middlemarch.* 2nd ed., 1874. Ed. Rosemary Ashton. Harmondsworth: Penguin, 2003.

———. *The Mill on the Floss.* Ed. A. S. Byatt. Harmondsworth: Penguin, 2003.

Eliot, Simon. "The Business of Victorian Publishing." *The Cambridge Companion to the Victorian Novel.* Ed. Deirdre David. Cambridge: Cambridge University Press, 2001. 37–60.

———. "Circulating Libraries in the Victorian Age and After." *The Cambridge History of Libraries in Britain and Ireland*. Ed. Peter Hoare et al. Vol. 3. Cambridge: Cambridge University Press, 2006. 125–46.

———. *Some Patterns and Trends in British Publishing, 1800–1919*. Occasional Papers of the Bibliographical Society, No. 8. London: Bibliographical Society, 1994.

Eliot, Simon, and Jonathan Rose. *A Companion to the History of the Book*. Blackwell Companions to Literature and Culture, 48. Malden, Mass.: Blackwell Pub., 2007.

Elkins, James. *Pictures & Tears: A History of People Who Have Cried in Front of Paintings*. New York London: Routledge, 2001.

Ellis, Sarah. "The Art of Reading Well, as Connected with Social Improvement." *Victorian Print Media: A Reader*. Ed. John Plunkett and Andrew King. Oxford: Oxford University Press, 2005. 250–54.

Emerson, Ralph Waldo. "Books." *Atlantic Monthly* 1 (1858): 343–54.

Emerson, Ralph Waldo, and John Lubbock. *In Praise of Books: A Vade Mecum for Book-Lovers*. New York: The Perkins Book Company, 1901.

"The Encyclopedia Britannica." *Quarterly Review* 70 (1842): 44–72.

Erikson, Lee. *The Economy of Literary Form*. Baltimore, Md.: Johns Hopkins University Press, 1996.

Erwin, Miles. "Pensioners Burn Books for Warmth." *Metro.co.uk,* 5 January 2010, sec. News, 1.

Evans, M.D.R., Jonathan Kelley, Joanna Sikora, and Donald J. Treiman. "Family Scholarly Culture and Educational Success: Books and Schooling in 27 Nations." *Research in Social Stratification and Mobility* 28 (2010): 171–97.

"Excerpt from *Hereford Times*." *Phonetic Journal*, 1868, 281.

"Excessive Reading." *Pall Mall Gazette*, 1869, 9.

Fabian, Ann. *The Unvarnished Truth: Personal Narratives in Nineteenth-Century America*. Berkeley: University of California Press, 2000.

Fadiman, Anne. *Rereadings*. 1st ed. New York: Farrar, Straus and Giroux, 2005.

Favret, Mary. *Romantic Correspondence: Women, Politics, and the Fiction of Letters*. Cambridge: Cambridge University Press, 1993.

Fergus, Jan S. "Provincial Servants' Reading in the Late 18th Century." *The Practice and Representation of Reading in England*. Ed. James Raven, Helen Small, and Naomi Tadmor. Cambridge: Cambridge University Press, 1996. 202–225.

Ferris, Ina. *The Achievement of Literary Authority: Gender, History, and the Waverley Novels*. Ithaca, N.Y.: Cornell University Press, 1991.

———. "Bibliographic Romance: Bibliophilia and the Book-Object." *Romantic Libraries*. Ed. Ina Ferris. College Park: University of Maryland, 2004.

———. "Introduction." *Romantic Libraries*. College Park: University of Maryland, 2004.

Festa, Lynn M. *Sentimental Figures of Empire in Eighteenth-Century Britain and France*. Baltimore, Md.: Johns Hopkins University Press, 2006.

"A Few Words About Reading." *Chambers's Journal of Popular Literature, Science, and Arts* 70 (1893): 225–27.

Fielding, Henry. *The History of Tom Jones*. Ed. R.P.C. Mutter. Baltimore, Md.: Penguin, 1966.

Fielding, Penny. "Reading Rooms: M. R. James and the Library of Modernity." *Modern Fiction Studies* 46.3 (2000): 749–71.

Fifty-Sixth Report of the British and Foreign Bible Society. London: Spottiswoode, 1860.

Finn, Margot. "Men's Things: Masculine Possession in the Consumer Revolution." *Social History* 25.2 (2000): 133–55.

Flaubert, Gustave. *Madame Bovary: Provincial Manners.* Trans. Margaret Mauldon. Oxford: Oxford University Press, 2004.

Flint, Christopher. "Speaking Objects: The Circulation of Stories in Eighteenth-Century Prose Fiction." *PMLA* 113.2 (1998): 212–26.

Flint, Kate. *The Feeling of Reading: Affective Experience and Victorian Literature.* Ed. Rachel Ablow. Ann Arbor: University of Michigan Press, 2010.

———. *The Woman Reader, 1837–1914.* Oxford: Clarendon Press, 1993.

"Foreign Missions at Home." *Indian Evangelical Review* 6 (1879).

"Frank" and I. New York: Grove Press, 1968.

Frankel, Oz. "Blue Books and the Victorian Reader." *Victorian Studies* 46.2 (2004): 308–18.

Fraser, Robert. *Book History through Postcolonial Eyes.* London: Routledge, 2008.

Freedgood, Elaine. *The Ideas in Things: Fugitive Meaning in the Victorian Novel.* Chicago: University of Chicago Press, 2006.

Freeland, Natalka. "Trash Fiction: The Victorian Novel and the Rise of Disposable Culture." Unpublished MS.

Fried, Michael. *Absorption and Theatricality: Painting and Beholder in the Age of Diderot.* Chicago: University of Chicago Press, 1980.

———. *Realism, Writing, Disfiguration: On Thomas Eakins and Stephen Crane.* Chicago: University of Chicago Press, 1987.

Friswell, Hain. "Circulating Libraries." *London Society* 20 (1871): 515–24.

Frith, Gail. "Women, Writing and Language: Making the Silences Speak." *Thinking Feminist: Key Concepts in Women's Studies.* New York: Guilford, 1993. 151–76.

Fritzsche, Peter. "1936, May 1." *A New History of German Literature.* Ed. David E. Wellbery, Judith Ryan, and Hans Ulrich Gumbrecht. Cambridge, Mass.: Belknap Press of Harvard University Press, 2004.

Fuller, Margaret. *Memoirs of Margaret Fuller Ossoli.* Vol. 1. 3 vols. Boston: Sampson, 1852.

"Furniture Books." *Fraser's Magazine,* January 1859, 95–102.

Fyfe, Aileen. "Commerce and Philanthropy: The Religious Tract Society and the Business of Publishing." *Journal of Victorian Culture* (2004): 164–88.

———. *Science and Salvation: Evangelical Popular Science Publishing in Victorian Britain.* Chicago: University of Chicago Press, 2004.

———. "Societies as Publishers: The Religious Tract Society in the Mid-Nineteenth Century." *Publishing History* 58 (2005): 5–41.

Gagnier, Regenia. *Subjectivities: A History of Self-Representation in Britain, 1832–1920.* New York: Oxford University Press, 1991.

Gallagher, Catherine. *Nobody's Story: The Vanishing Acts of Women Writers in the Marketplace, 1670–1820.* Berkeley: University of California Press, 1994.

Gamer, Michael. "*Waverley* and the Object of (Literary) History." *Modern Language Quarterly* 70.4 (2009): 495–525.

Garvey, Ellen Gruber. "Scissoring and Scrapbooks: Nineteenth-Century Reading, Remaking, and Recirculating." *New Media*. Ed. Lisa Gitelman and Geoffrey Pingree. Cambridge, Mass.: MIT Press, 2003. 207–27.

Gaskell, Elizabeth Cleghorn. *The Life of Charlotte Brontë*. Ed. Angus Easson. Oxford: Oxford University Press, 1996.

———. *My Lady Ludlow and Other Stories*. Ed. Edgar Wright. Oxford: Oxford University Press, 1989.

Gaskell, Elizabeth. *Mary Barton*. Ed. Macdonald Daly. New York: Penguin, 1996.

———. *North and South*. Ed. Patricia Ingham. New York: Penguin, 1995.

Gates, Henry Louis. *The Signifying Monkey: A Theory of Afro-American Literary Criticism*. New York: Oxford University Press, 1988.

Gee, Sophie. *Making Waste: Leftovers and the Eighteenth-Century Imagination*. Princeton, N.J.: Princeton University Press, 2010.

Gettelman, Debra. "Reverie, Reading, and the Victorian Novel." PhD diss., Harvard University, 2005.

Ghosh, Anindita. "The Many Worlds of the Vernacular Book." *Books without Borders: Perspectives from South Asia*. Ed. Robert Fraser and Mary Hammond. Vol. 2. Basingstoke: Palgrave Macmillan, 2008. 24–57.

Gigante, Denise. *The Great Age of the English Essay: An Anthology*. New Haven, Conn.: Yale University Press, 2008.

———. *Taste: A Literary History*. New Haven, Conn.: Yale University Press, 2005.

Gilmartin, Kevin. *Print Politics: The Press and Radical Opposition in Early Nineteenth-Century England*. Cambridge Studies in Romanticism, 21. Cambridge: Cambridge University Press, 1996.

———. "'Study to Be Quiet': Hannah More and the Invention of Conservative Culture in Britain." *ELH* 70.2 (2003): 493–540.

Ginsburg, Michal Peled. "The Case against Plot in *Bleak House* and *Our Mutual Friend*." *ELH* 59.1 (1992): 175–95.

Ginzburg, Carlo. "Clues: Morelli, Freud, and Sherlock Holmes." *The Sign of Three: Dupin, Holmes, Peirce*. Ed. Umberto Eco and Thomas A. Sebeok. Bloomington: Indiana University Press, 1988. 81–118.

Gissing, George. *New Grub Street*. Ed. Bernard Bergonzi. Harmondsworth: Penguin, 1985.

———. *Our Friend the Charlatan*. Society and the Victorians, no. 28. [New] ed. Hassocks, U.K.: Harvester Press, 1976.

———. *The Private Papers of Henry Ryecroft*. New York: Modern Library, n.d..

———. "The Prize Lodger." *Victorian Short Stories 2: The Trials of Love*. Ed. Harold Orel. London: Everyman, 1990. 143–55.

Gitelman, Lisa. *Always Already New: Media, History, and the Data of Culture*. Cambridge, Mass.: MIT Press, 2006.

———. *Scripts, Grooves, and Writing Machines: Representing Technology in the Edison Era*. Stanford, Calif.: Stanford University Press, 1999.

Gladstone, W. E. "On Books and the Housing of Them." *The Nineteenth Century* 27 (1890): 384–96.

Goffman, Erving. *Behavior in Public Places: Notes on the Social Organization of Gatherings*. New York: Free Press, 1963.

Goodell, Charles Le Roy. *My Mother's Bible: A Memorial Volume of Addresses for the Home*. Boston: Lee and Shepard, 1891.

Goody, Jack. *Literacy in Traditional Societies*. Cambridge: Cambridge University Press, 1968.

Goolsbee, Austan. "Are Durable Goods Consumers Forward looking? Evidence from College Textbooks." 2005. http://faculty.chicagobooth.edu/austan .goolsbee/research/texts.pdf.

Gosse, Edmund. *Father and Son: A Study of Two Temperaments*. Ed. A.O.J. Cockshut. Keele, Staffordshire: Ryburn Keele University Press, 1994.

"Government Waste-Paper." *Chambers's Journal* 13 (1896): 747–49.

Grand, Sarah. *The Beth Book*. New York: D. Appleton, 1897.

———. *The Heavenly Twins*. New York: Cassell Publishing Company, 1893.

Green, James N., and Peter Stallybrass. *Benjamin Franklin: Writer and Printer*. New Castle, Del.: Oak Knoll Press; London: British Library, 2006.

Green, Laura. *Educating Women: Cultural Conflict and Victorian Literature*. Athens: Ohio University Press, 2001.

———. "'I Recognized Myself in Her': Identifying with the Reader in George Eliot's *The Mill on the Floss* and Simone De Beauvoir's *Memoirs of a Dutiful Daughter*." *Tulsa Studies in Women's Literature* 24.1 (2005): 57–79.

Green, Samuel G. *The Story of the Religious Tract Society for One Hundred Years*. London: Religious Tract Society, 1899.

Greenwood, James. "Penny Awfuls." *St Paul's* 12 (1873): 161–68.

Greenwood, Thomas. *Public Libraries: A History of the Movement and a Manual for the Organization and Management of Rate-Supported Libraries*. London: Simpkin Marshall, 1890.

Greetham, David C. *Theories of the Text*. Oxford: Oxford University Press, 1999.

Greg, W. W. "What Is Bibliography?" *Transactions of the Bibliographical Society* 12 (1914): 39–53.

Grenby, M. O. "Chapbooks, Children, and Children's Literature." *The Library* 3 (2007): 277–303.

Griest, Guinevere. *Mudie's Circulating Library and the Victorian Novel*. Bloomington: Indiana University Press, 1970.

Griswold, Wendy. "Glamour and Honor: Going Online and Reading in West African Culture." *Information Technologies and International Development* 3.4 (2006): 37–52.

Groller, Balduin. "Anonymous Letters." *More Rivals of Sherlock Holmes: Cosmopolitan Crimes*. Ed. Hugh Greene. London: Bodley Head, 1971.

Hack, Daniel. *The Material Interests of the Victorian Novel*. Victorian Literature and Culture Series. Charlottesville: University of Virginia Press, 2005.

Hadley, Elaine. *Living Liberalism: Practical Citizenship in Mid-Victorian Britain*. Chicago: University of Chicago Press, 2010.

Hadley, Tessa. "Seated Alone with a Book . . . " *Henry James Review* 26 (2005): 229–36.

Haggard, H. Rider. *Allan's Wife: With Hunter Quatermain's Story; a Tale of Three Lions; and Long Odds*. London: Macdonald, 1969.

Hale, Dorothy J. "Fiction as Restriction: Self-Binding in New Ethical Theories of the Novel." *Narrative* 15 (2007): 187–206.

Hamilton, Elizabeth, and Claire Grogan. *Memoirs of Modern Philosophers*. Broadview Literary Texts. Peterborough, Ont: Broadview Press, 2000.

Hamilton, James. "Unearthing Broadcasting in the Anglophone World." *Residual Media*. Ed. Charles R. Acland. Minneapolis: University of Minnesota Press, 2007. 283–300.

Hammond, Mary. *Reading, Publishing, and the Formation of Literary Taste in England 1880–1914*. Aldershot: Ashgate, 2006.

Handed-On: Or, the Story of a Hymn Book. London: Society for Promoting Christian Knowledge, 1893.

Hands, Elizabeth. "A Poem, on the Supposition of the Book Having Been Published and Read (1789)." *British Women Poets of the Romantic Era: An Anthology*. Ed. Paula R. Feldman. Baltimore, MD: Johns Hopkins University Press, 1997. 258–60.

Hardy, Thomas. *The Mayor of Casterbridge*. New York: Bantam, 1981.

Hare, Augustus J. C. *The Story of My Life*. Vol. 1. 6 vols. London: G. Allen, 1896.

Harries, Patrick. "Missionaries, Marxists, and Magic: Power and the Politics of Literacy in South-East Africa." *Journal of Southern African Studies* 27.3 (2001): 405–27.

Harrison, Frederic. *The Choice of Books, and Other Literary Pieces*. London: Macmillan, 1886.

Hazlitt, William. *Lectures on the English Poets. Delivered at the Surrey Institution*. London: Printed for Taylor and Hessey, 1818.

Henkin, David. *City Reading: Written Words and Public Spaces in Antebellum New York*. New York: Columbia University Press, 1998.

Henkin, David M. *The Postal Age: The Emergence of Modern Communications in Nineteenth-Century America*. Chicago: University of Chicago Press, 2006.

Herbert, Christopher. *Culture and Anomie*. Chicago: University of Chicago Press, 1991.

Hill, Rowland, and George Birkbeck Norman Hill. *The Life of Sir Rowland Hill . . . And the History of Penny Postage*. 2 vols. London: Thos. De La Rue, 1880.

Hill, Sir Rowland. *Post Office Reform, Its Importance and Practicability*. London: C. Knight and Co., 1837.

Hilton, Boyd. *The Age of Atonement: The Influence of Evangelicalism on Social and Economic Thought, 1785–1865*. Oxford: Clarendon Press, 1991.

History of a Bible. Ballston Spa [N.Y.]: James Comstock, 1811.

The History of a Religious Tract Supposed to Be Related by Itself. London: Printed by W. Nicholson, for Williams & Smith, 1806.

"The History of an Old Pocket Bible." *Cottage Magazine; or, Plain Christian's Library,* March–December 1812.

Hitchman, Francis. "Penny Fiction." *Quarterly Review* 171 (1890): 150–71.

Hlubinka, Michelle. "The Datebook." *Evocative Objects: Things We Think With*. Ed. Sherry Turkle. Cambridge, Mass.: MIT Press, 2007. 78–85.

Hobsbawm, E. J. *Workers: Worlds of Labor*. 1st American ed. New York: Pantheon Books, 1984.

Hofmeyr, Isabel. "Metaphorical Books." *Current Writing* 13.2 (2001): 100–108.

Hofmeyr, Isabel, and Sarah Nuttall. "The Book in Africa." *Current Writing* 13.2 (2001): 1–8.

Holcroft, Thomas. *The Adventures of Hugh Trevor*. Ed. Seamus Deane. London: Oxford University Press, 1973.

Holsinger, Bruce. "Of Pigs and Parchment: Medieval Studies and the Coming of the Animal." *PMLA* 124.2 (2009): 616–23.

Homans, Margaret. *Bearing the Word: Language and Female Experience in Nineteenth-Century Women's Writing*. Chicago: University of Chicago Press, 1986.

Hood, Thomas, and Richard Herne Shepherd. *The Choice Works of Thomas Hood, in Prose and Verse, Including the Cream of the Comic Annuals*. Boston: DeWolfe, Fiske, n.d.

Horsburgh, Matilda. *The Story of a Red Velvet Bible*. Edinburgh: Johnstone, Hunter, & Co., 1862.

"How to Read Tracts." *The Children's Friend*, 1851, 164.

Howsam, Leslie. *Cheap Bibles: Nineteenth-Century Publishing and the British and Foreign Bible Society*. Cambridge Studies in Publishing and Printing History. Cambridge: Cambridge University Press, 1991.

Hughes, Linda K., and Michael Lund. *The Victorian Serial*. Charlottesville: University of Virginia Press, 1991.

Hughes, M. V. *A London Child of the 1870s*. London: Persephone Books, 2005.

Humpherys, Anne. *Travels into the Poor Man's Country: The Work of Henry Mayhew*. Athens: University of Georgia Press, 1977.

Humphreys, Arthur Lee. *The Private Library: What We Do Know, What We Don't Know, What We Ought to Know About Our Books*. 3rd ed. London: Strangeways & Sons, 1897.

Hunt, Lynn Avery. *Inventing Human Rights: A History*. New York: W. W. Norton & Co., 2007.

Hunter, Dard. *Papermaking: The History and Technique of an Ancient Craft*. 2nd ed. New York: A. A. Knopf, 1947.

Hunter, Ian. "The History of Theory." *Critical Inquiry* 33.1 (2006): 78–112.

———. "Literary Theory in Civil Life." *South Atlantic Quarterly* 95 (Fall 1996): 1099–134.

———. "Setting Limits to Culture." *New Formations* 2 (Spring 1988): 103–23.

Hutton, R. H. "The Reporter in Mr. Dickens." *A Victorian Spectator: Uncollected Writings of R. H. Hutton*. Ed. Robert Tener and Malcolm Woodfield. Bristol: Bristol Classical Press, 1865. 575–76.

Hyde, Lewis. *The Gift: Imagination and the Erotic Life of Property*. New York: Random House, 1983.

"Institution for the Evangelization of Gypsies." *The Town and Village Mission Record* (1855): 169–71.

Irving, Washington. *The Sketch-Book of Geoffrey Crayon, Gent. Legends of the Conquest of Spain*. New York: Co-operative Publication Society, 1881.

Jackson, H. J. *Marginalia: Readers Writing in Books*. New Haven, Conn.: Yale University Press, 2001.

———. *Romantic Readers: The Evidence of Marginalia*. New Haven, Conn.: Yale University Press, 2005.

Jackson, Holbrook, ed. *The Anatomy of Bibliomania*. Chicago: University of Illinois Press, 2001.

Jakobson, Roman. "The Metaphoric and Metonymic Poles." *Critical Theory since Plato*. Ed. Hazard Adams. New York: Harcourt, 1992. 1041–44.

James, Henry. *The Awkward Age. Novels 1896–1899*. Ed. Myra Jehlen. New York: Library of America, 1899.

———. "Brooksmith." *Tales of Henry James: The Texts of the Tales, the Author on His Craft, Criticism*. Ed. Christof Wegelin and Henry B. Wonham. 2nd ed. New York: W. W. Norton, 2003.

———. "Greville Fane." *Henry James: Collected Stories 1892–1898*. Ed. Denis Donoghue. New York: Library of America, 1996. 217–33.

———. *In the Cage*. Ed. Hortense Calisher. New York: Modern Library, 2001.

———. "The Middle Years." *Henry James: Collected Stories 1892–1898*. Ed. Denis Donoghue. New York: Library of America, 1996. 335–53.

———. "Miss Braddon." *Nation*, 1865, 593–94.

———. *What Maisie Knew*. 1897. *Novels 1896–1899*. Ed. Myra Jehlen. New York: Library of America, 1899.

James, Louis. *Fiction for the Working Man*. Harmondsworth: Penguin, 1973.

James, M. R. *Collected Ghost Stories*. 1931. Ware: Wordsworth Classics, 1992.

Jameson, Fredric. *The Political Unconscious: Narrative as a Socially Symbolic Act*. Ithaca, N.Y.: Cornell University Press, 1981.

Jeffrey, Francis. "Specimens of the British Poets: With Biographical and Critical Notices, and an Essay on English Poetry." *Edinburgh Review* 31 (March 1819): 471–72.

Jerrold, Douglas William. *The Best of Mr. Punch; the Humorous Writings of Douglas Jerrold*. Knoxville: University of Tennessee Press, 1970.

Jerrold, Douglas William, and Charles Keene. *Mrs. Caudle's Curtain Lectures*. New York: Hurd and Houghton, 1867.

Jerrold, Walter. *Thomas Hood: His Life and Times*. New York: John Lane, 1909.

Jevons, William Stanley. "The Rationale of Free Public Libraries." *Methods of Social Reform: And Other Papers*. London: Macmillan, 1882.

Jewsbury, Geraldine. *The Half-Sisters*. Oxford: Oxford University Press, 1994.

Johns, Adrian. "Changes in the World of Publishing." *The Cambridge History of English Romantic Literature*. Ed. James Chandler. Cambridge: Cambridge University Press, 2009. 377–402.

Johnson, Edgar. *Charles Dickens: His Tragedy and Triumph*. New York: Viking, 1977.

Johnstone, Charles. *Chrysal*. Ed. Malcolm J. Bosse. *The Novel, 1720–1805*. 4 vols. New York: Garland Pub., 1979.

Jones, William. *The Jubilee Memorial of the Religious Tract Society Containing a Record of Its Origin, Proceedings, and Results, A.D. 1799 to A.D. 1849.* London: Religious Tract Society, 1850.

Jordan, John, and Robert Patten. *Literature in the Marketplace: Nineteenth-Century British Publishing and the Circulation of Books.* Cambridge: Cambridge University Press, 1995.

Kaplan, Carla. "Girl Talk: *Jane Eyre* and the Romance of Women's Narration." *Novel: A Forum on Fiction* 30.1 (1996): 5–31.

Kearney, James. "The Book and the Fetish: The Materiality of Prospero's Text." *Journal of Medieval and Early Modern Studies* 32.3 (2002): 433–68.

———. *The Incarnate Text*. Material Texts. Philadelphia: University of Pennsylvania Press, 2009.

Keen, Paul. *Revolutions in Romantic Literature: An Anthology of Print Culture, 1780–1832.* Peterborough, Ont.: Broadview, 2004.

Keller, Helen. *The Story of My Life*. Riverside Literature Series, no. 253. School ed. Boston: Houghton Mifflin Co., 1905.

Kelly, Gary. "Revolution, Reaction, and the Expropriation of Popular Culture: Hannah More's *Cheap Repository*." *Man and Nature* 6 (1987): 147–59.

Kemble, Frances Ann. *Records of Later Life*. New York: Henry Holt and Company, 1882.

Kirschenbaum, Matthew G. *Mechanisms: New Media and the Forensic Imagination.* Cambridge, Mass.: MIT Press, 2008.

Kittler, Friedrich A. *Discourse Networks.* Trans. Michael Metteer and Chris Cullens. Stanford, Calif.: Stanford University Press, 1990.

Klancher, Jon P. "The Bibliographer's Tale, or the Rise and Fall of Book History in Britain (1797–1825)." Paper presented at the American Society for Eighteenth Century Studies, Montreal, 31 March 2006.

———. *The Making of English Reading Audiences, 1790–1832.* Madison: University of Wisconsin Press, 1987.

Knight, Charles. *The Old Printer and the Modern Press.* London: J. Murray, 1854.

———. *Passages of a Working Life During Half a Century: With a Prelude of Early Reminiscences.* London: Bradbury & Evans, 1864.

Knight, Jeffrey Todd. "'Furnished' for Action: Renaissance Books as Furniture." *Book History* 12 (2009).

Koops, Matthias. *Historical Account of the Substances Which Have Been Used to Describe Events and to Convey Ideas from the Earliest Date to the Invention of Paper.* London: T. Burton, 1800.

Kopytoff, Igor. "The Cultural Biography of Things." *The Social Life of Things: Commodities in Cultural Perspective.* Ed. Arjun Appadurai. Cambridge: Cambridge University Press, 1986. 64–94.

Krashen, Stephen D. *The Power of Reading: Insights from the Research.* 2nd ed. Westport, Conn.: Libraries Unlimited; Portsmouth, N.H.: Heinemann, 2004.

Kreilkamp, Ivan. "Speech on Paper: Charles Dickens, Victorian Phonography, and the Reform of Writing." *Literary Secretaries/Secretarial Culture.* Ed. Leah Price and Pamela Thurschwell. Aldershot: Ashgate, 2005. 13–31.

———. *Voice and the Victorian Storyteller*. Cambridge Studies in Nineteenth-Century Literature and Culture, 49. Cambridge: Cambridge University Press, 2005.

Kucich, John. "George Eliot and Objects: Meaning and Matter in the Mill on the Floss." *Dickens Studies Annual* 12 (1983): 319–40.

La Bruyère, Jean de. *Les caractères: ou, les mœurs de ce siècle*. Paris: Gallimard, 1975.

Lady Chatterley's Trial. Pocket Penguins 70's, 1960.

Lamb, Charles. "Detached Thoughts on Books and Reading." *The Romance of the Book*. Ed. Marshall Brooks. Delhi, N.Y.: Birch Books Press, 1995. 133–39.

———. "Readers against the Grain." *Works*. Ed. William MacDonald. Vol. 3. London: J. M. Dent & Sons; New York: E. P. Dutton & Co., 1914 [1903]. 237–40.

Lamb, Jonathan. "Modern Metamorphoses and Disgraceful Tales." *Critical Inquiry* 28 (2001): 133–66.

Lang, Andrew. *The Library*. London: Macmillan & Co., 1881.

A Lantern Lecture on Isaac Pitman and Pitman's Shorthand. London: Isaac Pitman, [1928?].

Laqueur, Thomas Walter. *Religion and Respectability: Sunday Schools and Working Class Culture, 1780–1850*. New Haven, Conn.: Yale University Press, 1976.

Latham, Sean, and Robert Scholes. "The Rise of Periodical Studies." *PMLA* 121 (2006): 517–31.

Latour, Bruno. *Aramis, or, the Love of Technology*. Cambridge, Mass.: Harvard University Press, 1996.

———. "Drawing Things Together." *Representation in Scientific Practice*. Cambridge, Mass.: MIT Press, 1990. 19–68.

———. "On Technical Mediation—Philosophy, Sociology, Genealogy." *Common Knowledge* 3.2 (1994): 29–64.

Law, Graham. *Serializing Fiction in the Victorian Press*. Houndsmills: Palgrave, 2000.

Leckie, Barbara. *Culture and Adultery: The Novel, the Newspaper, and the Law, 1857–1914*. Philadephia: University of Pennsylvania Press, 1999.

Ledger-Lomas, Michael. "Mass Markets: Religion." *The Cambridge History of the Book in Britain, 1830–1914*. Ed. David McKitterick. Cambridge: Cambridge University Press, 2009. 324–58.

Lee, Julia Sun-Joo. *The American Slave Narrative and the Victorian Novel*. New York: Oxford University Press, 2010.

Lerer, Seth. "Falling Asleep over the History of the Book." *PMLA* 21 (2006): 229–34.

———. *Inventing English: A Portable History of the Language*. New York: Columbia University Press, 2007.

Levine, Caroline. *The Serious Pleasures of Suspense: Victorian Realism and Narrative Doubt*. Victorian Literature and Culture Series. Charlottesville: University of Virginia Press, 2003.

Levine, George. *The Realistic Imagination: English Fiction from Frankenstein to Lady Chatterley*. Chicago: University of Chicago Press, 1981.

Levitt, Steven D., and Stephen J. Dubner. *Freakonomics: A Rogue Economist Explores the Hidden Side of Everything*. Rev. and expanded ed. New York: William Morrow, 2006.

Lewes, G. H. *Ranthorpe*. London: Chapman and Hall, 1847.

Lewins, William. *Her Majesty's Mails: A History of the Post-Office, and an Industrial Account of Its Present Condition*. London: S. Low, Son, and Marston, 1864.

Lewis, Monica. "Anthony Trollope among the Moderns: Reading Aloud in Britain 1850–1960." PhD diss., Harvard University, 2006.

"The Life and Adventures of a Number of Godey's Lady's Book. Addressed Particularly to Borrowers, Having Been Taken Down in Short-Hand from a Narration Made by Itself, When the Unfortunate Creature Was in a Dilapidated State, from the Treatment Received at the Hands of Cruel Oppressors." *Godey's Magazine,* November 1855, 425–27.

"Literary Voluptuaries." *Blackwood's* 142 (December 1887): 805–17.

"The Literature of the Rail." *Times,* 9 August 1851, 7.

"Little Jack of All Trades." *Flowers of Delight*. Ed. Leonard de Vries. 1965 ed. New York: Pantheon Books, 1806.

Little Wide-awake: An Anthology of Victorian Children's Books and Periodicals. Ed. Leonard DeVries. London: Arthur Barker, 1967.

Lloyd, Rosemary. "Reading As If for Life." *Journal of European Studies* 22.3 (1992): 259–72.

Logan, Peter Melville. *Victorian Fetishism: Intellectuals and Primitives*. SUNY Series, Studies in the Long Nineteenth Century. Albany: State University of New York Press, 2009.

Long, Elizabeth. *Book Clubs: Women and the Uses of Reading in Everyday Life*. Chicago: University of Chicago Press, 2003.

Losano, Antonia. "Reading Women/Reading Pictures: Textual and Visual Reading in Charlotte Brontë's Fiction and Nineteenth-Century Painting." *Reading Women: Literary and Cultural Icons from the Victorian Age to the Present*. Ed. Janet Badia and Jennifer Phegley. Toronto: University of Toronto Press, 2005. 27–52.

Lupton, Christina. "The Knowing Book: Authors, It-Narratives, and Objectification in the Eighteenth Century." *Novel* 39 (2006): 402–20.

———. "Theorizing Surfaces and Depths: Gaskell's Cranford." *Criticism* 50.2 (2008): 235–54.

Lynch, Deidre. "Canon's Clockwork: Novels for Everyday Use." *At Home in English: A Cultural History of the Love of Literature*. Chicago: University of Chicago Press, forthcoming.

———. *The Economy of Character: Novels, Market Culture, and the Business of Inner Meaning*. Chicago: University of Chicago Press, 1998.

———. "'Wedded to Books': Bibliomania and the Romantic Essayists." *Romantic Libraries*. Ed. Ina Ferris. College Park, Md.: University of Maryland, 2004.

Lyons, Martyn. "New Readers in the Nineteenth Century: Women, Children, Workers." *A History of Reading in the West*. Ed. Guglielmo Cavallo and Roger Chartier. Cambridge: Polity Press, 1999. 313–44.

Macaulay, Thomas Babington. "Minute on Indian Education." *The Great Indian Education Debate*. 1835. Ed. Lynn Zastoupil. Richmond: Curzon, 1999. 161–73.

———. "Mr. Robert Montgomery." April 1830. *Critical and Historical Essays, Contributed to the Edinburgh Review*. Vol. 1. London: Longman, Brown, Green, and Longmans, 1843. 270–94.

———. "On the Royal Society of Literature." *Miscellaneous Writings*. 1823. Vol. 1. London: Longman, Green, Longman, and Roberts, 1860. 20–28.

Macaulay, Zachary. *Life and Letters of Zachary Macaulay*. Ed. Margaret Jean Trevelyan Knutsford. Ann Arbor, Mich.: University Microfilms International, 1973.

Macray, William Dunn. *Annals of the Bodleian Library, Oxford; with a Notice of the Earlier Library of the University*. 2nd enl. ed., and continued from 1868 to 1880. Oxford: Clarendon Press, 1890.

Maidment, Brian. *Dusty Bob: A Cultural History of Dustmen, 1780–1870*. Manchester: Manchester University Press, 2007.

Manguel, Alberto. *A History of Reading*. London: HarperCollins, 1996.

———. *The Library at Night*. 1st ed. Toronto: Knopf Canada, 2006.

Mann, Gurinder Singh. "Scriptures and the Nature of Authority: The Case of the Guru Granth in Sikh Tradition." *Theorizing Scriptures: New Critical Orientations to a Cultural Phenomenon*. Ed. Vincent L. Wimbush. New Brunswick, N.J.: Rutgers University Press, 2008.

Mann, Thomas. *Buddenbrooks: The Decline of a Family*. Trans. John E. Woods. New York: Knopf, 1993.

Manning, Anne. *Claude the Colporteur, by the Author of 'Mary Powell'*. Religious Tract Society, 1880.

Marcus, Sharon. *Between Women: Friendship, Desire, and Marriage in Victorian England*. Princeton, N.J.: Princeton University Press, 2007.

———. "The Profession of the Author: Abstraction, Advertising, and *Jane Eyre*." *PMLA* 110.2 (1995): 206–19.

Marcus, Steven. "Language into Structure: Pickwick Revisited." *Daedalus* 101.1 (1972): 183–202.

Marshman, John Clark. *The Life and Times of Carey, Marshman, and Ward: Embracing the History of the Serampore Mission*. Vol. 1. 2 vols. London: Longman Brown Green Longmans & Roberts, 1859.

Martin, Roger. "Women and the Bible Society." *Sowing the Word: The Cultural Impact of the British and Foreign Bible Society, 1804–2004*. Ed. Stephen K. Batalden, Kathleen Cann, and John Dean. Sheffield: Sheffield Phoenix, 2004. 38–52.

Martineau, Harriet. *Harriet Martineau's Autobiography*. Ed. Maria Weston Chapman Vol. 1. 2 vols. Boston: J. R. Osgood and Company, 1877.

———. *Harriet Martineau—Selected Letters*. Ed. Valerie Sanders. Oxford: Clarendon Press, 1990.

Mathers, Helen. *Comin' Thro' the Rye: A Novel*. New York: A. L. Burt, 1876.

Maxwell, Herbert. "The Craving for Fiction." *The Nineteenth Century,* June 1893, 1046–61.

Mayhew, Henry. *The Criminal Prisons of London and Scenes of Prison Life*. Ed. John Binny. 1862. London: Frank Cass, 1968.

———. *The Essential Mayhew: Representing and Communicating the Poor*. Ed. Bertrand Taithe. London: Rivers Oram Press, 1996.

———. *London Labour and the London Poor*. Ed. Victor Neuburg. Harmondsworth: Penguin, 1985.

———. *London Labour and the London Poor; Cyclopaedia of the Condition and Earnings of Those That Will Work, Those That Cannot Work, and Those That Will Not Work*. London: Griffin Bohn, 1861–62.

———. *Voices of the Poor: Selections from the Morning Chronicle 'Labour and the Poor' (1849–1850)*. Ed. Anne Humpherys. Cass Library of Victorian Times, no. 10. London: Cass, 1971.

Mayhew, Horace. *Letters Left at the Pastrycook's: Being the Clandestine Correspondence between Kitty Clover at School and Her "Dear, Dear Friend" in Town*. London: Ingram Cooke and Co., 1853.

Mayhew, Robert J. "Materialist Hermeneutics, Textuality and the History of Geography: Print Spaces in British Geography, c. 1500–1900." *Journal of Historical Geography* 33 (2007): 466–88.

Mays, Kelly J. "The Disease of Reading and Victorian Periodicals." *Literature in the Marketplace: Nineteenth-Century British Reading and Publishing Practices*. Ed. John O. Jordan and Robert L. Patten. Cambridge: Cambridge University Press, 1995. 165–94.

McDonald, Peter. "Ideas of the Book and Histories of Literature: After Theory?" *PMLA* 121.1 (2006): 214–28.

McDonald, Peter D. *British Literary Culture and Publishing Practice, 1880–1914*. Cambridge Studies in Publishing and Printing History. Cambridge: Cambridge University Press, 1997.

McGill, Meredith. *American Literature and the Culture of Reprinting, 1834–1853*. Philadelphia: University of Pennsylvania Press, 2003.

McGurl, Mark. *The Novel Art: Elevations of American Fiction after Henry James*. Princeton, N.J.: Princeton University Press, 2001.

McKelvy, William R. *The English Cult of Literature: Devoted Readers, 1774–1880*. Victorian Literature and Culture Series. Charlottesville: University of Virginia Press, 2007.

McKitterick, David. *The Cambridge History of the Book in Britain, 1830–1914*. Cambridge: Cambridge University Press, 2009.

———. "Organizing Knowledge in Print." *The Cambridge History of the Book in Britain, 1830–1914*. Ed. David McKitterick. Cambridge: Cambridge University Press, 2009.

McLaughlin, Kevin. *Paperwork: Fiction and Mass Mediacy in the Paper Age*. Philadelphia: University of Pennsylvania Press, 2005.

McQuillan, Jeff. *The Literacy Crisis: False Claims, Real Solutions*. Portsmouth, N.H.: Heinemann, 1998.

Melville, Herman. "The Tartarus of Maids." *Complete Shorter Fiction*. New York: Everyman, 1987.

Menke, Richard. *Telegraphic Realism*. Stanford, Calif.: Stanford University Press, 2008.

Michie, Helena. *Victorian Honeymoons: Journeys to the Conjugal*. Cambridge Studies in Nineteenth-Century Literature and Culture, 53. Cambridge: Cambridge University Press, 2006.

"Midland District Conference of the National Federation of Shorthand Writers' Associations." *Phonetic Journal*, 1901, 102.

Miller, Andrew. *Novels behind Glass: Commodity Culture and Victorian Narrative*. Cambridge: Cambridge University Press, 1995.

Miller, D. A. *The Novel and the Police*. Berkeley: University of California Press, 1988.

Millington, Thomas Street. *Straight to the Mark*. London: Religious Tract Society, 1883.

Mills, John. *The English Fireside: A Tale of the Past*. 3 vols. London: Saunders and Otley, 1844.

Minnis, A. J., A. Brian Scott, and David Wallace. *Medieval Literary Theory and Criticism c.1100–c.1375: The Commentary Tradition*. Rev. ed. Oxford: Clarendon Press, 1991.

Miscellaneous Cabinet 1.23 (1823): 184.

Mitch, David. *The Rise of Popular Literacy in Victorian England*. Philadelphia: University of Pennsylvania Press, 1992.

Molesworth, Mrs. "On the Use and Abuse of Fiction." *Girl's Own Paper* 13 (1892): 452–54.

Moncrieff, W. T. *The March of Intellect; a Comic Poem*. London: William Kidd, 1830.

More, Hannah. *Cheap Repository Tracts; Entertaining, Moral, and Religious*, vol. 1. London: F. and C. Rivington, 1798.

———. *Tales for the Common People and Other Cheap Repository Tracts*. Ed. Clare MacDonald Shaw. Nottingham: Trent Editions, 2002.

———. *Works*. Vol. 1. New York: Harper & Brothers, 1846.

Moretti, Franco. *Graphs, Maps, Trees: Abstract Models for a Literary History*. London: Verso, 2005.

———. *The Novel*. 2001–2003. 2 vols. Princeton, N.J.: Princeton University Press, 2006.

———. *The Way of the World: The Bildungsroman in European Culture*. London: Verso, 1987.

Motherly, Mrs. *The Servant's Behavior Book: Or. Hints on Manners and Dress*. London: Bell and Daldy, 1859.

Mozley, Anne. "On Fiction as an Educator." *A Serious Occupation: Literary Criticism by Victorian Woman Writers* 108 (1870): 187–207.

Mudie, Charles Edward. "Mr. Mudie's Library." *The Athenaeum* 1719 (1860): 451.

Munro, Jeffrey. *Half Hours with Popular Authors, printed in the advanced stage of Pitman's shorthand*. London: Pitman,1927.

National Endowment for the Arts. *Reading at Risk: A Survey of Literary Reading in America*. 2004.

————. *Survey of Public Participation in the Arts*, 2002.

————. *To Read or Not to Read: A Question of National Consequence*, 2007.

Nead, Lynda. *Victorian Babylon: People, Streets and Images in Nineteenth-Century London*. New Haven, Conn.: Yale University Press, 2000.

Needham, Joseph, et al. *Science and Civilisation in China*. Vol. 5, *Chemistry and Chemical Technology*, pt. 1, *Paper and Printing*. Cambridge: Cambridge University Press, 1954.

Nelson, Craig. *Thomas Paine: Enlightenment, Revolution, and the Birth of Modern Nations*. New York: Viking, 2006.

Neuburg, Victor, ed. *The Invention of the Streets*. 2 vols. London: Routledge & Kegan Paul, 1973.

Newlyn, Lucy. *Reading, Writing, and Romanticism: The Anxiety of Reception*. Oxford: Oxford University Press, 2000.

Newton, John. *The Posthumous Works of the Late Rev. John Newton; Rector of the United Parishes of St. Mary Woolnorth, and St. Mary Woolchurch*. Vol. 2. Philadelphia: W. W. Woodward, 1809.

Nightingale, Florence. *Cassandra and Other Selections from Suggestions for Thought*. Ed. Mary Poovey. NYU Press Women's Classics. New York: New York University Press, 1992.

————. *Florence Nightingale on Mysticism and Eastern Religions*. Vol. 4 of *The Collected Works of Florence Nightingale*. Ed. Gerard Vallee. 16 vols. Waterloo, Ont.: Wilfrid Laurier University Press, 2001–.

Nissenbaum, Stephen. *The Battle for Christmas*. 1st ed. New York: Alfred A. Knopf, 1996.

Nixon, Edward John. *A Manual of District Visiting: With Hints and Directions to Visitors*. London: Seeleys, 1848.

Nunokawa, Jeff. *The Afterlife of Property: Domestic Security and the Victorian Novel*. Princeton, N.J.: Princeton University Press, 1994.

————. "Eros and Isolation: The Antisocial George Eliot." *ELH* 69.4 (2002): 835–60.

————. *Tame Passions of Wilde*. Princeton, N.J.: Princeton University Press, 2003.

Nussbaum, Martha. *Love's Knowledge: Essays on Literature and Philosophy*. Oxford: Oxford University Press, 1992.

O'Brien, Flann. *The Best of Myles*. Normal, Ill.: Dalkey Archive Press, 1999.

Ogden, R. "A Hundred Years of Tracts." *The Nation* 68.1769 (1899): 390.

Oliphant, Margaret. "The Byways of Literature." *Blackwood's Edinburgh Magazine* 84 (1858): 200–16.

————. *Kirsteen*. Ed. Merryn Williams. London: Everyman, 1984.

"One Thing at a Time." *Making the Best of It, and Other Picture Stories*. London: Religious Tract Society, n.d.

Ong, Walter J. *Rhetoric, Romance, and Technology: Studies in the Interaction of Expression and Culture*. Ithaca, N.Y.: Cornell University Press, 1971.

Orwell, George. *Animal Farm: A Fairy Story*. Signet Classic. New York: New American Library, 1960.

Paget, Francis Edward. *Lucretia; or, The Heroine of the Nineteenth Century. A Correspondence, Sensational and Sentimental*. London: Joseph Masters, 1868.

Park, Rev. Harrison G. *The Father's and Mother's Manual and Youth's Instructor.* Boston: Fitz, Hobbs & Co., 1850.

Parks, Lisa. "Falling Apart: Electronics Salvaging and the Global Media Economy." *Residual Media.* Ed. Charles R. Acland. Minneapolis: University of Minnesota Press, 2007. 32–47.

Patten, Robert. *Charles Dickens and His Publishers.* Oxford: Clarendon Press, 1978.

Pawnbroker's Gazette 865 (1855).

Pearson, David. "What Can We Learn by Tracking Multiple Copies of Books?" *Books on the Move: Tracking Copies through Collections and the Book Trade.* Ed. Robin Myers, Michael Harris, and Giles Mandelbrote. New Castle, Del.: Oak Knoll Press; London: British Library, 2007.

Pearson, Jacqueline. *Women's Reading in Britain, 1750–1835.* Cambridge: Cambridge University Press, 1999.

Pedersen, Susan. "Hannah More Meets Simple Simon: Tracts, Chapbooks, and Popular Culture in Late Eighteenth-Century England." *Journal of British Studies* 25 (1986): 84–113.

Pellegram, Andrea. "The Message in Paper." *Material Cultures: Why Some Things Matter.* Ed. Daniel Miller. Chicago: University of Chicago Press, 1998.

Peters, John Durham. *Speaking into the Air: A History of the Idea of Communication.* Chicago: University of Chicago Press, 1999.

Peterson, Carla. *The Determined Reader: Gender and Culture in the Novel from Napoleon to Victoria.* New Brunswick, N.J.: Rutgers University Press, 1986.

Petroski, Henry. *Small Things Considered: Why There Is No Perfect Design.* New York: Alfred A. Knopf, 2003.

Picker, John M. *Victorian Soundscapes.* New York: Oxford University Press, 2003.

Pickering, Samuel F., Jr. "The Old Curiosity Shop—a Religious Tract?" *Illinois Quarterly* 36.1 (1973): 5–20.

"Pioneer Work in China." *Wesleyan-Methodist magazine*, December 1885, 894–903.

Piozzi, Hester Lynch. *Anecdotes of the Late Samuel Johnson, LL.D.* London: T. Cadell, 1786.

Pitman, Benn. *Sir Isaac Pitman: His Life and Labors.* Cincinnati: Benn Pitman, 1902.

Pitman's Journal 17 (1905): 85.

Plotz, John. "Out of Circulation: For and Against Book Collecting." *Southwest Review* 84.4 (1999): 462–78.

———. *Portable Property: Victorian Culture on the Move.* Princeton, N.J.: Princeton University Press, 2008.

Polastron, Lucien X. *Books on Fire: The Destruction of Libraries throughout History.* 1st US ed. Rochester, Vt.: Inner Traditions, 2007.

Poovey, Mary. *Genres of the Credit Economy: Mediating Value in Eighteenth- and Nineteenth-Century Britain.* Chicago: University of Chicago Press, 2008.

———. "The Limits of the Universal Knowledge Project: British India and the East Indiamen." *Critical Inquiry* 31.1 (2004): 183–202.

———. *Uneven Developments: The Ideological Work of Gender in Mid-Victorian England*. Chicago: University of Chicago Press, 1988.

Pope, Alexander. *Poems: A One Volume Edition of the Twickenham Text, with Selected Annotations*. Ed. John Everett Butt. New Haven, Conn.: Yale University Press, 1963.

Popper, Karl Raimund. *The Philosophy of Karl Popper*. Ed. Paul Arthur Schilpp. 1st ed. La Salle, Ill.: Open Court, 1974.

Porter, George Richardson. *The Progress of the Nation in Its Various Social and Economic Relations from the Beginning of the Nineteenth Century*. 1843. A completely new ed. London: Methuen & Co., 1912.

"The Post-Office." *Fraser's Magazine,* February 1850, 224–32.

Pratt, Sarah G. S. *First Homes*. 1882. American Periodicals Series Online, Pro-Quest. 1 November 2009.

Price, Leah. *The Anthology and the Rise of the Novel*. Cambridge: Cambridge University Press, 2000.

———. "The Life of Charlotte Brontë and the Death of Miss Eyre." *SEL: Studies in English Literature, 1500–1900* 35.4 (1995): 757–68.

———. "Read a Book, Get out of Jail." *New York Times Book Review,* 26 February 2009, 23.

———. "Reading: The State of the Discipline." *Book History* 7 (2004): 303–20.

———. "Stenographic Masculinity." *Literary Secretaries/Secretarial Culture*. Ed. Leah Price and Pamela Thurschwell. Aldershot: Ashgate, 2005. 32–47.

"The Progress of Fiction as an Art." *Westminster Review,* 1853, 342–74.

"Prospectus for the Leisure Hour: A Family Journal of Instruction and Entertainment." *Publishers' Circular,* 1851.

Prosser, Sophie Amelia. *Susan Osgood's Prize: A New Story About an Old One*. Boston: Henry Hoyt, n.d.

Proust, Marcel. *A l'ombre des jeunes filles en fleurs*. Vol. 1. 3 vols. Paris: Gallimard, 1919.

Puccini, Giacomo, and Henri Murger. *La Bohème*. [Paris]: Calmann-Lévy, Erato, 1988.

Pugh, S. S. *Christian Home-Life: A Book of Examples and Principles*. London: Religious Tract Society, 1864.

Radway, Janice. "Reading Is Not Eating: Mass-Produced Literature and the Theoretical, Methodological, and Political Consequences of a Metaphor." *Book Research Quarterly* 2 (1986): 7–29.

———. *Reading the Romance: Women, Patriarchy and Popular Literature*. Chapel Hill: University of North Carolina Press, 1991.

Rae, W. Fraser. "Sensation Novelists: Miss Braddon." *North British Review* 43 (1865): 180–204.

The Railway Anecdote Book: A Collection of Anecdotes and Incidents of Travel by River and Rail. 1864. New York: D. Appleton & Co., 1871.

"Railway Literature." *Dublin University Magazine* 34 (1849): 280–91.

Rainey, Lawrence S. *Institutions of Modernism: Literary Elites and Public Culture*. The Henry McBride Series in Modernism and Modernity. New Haven, Conn.: Yale University Press, 1998.

Raitt, Suzanne. "Psychic Waste." *Culture and Waste*. Ed. Guy Hawkins and Stephen Muecke. Oxford: Rowan and Littlefield, 2002.

Rancière, Jacques. *The Philosopher and His Poor*. Trans. Andrew Parker. Durham, N.C.: Duke University Press, 2004.

Ransome, Arthur. *Bohemia in London*. New York: Dodd, Meade, 1907.

R[anyard], L. N. *The Missing Link, or, Bible-Women in the Homes of the London Poor*. New York: Robert Carter, 1860.

Rauch, Alan. *Useful Knowledge: The Victorians, Morality, and the March of Intellect*. Durham, N.C.: Duke University Press, 2001.

Raven, James. *The Business of Books: Booksellers and the English Book Trade, 1450–1850*. New Haven, Conn.: Yale University Press, 2007.

———. *Free Print and Non-Commercial Publishing since 1700*. Aldershot: Ashgate, 2000.

Raverat, Gwen. *Period Piece; a Cambridge Childhood*. London: Faber and Faber, 1952.

Reach, Angus B. "The Coffee Houses of London." *Victorian Print Media: A Reader*. Ed. John Plunkett and Andrew King. Oxford: Oxford University Press, 2005.

"Reading and Readers." *Sunday Magazine*, 22 March 1893, 189.

"Recent Novels: Their Moral and Religious Teaching." *London Quarterly Review* 27 (1866).

Reed, Charles. "The Bible Work of the World." *Proceedings of the General Conference on Foreign Missions,* 1879. 229–36.

Reed, Charles M. *Reading As If for Life: Preparing Young Women for the Real World*. Speech delivered at the Women's College at Brenau University, Gainesville, Georgia, 13 September 2001

Reeser, Todd W., and Steven D. Spalding. "Reading Literature/Culture: A Translation of 'Reading as a Cultural Practice.' " *Style* 36.4 (2002): 659–76.

Religious Tract Society. *Sarah Martin, of Great Yarmouth*. London: Religious Tract Society, n.d.

"Report of the Select Committee on Parliamentary Petitions." *Punch* 6 (1844): 89.

"The Republican Refuted; in a Series of Biographical, Critical, and Political Strictures on Thomas Paine's Rights of Man." *Monthly Review* 7 (1792): 82–84.

"Review of Castle Richmond." *Saturday Review* 9 (1860): 643–44.

"Review of Illustrations of Political Economy." *Spectator*, 4 August 1832.

"Review of the Twenty-Third Annual Report of the Bombay Tract and Book Society." *Bombay Quarterly Magazine and Review* 7 (1852): 497–99.

Reynolds, John Stewart, and W. H. Griffith Thomas. *Canon Christopher of St. Aldate's, Oxford*. Abingdon (Berks.): Abbey Press, 1967.

Reynolds, Kimberley. "Rewarding Reads? Giving, Receiving and Resisting Evangelical Reward and Prize Books." *Popular Children's Literature in Britain*. Ed. Julia Briggs, Dennis Butts, and M. O. Grenby. Aldershot: Ashgate, 2008.

Rich, Motoko. "With Kindle, the Best Sellers Don't Need to Sell." *New York Times*, 22 January 2010, Global Edition ed., sec. Books, 1–2.

Richards, Thomas. *The Imperial Archive: Knowledge and the Fantasy of Empire*. London: Verso, 1993.

Richardson, Alan. *Literature, Education, and Romanticism: Reading as Social Practice, 1780–1832.* Cambridge: Cambridge University Press, 1994.

Richardson, Samuel. *Pamela, or, Virtue Rewarded.* Boston: Houghton Mifflin, 1971.

———. *Selected Letters of Samuel Richardson.* Ed. John Carroll. Oxford: Clarendon Press, 1964.

Rickards, Maurice, and Michael Twyman. *The Encyclopedia of Ephemera: A Guide to the Fragmentary Documents of Everyday Life for the Collector, Curator, and Historian.* New York: Routledge, 2000.

Rindisbacher, Hans J. *The Smell of Books: A Cultural-Historical Study of Olfactory Perception in Literature.* Ann Arbor: University of Michigan Press, 1992.

Robb, Graham. *The Discovery of France.* London: Picador, 2007.

Robbins, Bruce. *The Servant's Hand: English Fiction from Below.* New York: Columbia University Press, 1986.

Roberts, Lewis C. "Disciplining and Disinfecting Working-Class Readers in the Victorian Public Library." *Victorian Literature and Culture* 26.1 (1998): 105–32.

Robinson, Howard. *Britain's Post Office: A History of Development from the Beginnings to the Present Day.* Princeton, N.J.: Princeton University Press, 1953.

Robson, Catherine. *Heart Beats: Everyday Life and the Memorized Poem.* Princeton, N.J.: Princeton University Press, 2011.

Roche, Regina Maria. *The Children of the Abbey; a Tale.* Chicago: Belford Clarke & Co., 1887.

Rose, Jonathan. "How to Do Things with Book History." *Victorian Studies* 37.3 (1994): 461–71.

———. *The Intellectual Life of the British Working Classes.* New Haven, Conn.: Yale University Press, 2001.

Rosman, Doreen M. *Evangelicals and Culture.* London: Croom Helm, 1984.

Rubery, Matthew. *The Novelty of Newspapers: Victorian Fiction after the Invention of the News.* Oxford: Oxford University Press, 2009.

Ruskin, John. *The Political Economy of Art.* London: Dent, 1968.

———. *Stones of Venice.* New York: John Alden, 1885.

———. *The Works of John Ruskin.* Ed. Edward Tyas Cook and Alexander Dundas Oligvy Wedderburn. Library ed. Vol. 11. 39 vols. London: G. Allen; New York: Longmans Green and Co., 1903.

Ruth, Jennifer. "Mental Capital, Industrial Time, and the Professional in *David Copperfield.*" *Novel* 32 (1999): 303–30.

Rymer, James Malcom. *The White Slave: A Romance for the Nineteenth Century. By the Author of "Ada."* London, 1844.

"The Sailors and Their Hardships on Shore." *The sailors' magazine and seamen's friend* 41 (1869): 97–104.

Sala, G. A. "A Journey Due North." *Household Words,* 1856, 449.

[Sargent, George]. "The Story of a Pocket Bible, 1st Series." *The Sunday at Home: A Family Magazine for Sabbath Reading* 28 (1856).

[Sargent, George]. "The Story of a Pocket Bible, 2nd Series." *The Sunday at Home: A Family Magazine for Sabbath Reading* 90 (1856).

[Sargent, George Etell]. *The Story of a Pocket Bible.* New York: Carlton & Lanahan, n.d.

Sargent, George Etell. *The Story of a Pocket Bible*. London: The Religious Tract Society, [1859].

Sartre, Jean Paul. *Les mots*. [Paris]: Gallimard, 1964.

Scarry, Elaine. *Dreaming by the Book*. Princeton, N.J.: Princeton University Press, 2001.

Schaffer, Talia. "Craft, Authorial Anxiety, and 'the Cranford Papers.'" *Victorian Periodicals Review* 38.2 (2005): 221–39.

Schneider, Mark A. "Culture-as-Text in the Work of Clifford Geertz." *Theory and Society* 16.6 (1987): 809–39.

Schreiner, Olive. *The Story of an African Farm: A Novel*. 1883. New York: Bantam, 1993.

Scott, Patrick. "The Business of Belief." *Sanctity and Secularity; the Church and the World. Papers Read at the Eleventh Summer Meeting and the Twelfth Winter Meeting of the Ecclesiastical History Society*. Oxford: Published for the Ecclesiastical History Society by B. Blackwell, 1973.

Scott, Rosemary. "The Sunday Periodical: *Sunday at Home*." *Victorian Periodicals* 25.4 (1992): 158–62.

Scott, Walter. *Waverley, or, 'Tis Sixty Years Since*. Ed. Claire Lamont. The World's Classics. Oxford: Oxford University Press, 1986.

Secord, James A. *Victorian Sensation: The Extraordinary Publication, Reception, and Secret Authorship of Vestiges of the Natural History of Creation*. Chicago: University of Chicago Press, 2000.

Sedgwick, Eve Kosofsky. *Touching Feeling: Affect, Pedagogy, Performativity*. Durham, N.C.: Duke University Press, 2003.

Seed, David. "The Flight from the Good Life: *Fahrenheit 451* in the Context of Postwar American Dystopias." *Journal of American Studies* 28.2 (1994): 225–40.

Sewell, Elizabeth. *Gertrude*. London: Longmans Green, n.d.

Sewell, Elizabeth Missing. *Cleve Hall*. New York, 1855.

———. *Margaret Percival*. New York: D. Appleton & Company, 1847.

Shaw, Graham. "South Asia." *A Companion to the History of the Book*. Ed. Simon Eliot and Jonathan Rose. Malden, Mass.: Blackwell Pub., 2007.

She Would Be a Governess: A Tale. London: Routledge, 1861.

Sheridan, Richard. *The Rivals. Twelve Famous Plays of the Restoration and Eighteenth Century*. New York: Modern Library, 1933. 793–872.

Sherman, William H. *Used Books: Marking Readers in Renaissance England*. Material Texts. Philadelphia: University of Pennsylvania Press, 2008.

Sherwood, Mrs. *A Drive in the Coach through the Streets of London: A Story Founded on Fact*. London: F. Houlston and Son, 1824.

Sherwood, Mary. *The Works of Mrs. Sherwood. Being the Only Uniform Edition Ever Published in the United States*. 16 vols. New York: Harper & Brothers, 1834.

Shiv, Baba, Ziv Carmon, and Dan Ariely. "Placebo Effects of Marketing Actions: Consumers May Get What They Pay For." *Journal of Marketing Research* 42 (November 2005): 383–93.

Sicherman, Barbara. "Sense and Sensibility: A Case Study of Women's Reading in Late-Victorian America." *Reading in America: Literature and Social History*.

Ed. Cathy Davidson. Baltimore, Md.: Johns Hopkins University Press, 1989. 201–25.

Silverman, Willa. "The Enemies of Books? Women and the Male Bibliophilic Imagination in *Fin-de-Siècle* France." *Contemporary French Civilization* 30.1 (Winter 2005/Spring 2006): 47–74.

Simmel, Georg. *The Sociology of Georg Simmel*. Ed. Kurt Wolff. New York: Free Press, 1950.

Simpson, James. "Bonjour Paresse: Literary Waste and Recycling in Book 4 of Gower's *Confessio Amantis*." *Proceedings of the British Academy* 151 (2006): 257–84.

———. "'Faith and Hermeneutics.'" *Journal of Medieval and Early Modern Studies* 33 (2003): 215–39.

Skallerup, Harry Robert. *Books Afloat & Ashore; a History of Books, Libraries, and Reading among Seamen During the Age of Sail*. Hamden, Conn.: Archon Books, 1974.

Slaughter, Joseph R. *Human Rights, Inc.: The World Novel, Narrative Form, and International Law*. New York: Fordham University Press, 2007.

Small, Helen. "A Pulse of 124: Charles Dickens and a Pathology of the Mid-Victorian Reading Public." *The Practice and Representation of Reading in England*. Ed. James Raven, Helen Small, and Naomi Tadmor. Cambridge: Cambridge University Press, 1996. 263–90.

Smith, Benjamin. *Sunshine in the Kitchen; or, Chapters for Maid-Servants*. Wesleyan Conference Office, 1872.

Southey, Robert. *Letters from England*. The Cresset Library. London: Cresset Press, 1951.

———. "State and Prospects of the Country." *The Emergence of Victorian Consciousness, the Spirit of the Age*. Ed. George Lewis Levine. New York: Free Press, 1967.

Spivak, Gayatri Chakravorty. "Three Women's Texts and a Critique of Imperialism." *Critical Inquiry* 12.1 (1985): 243–61.

Spufford, Francis. *The Child That Books Built*. London: Faber and Faber, 2002.

St Clair, William. *The Reading Nation in the Romantic Period*. Cambridge: Cambridge University Press, 2004.

Stallybrass, Peter. "Books and Scrolls: Navigating the Bible." *Books and Readers in Early Modern England*. Ed. Jennifer Andersen and Elizabeth Sauer. Philadelphia: University of Pennsylvania Press, 2002. 42–79.

Stallybrass, Peter, and Ann Rosalind Jones. *Renaissance Clothing and the Materials of Memory*. Cambridge: Cambridge University Press, 2000.

Steedman, Carolyn. *Labours Lost: Domestic Service and the Making of Modern England*. Cambridge: Cambridge University Press, 2009.

Stephen, James Fitzjames. "The Relation of Novels to Life." *Cambridge Essays*. London: John W. Parker, 1855. 148–92.

[Stephen, Leslie]. "Journalism." *Cornhill Magazine*, July 1862, 52–63.

Stephenson, Neal. *The Diamond Age; or, a Young Lady's Illustrated Primer*. New York: Bantam, 1995.

Sterne, Jonathan. "Out with the Trash." *Residual Media*. Ed. Charles R. Acland. Minneapolis: University of Minnesota Press, 2007. 16–31.

Stetz, Margaret Diane. "Life's 'Half-Profits': Writers and Their Readers in Fiction of the 1890s." *Nineteenth-Century Lives*. Ed. Laurence Lockridge et al. Cambridge: Cambridge University Press, 1989. 169–87.

Stewart, Garrett. *Dear Reader: The Conscripted Audience in Nineteenth-Century British Fiction*. Baltimore, Md.: Johns Hopkins University Press, 1996.

———. *The Look of Reading: Book, Painting, Text*. Chicago: University of Chicago Press, 2006.

———. "The Mind's Sigh: Pictured Reading in Nineteenth-Century Painting." *Victorian Studies* 46.2 (2004): 217–30.

———. "Painted Readers, Narrative Regress." *Narrative* 11 (2003): 126–75.

Stewart, Rev. R. "A Piece of Waste Paper." *Gleanings for the Young* 11 (1885): 124.

Stewart, Susan. *Crimes of Writing: Problems in the Containment of Representation*. Oxford: Oxford University Press, 1991.

Stimpson, Felicity. "Servants' Reading: An Examination of the Servants' Library at Cragside." *Library History* 19.1 (2003): 3–11.

Stott, Anne. *Hannah More: The First Victorian*. Oxford: Oxford University Press, 2003.

Sturgis, Howard Overing. *Belchamber*. New York Review Books Classics. New York: New York Review Books, 2008.

Suarez, Michael Felix, and H. R. Woudhuysen. "Introduction." *The Oxford Companion to the Book*. Ed. Michael Felix Suarez and H. R. Woudhuysen. Oxford: Oxford University Press, 2010.

Sutherland, John. *Victorian Novelists and Publishers*. Chicago: University of Chicago Press, 1976.

Symon, James David. *The Press and Its Story; an Account of the Birth and Development of Journalism up to the Present Day, with the History of All the Leading Newspapers: Daily, Weekly, or Monthly, Secular and Religious, Past and Present; Also the Story of Their Production from Wood-Pulp to the Printed Sheet*. London: Seeley Service & Co., 1914.

Tanselle, G. Thomas. "Libraries, Museums, and Reading." *Literature and Artifacts*. Charlottesville: Bibliographical Society of the University of Virginia, 1998. 3–23.

Tatar, Maria. *Enchanted Hunters: The Power of Stories in Childhood*. 1st ed. New York: W. W. Norton & Co., 2009.

Taylor, John Tinnon. *Early Opposition to the English Novel: The Popular Reaction from 1760 to 1830*. New York: King's Crown Press, 1943.

Terdiman, Richard. *Discourse/Counter-Discourse: The Theory and Practice of Symbolic Resistance in Nineteenth-Century France*. Ithaca, N.Y.: Cornell University Press, 1985.

[Thackeray, William Makepeace]. "Singular Letter from the Regent of Spain." *Punch,* 16 December 1843.

———. "Waiting at the Station." *Punch,* 9 March 1850, 92–93.

———. "Why Can't They Leave Us Alone in the Holidays?" *Punch's Almanack for 1851*. London: Bradbury and Evans. 1851. 23.

Thackeray, William Makepeace. "George de Barnwell." *Burlesques*. New York: A. L. Burt, n.d. 3–13.

———. "The Memoirs of Mr. C. J. Yellowplush." *Christmas Books; Snobs and Ballads*. New York: Metropolitan Publishing Company. 195–304.

———. *The Newcomes*. Ed. Andrew Sanders. Oxford: Oxford University Press, 1995.

———. *Roundabout Papers*. New York: Scribner's, 1904.

———. *Sketches and Travels in London*. New York: Scribner's, 1904.

———. *Vanity Fair*. Ed. Peter L. Shillingsburg. New York: Norton, 1994.

"Things It Is Better Not to Do." *Judy, or the London Serio-Comic Journal* 15 (1874): 255.

Thompson, E. P. "The Political Education of Henry Mayhew." *Victorian Studies* 11 (1967–68): 41–62.

Thompson, Michael. *Rubbish Theory: The Creation and Destruction of Value*. Oxford: Oxford University Press, 1979.

Thornton, Dora. *The Scholar in His Study: Ownership and Experience in Renaissance Italy*. New Haven, Conn.: Yale University Press, 1997.

Thornton, Sara. *Advertising, Subjectivity and the Nineteenth-Century Novel: Dickens, Balzac and the Language of the Walls*. Palgrave Studies in Nineteenth-Century Writing and Culture. Basingstoke: Palgrave Macmillan, 2009.

Thurschwell, Pamela. "Henry James and Theodora Bosanquet: On the Typewriter, in the Cage, at the Ouija Board." *Textual Practice* 13 (1999): 5–23.

Tobin, Beth Fowkes. *Superintending the Poor: Charitable Ladies and Paternal Landlords in British Fiction, 1770–1860*. New Haven, Conn.: Yale University Press, 1993.

Tonna, Charlotte Elizabeth. *The Wrongs of Woman*. 1844. J. S. Taylor.

"Traffic in Waste Paper." *Municipal Affairs* 2 (June 1898).

Trivedi, Harish. "The 'Book' in India." *Books without Borders: Perspectives from South Asia*. Ed. Robert Fraser and Mary Hammond. Vol. 2. Basingstoke: Palgrave Macmillan, 2008. 12–33.

Trollope, Anthony. *An Autobiography*. 1883. Ed. P. D. Edwards. Oxford: Oxford University Press, 1980.

———. *Ayala's Angel*. Ed. Julian Thompson. The World's Classics. Oxford: Oxford University Press, 1986.

———. *Can You Forgive Her?* Ed. Stephen Wall. Harmondsworth: Penguin, 1972.

———. *The Claverings*. Oxford: Oxford University Press, 1959.

———. *The Eustace Diamonds*. Ed. W. J. McCormack. Oxford: Oxford University Press, 1983.

———. *He Knew He Was Right*. Oxford: Oxford University Press, 1978.

———. "The Higher Education of Women." *Four Lectures*. Ed. Morris L. Parrish. London: Constable and Co., 1938.

———. "Novel-Reading." *Nineteenth Century, a Monthly Review* 5 (1879): 24–43.

———. "On English Prose Fiction as a Rational Amusement." *Four Lectures*. Ed. Morris L. Parrish. London: Constable and Co., 1938.

———. *The Prime Minister*. Ed. David Skilton. Harmondsworth: Penguin, 1994.

———. *The Small House at Allington*. Ed. Julian Thompson. Harmondsworth: Penguin, 1991.

———. *The Struggles of Brown, Jones and Robinson*. Harmondsworth: Penguin, 1993.

Trollope, Frances Milton. *The Vicar of Wrexhill*. London: R. Bentley, 1837.

Troubridge, Laura. *Life amongst the Troubridges: Journal of a Young Victorian, 1873–1884*. Ed. Jacqueline Hope-Nicholson. London: Tite Street Press, 1999.

Trumpener, Katie. *Bardic Nationalism: The Romantic Novel and the British Empire*. Princeton, N.J.: Princeton University Press, 1997.

———. "The Making of Child Readers." *The Cambridge History of English Romantic Literature*. Ed. James Chandler. New Cambridge History of English Literature. Cambridge: Cambridge University Press, 2009.

Turkle, Sherry. "Introduction." *Evocative Objects: Things We Think With*. Ed. Sherry Turkle. Cambridge, Mass.: MIT Press, 2007.

"The Two Bibles." *The Children's Friend* 129 (1871): 142.

Tyndale, William. *The Obedience of a Christian Man*. Ed. David Daniell. Antwerp: Martin de Keyser, 1528. Penguin Classics. London: Penguin, 2000.

The Useful and the Beautiful: Or, Domestic and Moral Duties Necessary to Social Happiness. By the Author of "a Week at Glenville." Philadelphia: Lippincott, 1850.

Vanden Bossche, Chris. "Cookery, not Rookery: Family and Class in *David Copperfield*." *David Copperfield and Hard Times*. Ed. John Peck. London: St. Martin's, 1995. 31–57.

Vickery, Amanda. *Behind Closed Doors: At Home in Georgian England*. New Haven, Conn.: Yale University Press, 2009.

Vincent, David. *The Culture of Secrecy in Britain, 1832–1998*. Oxford: Oxford University Press, 1998.

———. *Literacy and Popular Culture: England 1750–1914*. Cambridge: Cambridge University Press, 1989.

———. *The Rise of Mass Literacy: Reading and Writing in Modern Europe*. Cambridge: Polity, 2000.

Viswanathan, Gauri. *Masks of Conquest: Literary Study and British Rule in India*. New York: Columbia University Press, 1989.

Waller, A. R., and Adolphus William Ward. *The Cambridge History of English Literature*. Cambridge: Cambridge University Press, 1932.

Waller, P. J. *Writers, Readers, and Reputations: Literary Life in Britain, 1870–1918*. Oxford: Oxford University Press, 2006.

Wallis, Alfred. "Another Chapter on Book-Plates." *The Antiquary* 1 (1880): 256–59.

Walsh, William Pakenham. *Modern Heroes of the Mission Field*. 4th ed. New York: T. Whittaker, n.d.

Ward and Lock's Home Book. London: Ward, Lock, 1882.

Ward, Mary Arnold. *David Grieve*. London: Macmillan, 1891.

Warner, Michael. *The Letters of the Republic: Publication and the Public Sphere in Eighteenth-Century America*. Cambridge, Mass.: Harvard University Press, 1990.

———. "Uncritical Reading." *Polemic: Critical or Uncritical*. Ed. Jane Gallop. New York: Routledge, 2004. 13–38.

"Waste Paper." *Leisure Hour* 30 (1881): 419–20.

Waters, Catherine. *Commodity Culture in Dickens's Household Words: The Social Life of Goods*. Aldershot: Ashgate, 2008.

Watkins, M. G. "The Library." *Gentleman's Magazine* 252 (1882): 89–103.

Watson, Rowan "Some Non-Textual Uses of Books." *A Companion to the History of the Book*. Ed. Simon Eliot and Jonathan Rose. Malden, Mass.: Blackwell Pub., 2007.

Watts, Isaac. *Divine Songs Attempted in Easy Language for the Use of Children*. Boston: Printed by S. Kneeland and T. Green, for D. Henchman, in Cornhill, 1730.

Watts, Newman. *The Romance of Tract Distribution*. London: Religious Tract Society, 1934.

Waugh, Evelyn. *A Handful of Dust*. Boston: Back Bay Books, 1999.

Webb, Robert K. *The British Working Class Reader, 1790–1848: Literacy and Social Tension*. London: Allen & Unwin, 1955.

Weedon, Alexis. *Victorian Publishing: The Economics of Book Production for a Mass Market, 1836–1916*. The Nineteenth Century Series. Aldershot: Ashgate, 2003.

Welsh, Alexander. *From Copyright to Copperfield: The Identity of Dickens*. Cambridge, Mass.: Harvard University Press, 1987.

———. *George Eliot and Blackmail*. Cambridge, Mass.: Harvard University Press, 1985.

———. *The Hero of the Waverley Novels*. New Haven, Conn.: Yale University Press, 1963.

———. "Writing and Copying in the Age of Steam." *Victorian Literature and Society*. Ed. James Kincaid and Albert Kuhn. Columbus: Ohio State University Press, 1984. 30–45.

West, William. "Less Well-Wrought Urns: Henry Vaughan and the Decay of the Poetic Monument." *ELH* 75.1 (2008): 197–217.

Weylland, John Matthias. *These Fifty Years: Being the Jubilee Volume of the London City Mission*. London: Partridge, 1884.

Wharton, Edith. *A Backward Glance*. New York: Scribner, 1964.

———. *The Collected Short Stories of Edith Wharton*. Ed. R.W.B. Lewis. New York: Scribner, 1968.

White, Gleeson. *Book-Song; an Anthology of Poems of Books and Bookmen from Modern Authors*. The Book-Lover's Library. London: E. Stock, 1893.

White, Hayden. *Tropics of Discourse: Essays in Cultural Criticism*. Baltimore, Md.: Johns Hopkins University Press, 1978.

Wiener, Joel H. *The War of the Unstamped: The Movement to Repeal the British Newspaper Tax, 1830–1836*. Ithaca, N.Y.: Cornell University Press, 1969.

Wike, Jonathan. "The World as Text in Hardy's Fiction." *Nineteenth-Century Literature* 47.4 (1993): 455–71.

Wilde, Oscar. *The Picture of Dorian Gray*. Harmondsworth: Penguin, 1981.

Willemen, Paul, and British Film Institute. *Looks and Frictions: Essays in Cultural Studies and Film Theory*. Bloomington: Indiana University Press; London: British Film Institute, 1994.

Williams, Raymond. *Culture and Society, 1780–1950*. New York: Columbia University Press, 1983.

———. *The Long Revolution*. Peterborough, Ont.: Broadview, 1961.

Williams, Ronald Earnest. *A Century of Punch*. London: W. Heinemann, 1956.

Williams, William Proctor, and Craig S. Abbott. *An Introduction to Bibliographical and Textual Studies*. 4th ed. New York: Modern Language Association of America, 2009.

Wills, W. H. "The Appetite for News." *Household Words,* 1850, 238–40.

Windscheffel, Ruth Clayton. *Reading Gladstone*. Basingstoke: Palgrave Macmillan, 2008.

Wogan, Peter. "Perceptions of European Literacy in Early Contact Situations." *Ethnohistory* 41.3 (1994): 407–29.

Wood, Mrs. Henry. *The Earl's Heirs: A Tale of Domestic Life*. Philadelphia: T. B. Peterson, 1862.

Woodburn, James. *Rules of Etiquette*. Chicago: Rand McNally, 1893.

Woodward, Donald. "Swords into Ploughshares: Recycling in Pre-Industrial England." *Economic History Review* 38 (1985): 175–91.

Woolf, Leonard and Virginia. "Are Too Many Books Written and Published?" 1927. *PMLA* 121 (2006): 235–44.

Woolf, Virginia. "How Should One Read a Book?" *The Common Reader. Second Series*. London: Hogarth Press, 1932. Ed. Andrew McNeillie. London: Hogarth Press, 1986.

———. *The Voyage Out*. 1915. New York: Modern Library, 2001.

Worboise, Emma. *Thornycroft Hall: Its Owners and Its Heirs*. London: J. Clarke, 1886.

Wordsworth, William. *The Prelude, 1799, 1805, 1850: Authoritative Texts, Context and Reception, Recent Critical Essays*. Ed. Jonathan Wordsworth, M. H. Abrams, and Stephen Charles Gill. 1st ed. New York: Norton, 1979.

Wynter, Andrew. "Mudie's Circulating Library." *Victorian Print Media: A Reader*. Ed. John Plunkett and Andrew King. Oxford: Oxford University Press, 2005.

Yeames, James. *Gilbert Guestling, or, the Story of a Hymn-Book*. London: Wesleyan Conference Office, 1881.

Yonge, Charlotte. "Children's Literature of the Last Century: Part II—Didactic Fiction." *Macmillan's* 20.118 (1869): 302–10.

———. "Children's Literature: Part III—Class Literature of the Last Thirty Years." *Macmillan's* 20.119 (1869): 448–56.

———. *P's and Q's, or, the Question of Putting Upon*. London: Macmillan, 1872.

Yonge, Charlotte Mary. *The Pillars of the House; or, under Wode, under Rode*. London: Macmillan and Co., 1874.

———. *What Books to Lend and What to Give*. London: National Society's Depository, 1887.

Zeitlin, Judith. "Xiaoshuo." *The Novel*. 2001–2003. Ed. Franco Moretti. Vol. 1. Princeton, N.J.: Princeton University Press, 2006. 249–61.

Zemka, Sue. "The Holy Books of Empire: Translations of the British and Foreign Bible Society." *Macropolitics of Nineteenth-Century Literature: Nationalism, Exoticism, Imperialism*. Ed. Jonathan Arac and Harriet Ritvo. Philadelphia: University of Pennsylvania Press, 1991.

Index

Milton Keynes UK
Ingram Content Group UK Ltd.
UKHW010309011223
433577UK00004B/275